NEW APPROACHES IN
EUKARYOTIC DNA REPLICATION

NEW APPROACHES IN EUKARYOTIC DNA REPLICATION

Edited by

A. M. de Recondo

Institute of Scientific Research on Cancer
Villejuif, France

PLENUM PRESS • NEW YORK AND LONDON

Library of Congress Cataloging in Publication Data

Main entry under title:

New approaches in Eukaryotic DNA replication.

"Proceedings of a meeting on new approaches in Eukaryotic DNA replication, held June 13-17, 1980, in Cargèse, Corsica, France"—Verso t.p.
Includes bibliographical references and index.
1. DNA replication—Congresses. I. Recondo, A. M. de (Anne-Marie de) [DNLM: replication—Congresses. 2. DNA repair—Congresses. QH 462.D8 N532 1980]
QH450.N48 1982 574.87′3282 82-18550
ISBN 0-306-41182-2

Proceedings of a meeting on New Approaches in Eukaryotic DNA Replication, held June 13–17, 1980, in Cargèse, Corsica, France

A Division of Plenum Publishing Corporation
233 Spring Street, New York, N.Y. 10013

Printed in the United States of America

PREFACE

DNA replication in eukaryotes is an important field, particularly because of its direct impact on the study of cancer. The understanding of molecular mechanisms of replication and their regulation should allow a better comprehension of the alterations that lead to the proliferation of tumor cells and to error-prone repair in cells exposed to radiation or chemical carcinogens.

During the last several years, many enzymes and proteins which participate in replication of DNA in eukaryotic cells have been identified, isolated and characterized. New concepts in chromatin structure have refocused attention on the study of replication of DNA complexed with histones and non-histone chromosomal proteins. However, progress has been noticeably slower than for prokaryotes, essentially because of the difficulty in genetic analysis of eukaryotic DNA replication.

In June 1980, a workshop was organized in Cargèse, Corsica (France) to facilitate exchanges of information between workers specializing in prokaryotes and those specializing in eukaryotes, and to allow discussion of new experimental approaches. With this in mind, special interest has been taken in the origin and termination of chromosome cycles and how they are controlled.

To convey the main contributions and new ideas discussed at this meeting, some of the lectures given at Cargèse and reviews concerning problems remaining to be solved have been compiled in this book. The authors have included in their articles the most recent results and their opinions on the most controversial areas; this book thus provides an up to date summary of DNA replication and reparation which can be used as a source of information and orientation for future research.

I am indebted to the Centre National de la Recherche Scientifique and to the Association pour le Developpement de la Recherche sur le Cancer for the financial support which made this workshop possible. I am also grateful to Drs. M. Kohiyama and D. Korn, who helped me to organize this meeting. Finally, I express my gratitude to Mrs. Dee Le Roy, Mrs. Jerri Bram, and Mrs. Josette Bannelier for their helpful editorial assistance.

Anne-Marie de Recondo

CONTENTS

REPLICATION MACHINERY

REGULATION OF DNA REPLICATION

REPAIR OF DNA DAMAGE

PART I

REPLICATION MACHINERY

ENZYMATIC MECHANISMS IN DNA REPLICATION: INITIATION OF STRANDS*

Arthur Kornberg

Department of Biochemistry
Stanford University School of Medicine
Stanford, California

ABSTRACT

A model for the initiation of nascent (Okazaki) fragments in the discontinuous phase of replication is the conversion of ϕX174 viral single-stranded (SS) DNA to the parental duplex replicative form (RF). The conversion *in vitro* requires at least eleven *Escherichia coli* proteins. Among these protein n', a highly specific DNA-dependent ATPase (dATPase), guides proteins n, n'', i, *dnaC dnaB* and primase to the intragenic locus between structural genes F and G to assemble a mobile priming system, the *primosome*. Remarkably, the primosome, having served in the SS→RF reaction, is still physically and functionally present in nearly intact form following isolation by sucrose gradient centrifugation. During the next replicative stage (RF→RF), the conserved primosome facilitates cleavage, by the ϕX-encoded gene A protein, of the viral strand, thereby initiating continuous elongation of this strand. When fully restored by the addition of protein i, the primosome repeatedly primes the discontinuous elongation of the complementary strand through several cycles of progeny RF production.

*This work was supported in part by grants from the National Institutes of Health and the National Science Foundation. I wish to acknowledge the major contributions to this work by K. Arai, N. Arai, J. Shlomai, P. M. J. Burgers, R. L. Low, J. Kobori, R. Fuller, L. Bertsch, J. Kanguni, M. M. Stayton and K. Taylor.

INTRODUCTION

Replication of the *E. coli* chromosome requires a large number of replication proteins as predicted by the numerous genes necessary for this process[1]. In fact the task of reconstituting the replication machinery *in vitro* is so formidable that a simpler approach to dissect the basic features of *E. coli* replication had to be sought. Our approach lies in the study of the *in vitro* DNA replication of single-stranded phages M13, G4 and ϕX174 that depend heavily upon the host *E. coli* for their DNA replication proteins. ϕX174 replication has been especially useful since it requires most, if not all, the replication proteins that operate at the *E. coli* replication fork.

The replication of ϕX174 DNA can be divided into three stages[2]. In the first stage, eleven purified *E. coli* replicative proteins serve to convert the single-stranded viral circle to the parental duplex, replicative form (SS→RF). In the second stage, multiple copies of progeny RF are produced using the parental RF as a template (RF→RF); in addition to the eleven proteins required in the first stage, this process requires the phage-encoded gene A protein and host *rep* protein[3,4]. In the final stage, production of progeny RF ceases and only the viral strand circles are synthesized for encapsulation into progeny phage (RF→SS)[5].

In this paper two key aspects of this process will be discussed. First, we report that the prepriming proteins n, n', n" i, *dnaC* and *dnaB* proteins and primase necessary for stage I (SS→RF) and required in stage II (RF→RF) for strand synthesis assemble into a mobile, functional priming complex, the primosome, on the viral single-stranded DNA coated with single strand binding protein (SSB). Secondly, this primosome assembled in stage I is remarkably conserved and active in stage II to help produce multiple copies of progeny RF.

RESULTS AND DISCUSSION

Proteins Required for the Conversion of SS ϕX DNA into the Duplex Form

Synthesis of the complementary strand on the single-stranded circle of small DNA phages (M13, G4 and ϕX174) is an example of discontinuous replication. The various stages of this process are: prepriming, priming, elongation, gap filling and ligation. These stages have been elucidated and reconstituted *in vitro* for the three phages (Table 1). Although all three use DNA polymerase III holoenzyme for elongation[6], they differ in the mechanism used to synthesize the primer RNA that holoenzyme can elongate[7].

RNA polymerase recognizes and primes SSB-covered M13 DNA[8]; primase recognizes and primes SSB-covered G4 DNA[9]. SSB-covered ϕX

Table 1. Protein Requirements for in vitro Conversion of Phage Single-stranded DNA to Duplex Closed Circular Form

Stage	M13	G4	φX174
Prepriming	SSB	SSB	SSB protein n protein n' protein n" protein i dnaB protein dnaC protein
Priming	RNA polymerase	Primase	Primase
Elongation	DNA polymerase III holoenzyme	DNA polymerase III holoenzyme	DNA polymerase III holoenzyme
Gap filling & Ligation	DNA polymerase I ligase	DNA polymerase I ligase	DNA polymerase I ligase

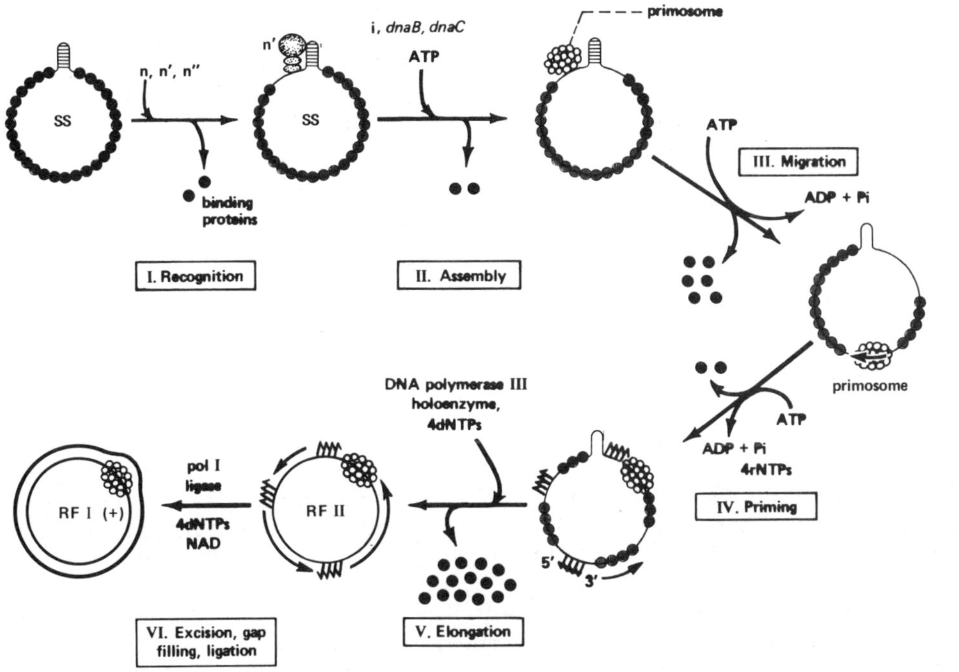

Fig. 1. Various steps in the conversion of φX174 SS DNA into the duplex replicative form.

DNA, however, is not recognized by primase, unless the DNA is preprimed by six prepriming proteins: proteins n, n', n", i, dnaB and dnaC[10] (Fig. 1).

Proteins n*, n'[11], dnaC** and dnaB[12] have each been purified to near homogeneity and protein n" has been partially purified through six fractionating steps***. The dnaB[12,13] and dnaC** genes, as well as the dnaG (primase) gene****[13] have been cloned into high-copy-number plasmids to overproduce the proteins 20- to 200-fold, considerably simplifying the task of purification and providing material needed for more extensive characterization. The genes coding for proteins n, n', n" and i have not yet been identified.

Irrespective of the mechanism of primer formation used by the different single-stranded phage DNA's, chain elongation of the primer in each case is carried out by DNA polymerase III holoenzyme, a multipolypeptide replication complex[6]. Although the exact subunit structure of holoenzyme is not yet known, several subunits have been extensively characterized and their structural genes determined[14-20] (Table 2). Reconstitution experiments have shown that the form of holoenzyme used in the replication of each phage DNA may differ[18] (Table 2). A new subunit, ζ, has been recently identified as necessary for G4 and M13 replication*****. The requirement for this subunit had not been appreciated before because of its presence as a minor impurity in primase[18]. Overproduction of primase (dnaG) eliminated this impurity**** and thus provided an assay for ζ activity. From SDS polyacrylamide gel electrophoresis, a molecular weight of 48,000 D is tentatively assigned to the ζ subunit.

Roles of Protein n' and dnaB in the Primosome

The ϕX174 primosome, defined as the active priming complex of the prepriming proteins and primase, contains the two DNA-dependent ATPases, n' protein and dnaB protein, that have been studied in detail.

Protein n' (factor Y of Wickner and Hurwitz[21]) contains an intrinsic, DNA-dependent ATPase (dATPase) activated by a locus on ϕX DNA between structural genes F and G[11]. This site is preserved in an exonuclease VII digest of the HaeIII Z-1 fragment, a 55-nucleotide long fragment from residue 2301 to 2354 that contains the potential for a 44-nucleotide hairpin structure[22]. Unlike the SSB-coated SS

*R. L. Low, J. Shlomai and A. Kornberg, manuscript in preparation.
**J. Kobori and A. Kornberg, manuscrip in preparation.
***R. L. Low, J. Shlomai, and A. Kornberg, unpublished results.
****M. Stayton, S. Yasuda and A. Kornberg, unpublished results.
*****P. M. J. Burgers and A. Kornberg, unpublished results.

Table 2. DNA Polymerase III Holoenzyme Subunits

Subunit	Mw (x 10^3)	Locus	Required for SS→RF G4	ϕX	M13	Reference
α	140	dnaE				(14,15)
core ε	25					
θ	10		+	+	+	
β	37	dnaN				(16,17)
γ	52	dnaZ				(18)
δ	32	dnaX	−	+	−	(19)
ζ	(48)[a]		+	n.d.	+	
τ	83		−	−	−	(20)

[a]P.M.J. Burgers and A. Kornberg, unpublished results; n.d. = not determined.

ϕX DNA, SSB-coated SS G4 and M13 DNA show negligible ATPase (dATPase) effector activity[23]. Furthermore, this HaeIII Z1 fragment of ϕX174 uniquely promotes assembly of the ϕX priming system complex, the primosome, indicating that protein n' initiates the assembly of proteins n, n", i, dnaC, dnaB, and primase at this site[24].

Protein n' effector sites have also been identified at the pBR322[25] and ColE1* lagging-strand origin sites. A separate site on ColE1 is also present upstream on the opposite (L) strand*. The ColE1 lagging-strand origin site (H strand of the HaeII E fragment) when cloned into M13 SS DNA supports an in vivo rifampicin-resistant, dnaG-dnaB dependent conversion of the chimeric SS DNA to the RF[26]. Recently, in collaboration with Nomura and Ray, we have found that this chimeric SS DNA in vitro promotes a rifampicin-resistant SS→RF conversion dependent on each of the purified ϕX replicative proteins*.

The dnaB protein DNA-dependent ATPase (GTPase) differs significantly from the protein n' activity[27-30]. The dnaB protein ATPase (GTPase) can use a wide variety of single-stranded DNAs, but ATP hydrolysis is completely suppressed by SSB. Studies with the dnaB-primase general priming system[31] and experiments with nonhydrolyzable ATP analogs suggest that the dnaB protein complexed with ATP helps the primosome identify sites where primers can be started; the dnaB protein:ATP complex induces DNA conformational changes required for primase initiation[30].

*R. L. Low, A. Kornberg, N. Nomura and D. S. Ray, unpublished results.

Protein n' and dnaB are Stably Bound to ϕX174 DNA

To determine the stoichiometry of proteins n' and dnaB in the prepriming intermediates and their fate at the different stages of the conversion of the single-stranded circle into the replicative form, these two proteins were isotopically labeled by reductive methylation with tritiated sodium borohydride[31,32]. At any given stage, proteins bound to the DNA can be separated from unbound proteins by Bio-Gel A-5m gel filtration or by sucrose density centrifugation. The stoichiometry of labeled, bound protein can be determined by the radioactivity associated with the DNA.

These experiments show that in the different stages of the ϕX SS→RF reaction, even up to the covalently closed duplex form, proteins n' and dnaB (expressed as a hexamer of the 50,000 D monomer) each remain bound in a stoichiometry of one molecule per replicated circle[33] (Table 3). Thus, these key primosomal proteins are fully retained in the completed parental RF. Moreover, the dnaB binding is highly specific as indicated by isotope exchange studies[30]. A stable non-exchangeable dnaB-DNA complex can be obtained only if all the prepriming proteins are present in the prepriming intermediate. Furthermore, only ϕX, not G4 or M13 DNA, can support this specific complex[30] (Table 4).

Functional Retention of the Primosome on Covalently Closed Duplex ϕX DNA

Recently, we have been able to reconstitute replication of the ϕX duplex form (RF→RF) from purified enzymes by combining the SS→RF with the RF→SS system[34]. This reaction requires all the prepriming proteins, primase, gene A protein, rep protein and DNA polymerase III holoenzyme. Many RFII molecules are made from one parental gene A-RFII molecule. Since proteins n' and dnaB are physically retained on the ϕX duplex form, we determined whether other prepriming proteins are functionally present. The RF→RF reaction provides such a functional test for the presence of an active primosome. The synthetic RFI isolated by sucrose density centrifugation, when supplemented with the RF→SS enzymes (gene A, rep, holoenzyme), produced no RFII, but only SS ϕX DNA[33]. However, when the reaction was further supplemented with protein i, the primosome became fully functional and many copies of RFII DNA were produced per input parental RFI. On the other hand, when the synthetic RFI was isolated by gel filtraton, the primosome had to be supplemented with prepriming proteins n + n", i and dnaC[33]. These results indicate that the primosome is unstable and gentler methods are needed to isolate ϕX RFI DNA in an intact and stable form.

Retention of the primosome renders the synthetic RFI a far more efficient substrate for gene A cleavage than supercoiled RFI extracted from infected cells. The initial rate of DNA synthesis in the RF→SS

Table 3. Retention of Proteins n' and dnaB with Synthetic RFI

Product	Molecules per replicated circle	
	Protein n'	dnaB protein
Prepriming intermediate	1.0	1.2
RFII (priming, DNA synthesis)	1.0	1.1
RFI (gap-filling, ligation)	0.9	0.9

The prepriming intermediate was formed in a 150μl reaction mixture containing: 90μl of buffer A (20mM Tris-HCl pH 7.5/20mM KCl/5% sucrose/8mM $MgCl_2$/0.1mM EDTA/5mM dithiothreitol/0.2mM ATP/0.1mg/ml bovine serum albumin), 2.2nmol (as nucleotide) of ϕX SS DNA, 6.6μg of SSB, 0.15μmol of ATP, 4.7μg of dnaC protein, 0.4μg of protein i, 0.5μg of proteins n + n" mixture, and either 0.7μg of ^{3}H-labeled dnaB protein (1.5 x 10^5cpm/μg) and43ng of protein n' or 0.7μg of dnaB protein and 43ng of ^{3}H-labeled protein n' (5 x 10^4cpm/μg). After a 30 min incubation at 30°C, the intermediate was filtered through a 5ml Bio-Gel A-5m column (0.5 x 26cm) equilibrated at room temperature with buffer A. Void-volume fractions (25μl aliquots) were assayed for DNA replication by using [α-^{32}P]dTTP (10cpm/pmol). The remainder of each fraction was assayed for labeled n' or dnaB protein. For synthesis of RFII, the reaction mixture was supplemented with rNTPs, unlabeled dNTPs, [α-^{32}P]dTTP (20cpm/pmol), primase (0.3μg), and pol III holoenzyme (0.25μg) and incubated an additional 20 min; for RFI, DNA polymerase I (10ng), NAD (7.5nmol), and DNA ligase (1μg) were also added and the incubation was extended 40 min.

reaction with the synthetic RFI is 20 times faster than that obtained with *in vivo* supercoiled RFI[33] Fig. 2. The need for gyrase catalyzed supercoiling, previously considered essential, can be dispensed with[35].

The Primosome Requires ATP

The stages in which the primosome requires the energy of ATP hydrolysis have been analyzed. Protein n' and dnaB protein ATPases can easily be distinguished since the former has dATPase and very low GTPase activity[11]., while the latter has GTPase and negligible dATPase activity[29]. A stable prepriming replication intermediate (primosome minus primase) was assembled on the SSB-coated ϕX circle and the complex isolated by gel filtration in the presence of ATP[36]. In the absence of ATP, no stable primosome could be retained on the DNA.

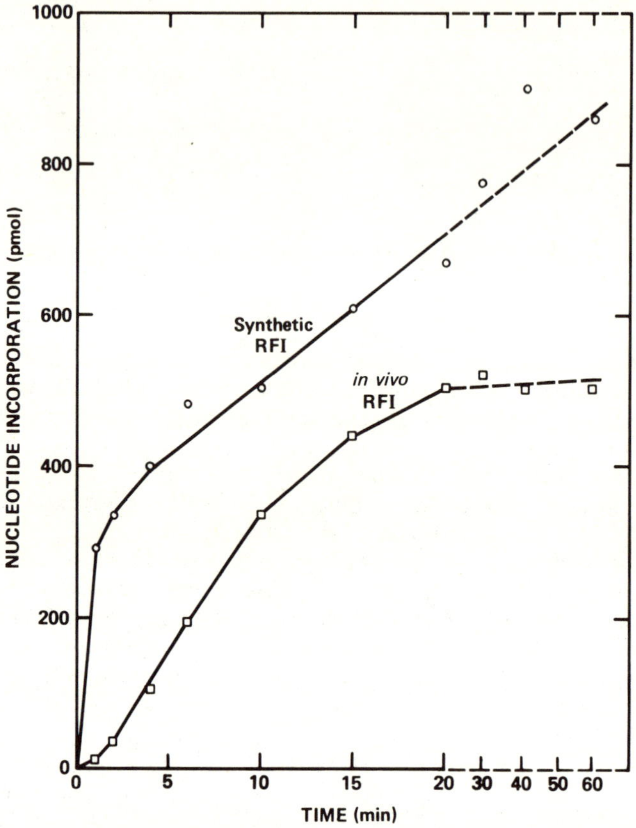

Fig. 2. Kinetics of the RF→SS DNA reaction. The reaction mixture (120μl) contained 360pmol (as nucleotide) of either Bio-Gel-purified synthetic RFI (o) or in vivo RFI (□) and the RF→SS DNA system (Fig. 1). Incorporation of |^{32}P|dTTP into DNA was determined by precipitation of 5μl aliquots with 10% trichloroacetic acid.

Studies with nonhydrolyzable analogs of ATP indicate that ATP hydrolysis is not necessary for stable binding of the primosome[36]. Although the primosome assembles at or near the n' site, movement of the primosome around the chromosome in the anti-elongation direction[37] requires the energy of ATP hydrolysis[24,36]. A comparison of the ATPase, dATPase and GTPase activities of the prepriming replication intermediate (10nmol, 6nmol, and 1nmol, respectively, of triphosphate hydrolyzed per pmol of complex per min at 30°C) is one indication that the n' ATPase likely fuels movement along the DNA[36].

Table 4. Requirement for the Formation of dnaB protein·DNA complex

DNA	Additions: SSB	n + n"	n'	i	dnaC	dnaB complexed with DNA (molecule/circle
φX174	+	−	−	−	−	<0.05
	+	+	−	−	−	<0.05
	+	−	+	−	−	<0.05
	+	−	−	+	−	<0.05
	+	−	−	−	+	<0.05
	+	−	+	+	+	<0.05
	+	+	−	+	+	0.07
	+	+	+	−	+	<0.05
	+	+	+	+	−	<0.05
	+	+	+	+	+	1.13
	−	+	+	+	+	0.73
M13	+	−	−	−	−	<0.05
	+	+	+	+	+	<0.05
G4	+	−	−	−	−	<0.05
	+	+	+	+	+	<0.05

The reaction mixture contained in 25μl: 5μl of buffer A (Table 2), 15nmol of ATP, 750pmol (as nucleotide) of φX, G4 or M13 SS DNA, 0.5μg of [^{3}H]dnaB protein, and, when present, 90ng of protein n', 0.3μg of proteins n + n" mixture, 90ng of protein i, 1.2μg of dnaC protein, and 2μg of SSB. Incubation was at 30°C for 20 min. The reaction mixtures were filtered on a Bio-Gel A-5m agarose column (0.2 x 20cm) equilibrated with buffer A. The [^{3}H]dnaB protein associated with DNA in the void-volume was measured in a scintillation spectrometer.

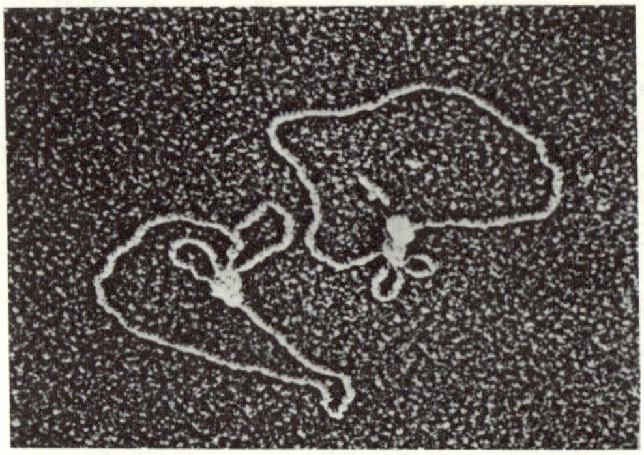

Fig. 3. Electron micrograph of synthetic RFI DNA.

The Primosome Freely Moves on the φX RFI DNA

SS φX DNA was converted to RFI DNA by the action of the SS→RF enzymes and the RFI DNA was isolated by sucrose density centrifugation in the presence of Mg-ATP[33]. Electron microscopy of the DNA showed that a large protein complex was present on the DNA* (Fig. 3). The size of the complex agrees with the expected mass of the primosome (ca. 0.5×10^6 D). To determine if the primosome was located at a specific site, the synthetic RFI was cut by the restriction endonuclease Pstl and the linear molecules examined by electron microscopy. The primosome was randomly distributed on the DNA (Fig. 4). However, when the synthetic RFI was first cut by gene A protein and then by Pstl endonuclease, most of the molecules had the primosome located at 20% from the nearest end of the linear molecule, exactly the distance between the gene A and Pstl sites[38] (Fig. 4). Thus the primosome appears to move freely along the DNA, using the n' ATPase (dATPase) as an energy source, until it encounters the site of gene A cleavage where further movement is restricted. Indeed, the ATPase (dATPase) activity associated with the primosome-RFI complex is abolished by gene A cleavage*.

During the φX DNA replicative cycle, conservation of the primosome has important advantages. First, retention of the primosome provides a DNA topology that favors efficient cleavage by gene A protein, independent of supercoiling by gyrase. Secondly, conservation of an intact primosome on the DNA may provide a more efficient mechanism for fork movement in duplex DNA replication (Fig. 5). Traveling along the lagging strand of the replication fork in the direction of fork movement[37] without the necessity for repeated dissociation and reassociation, the primosome may, together with holoenzyme and rep protein, constitute part of a larger replisome and promote efficient duplex replication (Fig. 5).

*R. L. Low, A. Kornberg and J. Griffith, unpublished results.

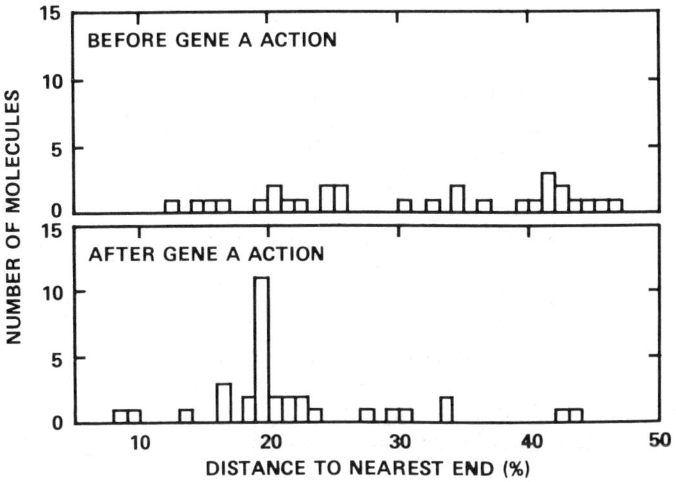

Fig. 4. Length histogram of primosome location on synthetic RFI DNA cut by Pstl without (upper) and with (lower) prior cleavage by gene A protein.

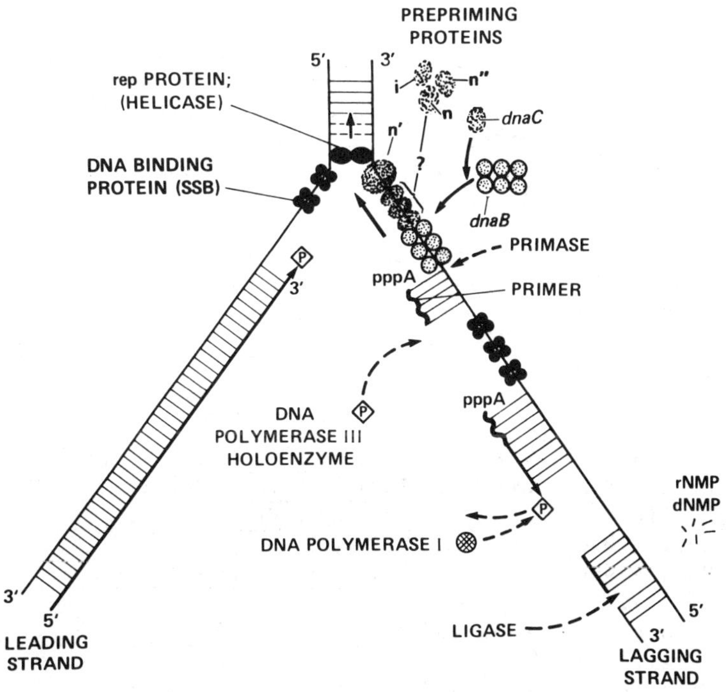

Fig. 5. A scheme of the structure of the replication fork of the E. coli chromosome.

REFERENCES

1. A. Kornberg, "DNA Replication," W. H. Freeman and Co., San Francisco, p.376 (1980).
2. D. T. Denhardt, CRC Crit. Rev. Microbiol. 4:1611 (1975).
3. E. S. Tessmar, J. Mol. Biol. 17:218 (1966).
4. D. T. Denhardt, M. Iwaya and L. L. Larison, Virology 49:486 (1972).
5. H. Fujisawa and M. Hayashi, J. Virol. 19:416 (1976).
6. C. McHenry and A. Kornberg, J. Biol. Chem. 252:6478 (1977).
7. R. McMacken, J. P. Bouché, S. L. Rowen, J. H. Weiner, K. Ueda, L. Thelander, C. McHenry and A. Kornberg, RNA Priming of DNA Replication, in:"Nucleic Acid-Protein Recognition", H. J. Vogel, ed., p.15, Academic Press, New York (1977).
8. K. Geider and A. Kornberg, J. Biol. Chem. 249:3999 (1974).
9. K. Zechel, J. P. Bouché,and A. Kornberg, J. Biol. Chem. 250:4684 (1974).
10. R. R. Meyer, J. Shlomai, J. Kobori, D. L. Bates, L. Rowen, R. McMacken, K. Ueda and A. Kornberg, Cold Spring Harbor Symp. Quant. Biol. 43:289 (1978).
11. J. Shlomai and A. Kornberg, J. Biol. Chem. 255:6789 (1980).
12. K. Arai, S. Yasuda and A. Kornberg, submitted for publication.
13. R. McMacken, J. Supramol. Str. Cell. Biochem. Supplement 5:343 (1981).
14. C. S. McHenry and W. Crow, J. Biol. Chem. 254:1748 (1979).
15. M. L. Gefter, Y. Hirota, T. Kornberg, J. A. Wechsler and C. Barnoux, Proc. Natl. Acad. Sci. USA 60:3150 (1971).
16. K. O. Johanson and C. S. McHenry, J. Biol. Chem. 255:10984 (1980).
17. P. M. J. Burgers, A. Kornberg and Y. Sakakibara, Proc. Natl. Acad. Sci. USA, in press.
18. U. Hübscher and A. Kornberg, J. Biol. Chem. 255:11698 (1980).
19. U. Hübscher and A. Kornberg, Proc. Natl. Acad. Sci. USA 77:6284 (1980).
20. C. S. McHenry, in:"Mechanistic Studies of DNA Replication and Genetic Recombination", B. Alberts and C. F. Fox, eds., Academic Press, New York, p.569 (1980).
21. S. Wickner and J. Hurwitz, Proc. Natl. Acad. Sci. USA 72:3342 (1975).
22. J. Shlomai and A. Kornberg, Proc. Natl. Acad. Sci. USA 77:799 (1980).
23. J. Shlomai and A. Kornberg, J. Biol. Chem. 255:6794 (1980).
24. K. Arai and A. Kornberg, Proc. Natl. Acad. Sci. USA 78:69 (1981).
25. S. L. Zipursky and K. J. Marians, Proc. Natl. Acad, Sci. USA 77: 6521 (1980).
26. N. Nomura and D. S. Ray, Proc. Natl. Acad. Sci. USA 77:6566 (1980).
27. R. McMacken, K. Ueda and A. Kornberg, Proc. Natl. Acad. Sci. USA 47:4190 (1977).
28. K. Ueda, R. McMacken and A. Kornberg, J. Biol. Chem. 253:4051 (1978).

29. L. Reha-Krantz and J. Hurwitz, J. Biol. Chem. 253:4051 (1978).
30. K. Arai and A. Kornberg, J. Biol Chem. 256:5247; and subsequent papers pp. 5247-5280 (1981).
31. K. Arai and A. Kornberg, Proc. Natl. Acad. Sci. USA 76:4308 (1979).
32. R. H. Rice and G. E. Means, J. Biol. Chem. 246:831 (1971).
33. R. L. Low, K. Arai and A. Kornberg, Proc Natl. Acad. Sci. USA 78:1436 (1981).
34. K. Arai, N. Arai, J. Shlomai and A. Kornberg, Proc. Natl. Acad. Sci. USA 77:3322 (1980).
35. T. J. Henry and R. Knippers, Proc. Natl. Acad. Sci. USA 71:1549 (1974).
36. K. Arai, R. L. Low and A. Kornberg, Proc. Natl. Acad. Sci. USA 78:707 (1981).
37. K. Arai and A. Kornberg, Proc. Natl. Acad. Sci. USA 78:69 (1981).
38. S. Eisenberg, J. Griffith and A. Kornberg, Proc. Natl. Acad. Sci. USA 74:3198 (1977).

ENZYMOLOGICAL CHARACTERIZATION OF HUMAN DNA POLYMERASES-α and β *

David Korn, Paul A. Fisher and Teresa Shu-Fong Wang

The Laboratory of Experimental Oncology
Department of Pathology
Stanford University School of Medicine
Stanford, CA. 94305

INTRODUCTION

During the past few years, we have described the results of an extensive series of investigations of the enzymological properties of essentially homogeneous preparations of human DNA polymerases-α and β[1-3] that are devoid of contaminating or associated endo- or exodeoxyribonuclease activities[4,5]. These studies have demonstrated a number of striking differences between the two enzymes with respect to their ability to catalyze deoxynucleotide incorporation on a variety of defined, natural and synthetic DNA primer-templates[4-8] and have suggested that there might be equally profound differences, possibly of physiological significance, in the nature of the specific molecular signals that regulate their catalytic interactions with nucleic acids. To pursue these observations in the face of exceedingly limited quantities of purified enzyme protein, we have successfully exploited the power of classical steady-state kinetics methodology to illuminate many of the key features of the polymerase-primer-template interaction and to gain substantial new insights into the fundamental mechanisms of polymerase catalysis. In performing these studies, we have employed direct sedimentation binding assays and novel methods of analysis of polymerase products synthesized on DNA molecules of known sequence to obtain, in every instance, direct physical corroboration of the principal conclusions derived from the kinetics experiments. We have thus been able to provide important

*The studies described in this review were supported by Grant CA-14835 and Training Grants GM-01922 and CA-09151 from the National Institutes of Health

reassurance of the validity of the kinetics interpretations and obviate much of the concern that might appropriately arise from the indirect nature of kinetics analyses, particularly in complex systems.

1. Primer-Template Recognition by DNA Polymerase-α

The first question to be addressed is the specification of the structural elements in DNA substrates that are required for catalytically significant interaction of DNA polymerase-α with primer-templates. To approach this problem, we employed competitive inhibition assays to evaluate the specific interaction of polymerase-α with a variety of defined DNA molecules that are essentially unable to support deoxynucleotide incorporation under standard[1,4] reaction conditions (Fig.1). Polymerase-α is incapable of kinetically detectable interaction with intact duplex circular (PM2) DNA, whether super-coiled (Form 1) or relaxed (Form IV)[9]; nor can it interact with multiply-nicked PM2 DNA molecules (Fig. 1D) or with blunt-ended linear fragments of PM2 DNA (Fig. 1C). In contrast, single-stranded DNA molecules, whether intact, closed-circular (M13) (Fig. 1A, B, D, broken lines), or linear fragments bearing 3'-PO_4(Fig. 1A) or 3'-OH (Fig. 1B) terminal residues, prove to be potent inhibitors that are fully competitive with the activated DNA substrate. These kinetics results, indicating that polymerase-α has measurable affinity only for single-stranded, but not for duplex, DNA were confirmed by direct sedimentation binding assays (Fig. 2) that readily demonstrated the binding of the polymerase to M13 DNA (Fig. 2B) or heat-denatured T7 DNA (Fig. 2D) and corroborated the complete absence of detectable binding to relaxed PM2 (Fig. 2A) or native T7 (Fig. 2C) duplex DNA species.

The results in Fig. 1 illustrate three other very important points. First, the identical affinity of polymerase-α for closed-circular M13 DNA molecules and linear M13 DNA fragments containing 3'-terminal phosphate residues (Fig. 1A) indicates that the enzyme can not recognize 3'-PO_4 termini. Second, the significantly enhanced affinity of the enzyme for linear M13 DNA fragments with 3'-terminal hydroxyl residues (Fig. 1B), together with the complete indifference of the polymerase to 3'-OH termini in duplex DNA (whether internally at nicks (Fig. 1D) or at flush ends (Fig. 1C)), strongly infer that polymerase-α is capable of recognizing 3'-OH (primer) termini only in concert with, or subsequent to, its binding to single-stranded DNA (template). Third, the non-linearity of the slope replots (not shown) obtained from Lineweaver-Burk data like those in Figures 1A, B and D suggested that the interaction of the polymerase with single-stranded DNA was cooperative, an interpretation that was strengthened by further kinetic analyses[9] of the interaction of the enzyme with both closed-circular and linear 3'-OH- and 3'-PO_4-terminated single-stranded molecules of M13 and ϕX 174 DNA. In every instance, those analyses generated linear Hill plots with slopes that ranged between

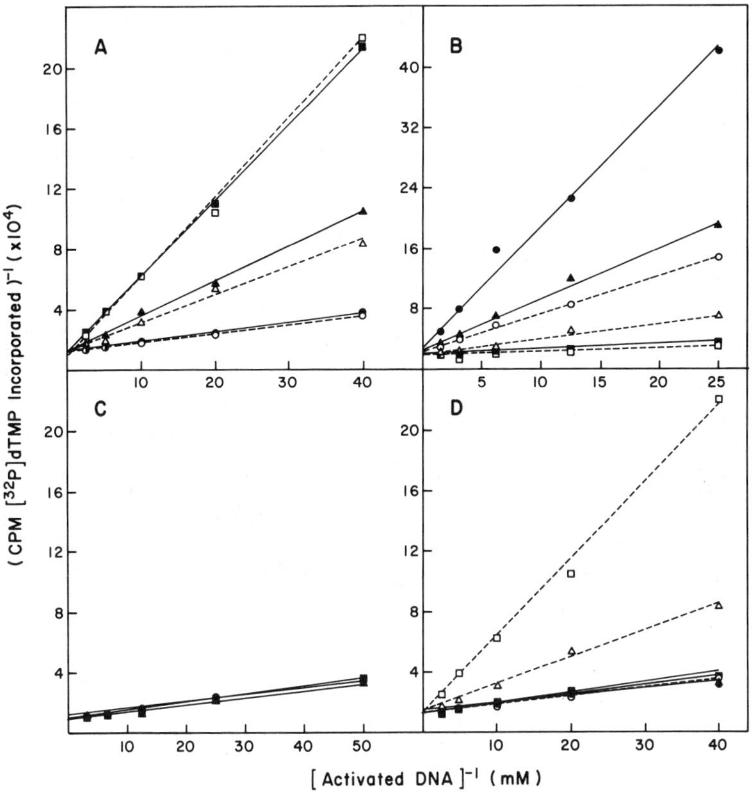

Fig. 1. Effects of duplex DNA, single-stranded DNA, and DNA termini on the reaction of DNA polymerase-α with activated DNA[9]. The data are presented as Lineweaver-Burk plots. The results obtained with intact circular M13 DNA are shown by the open symbols and dashed lines in panels A, B and D. In all panels, concentrations of substrate DNA are indicated (in reciprocal form) on the abscissa, and quantities of inhibitor DNA added to the reactions were none (control), 1X and 2X. Panel A: Effect of 3'-PO_4 fragments (300-400 nucleotides in average length) of M13 DNA. Panel B: Effect of 3'-OH fragments (200 nucleotides in average length) of M13 DNA. Panel C: Effect of flush-end duplex restriction fragments of PM2 DNA (form I and form IV PM2 DNA molecules are identically inert under these assay conditions). Panel D: Effect of duplex PM2 DNA molecules that contain about 10 nicks per molecule. In this and all other Figures, straight lines were generated by the method of least squares analysis.

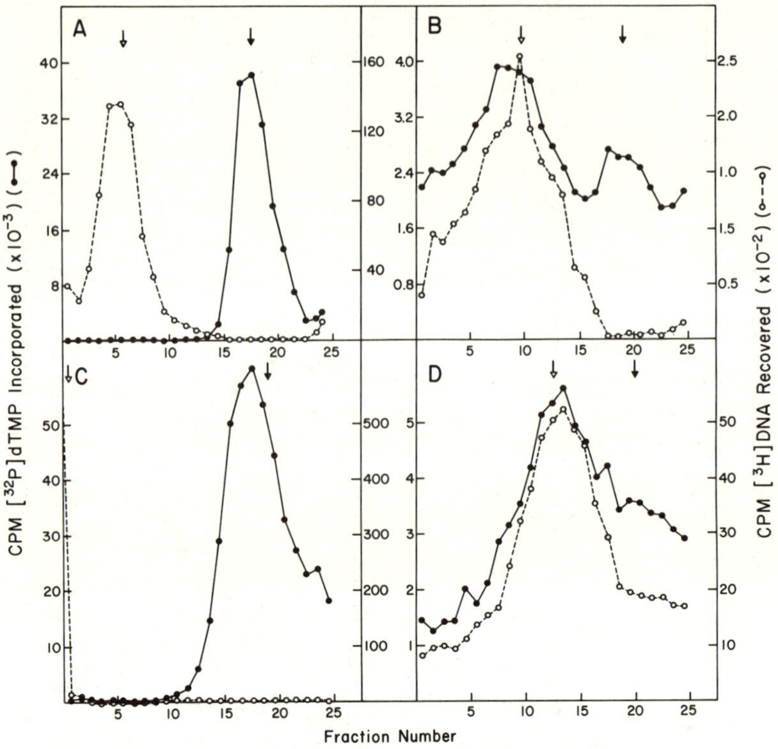

Fig. 2. Glycerol gradient sedimentation analysis of the binding of polymerase-α to DNA[9]. Glycerol gradients (20 to 40% (v/v) glycerol) were formulated as described[9]; identical results were obtained in the presence and absence of 4mM $MgCl_2$ and 20μM dNTPs. In all panels, the position of DNA sedimented alone is indicated by the open arrow; the position of polymerase-α alone, by the closed arrow. Panel A: Co-sedimentation of 114μg Form IV PM2 [^{3}H]DNA with 5 units of KB polymerase-α Fraction VIII[1]. Panel B: Co-sedimentation of 8.4μg M13 [^{3}H]DNA with 2.5 units of polymerase-α Fraction VIII. Panel C: Co-sedimentation of 9μg duplex T7 [^{3}H]DNA with 5 units of polymerase-α Fraction VIII. (The T7 DNA pelleted under these conditions and was recovered at the bottom of the tube). Panel D: Co-sedimentation of 9μg of heat-denatured T7 [^{3}H]DNA with 2.5 units of polymerase-α Fraction VIII. Recovery of loaded enzyme activity in all panels was 50 to 90%.

1.5 and 2.0. On the basis of these results, we concluded[9] that the interaction of polymerase-α with DNA is absolutely dependent on an initial binding reaction with single-stranded polydeoxynucleotide (template), and we posited[9] that each catalytically active molecule of KB cell polymerase-α must possess at least 2 strongly interactive (positively cooperative) single-stranded DNA binding sites. We further noted that both the cooperativity of single-stranded DNA binding and the apparent dependence of primer-terminus recognition on antecedent single-strand (template) binding were formally compatible with the behavior of a conformationally active protein, with which the initial binding of template at one site might be required for the (allosteric) activation of both a second template binding site as well as a primer binding site.

During the course of these experiments, we became aware of the ability of polymerase-α to catalyze deoxynucleotide incorporation on single-stranded DNA fragments by a self-priming, snap-back mechan-

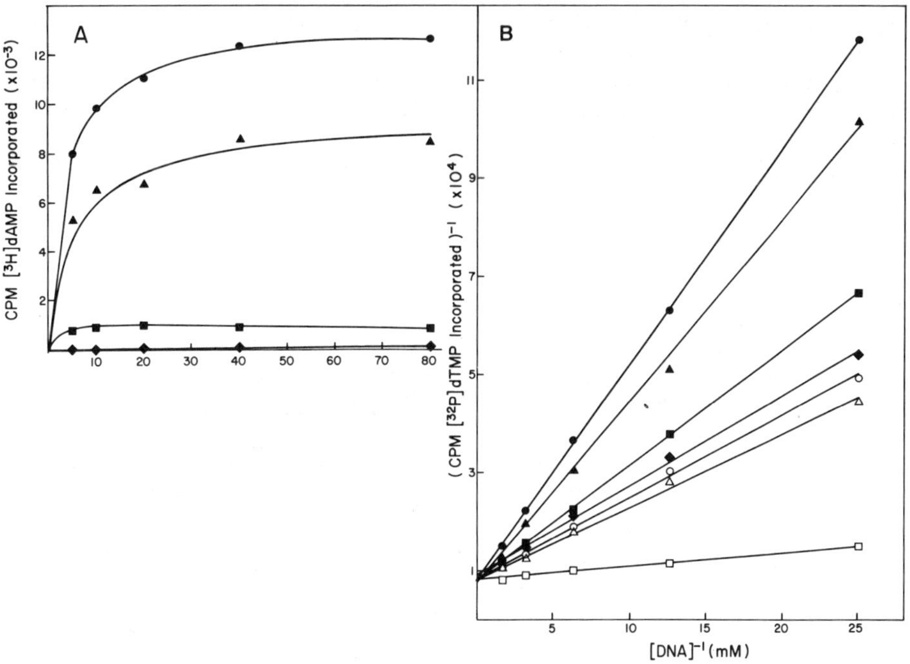

Fig. 3. Reaction of DNA polymerase-α with synthetic hook polymers, dT_n-dA_m[10]. Panel A: Polymerization reaction with $(dT)_{\overline{200}}$-$(dA)_{\overline{0.24}}$ (♦); $(dT)_{\overline{200}}$-$(dA)_{\overline{0.85}}$ (■); $(dT)_{\overline{200}}$ $(dA)_{\overline{3.3}}$ (▲); and $(dT)_{\overline{200}}$-$dA)_{\overline{6.8}}$ (●). Panel B: Effect of the synthetic polymers as inhibitors of the polymerization reaction with activated DNA. The data are presented as Lineweaver-Burk plots; the several synthetic polymer inhibitors were added to each series of reactions at final concentrations of 20μM (nucleotide). The filled symbols are used as in panel A. The other symbols are: (□), control; (△), $(dT)_{\overline{200}}$; (○), $(dT)_{\overline{200}}$-$(ddT)_1$ [(ddT) : dideoxythymidylate].

ism that we have characterized in detail[10]. Two observations reported in that study are particularly relevant to the present discussion because they revealed important additional features of the polymerase-nucleic acid binding reaction. First, by examining the kinetics of the interaction between polymerase-α and a set of synthetic hook polymers of the structure dT_n-dA_m (which could be tested either as polymerase substrates or as alternate product inhibitors of the reaction with activated DNA) (Fig. 3), we showed that the requirement for primer recognition by the template-bound polymerase could be satisfied by only a minimal degree of potential 3'-terminal base-pairing with template (for A-T base-pairs at 37°C, $2 \leq m \leq 5$); preformed, stably base-paired primer termini were clearly not necessary. Second, we observed that dT_n was itself an unusually potent inhibitor of polymerase-α, with significantly higher apparent affinity for the enzyme than, for example, was exhibited by single-stranded hetero-polymeric DNA. Preliminary investigation indicated that this enhanced affinity of dT_n was not a general property of homopolymers, and we thus considered for the first time the possibility that the polymerase-α-template binding reaction might in some way be modulated by signals that were template sequence-determined.

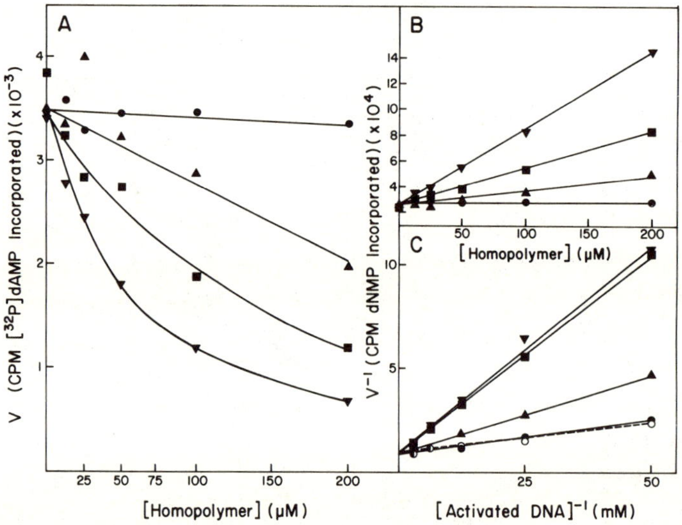

Fig. 4. Effect of unprimed homopolymers on the reaction of DNA polymerase-α with activated DNA[11]. Panel A: Reactions were carried out under standard conditions, with 500μM activated salmon sperm DNA and homopolymer at the concentrations indicated. (All polymer concentrations are expressed in terms of nucleotide.) (●), poly(dA); (▲), poly(dC); (■), poly(dG); (▼), poly(dT). Panel B: Dixon plot of the data in panel A. Panel C: Lineweaver-Burk plot of the data in panel A; (○), control reaction without homopolymers.

To examine this hypothesis, we carried out an extensive, quantitative analysis[11] of the interaction of DNA polymerase-α with a variety of synthetic polymers that we synthesized ourselves to ensure that they were of comparable average lengths and defined base composition. Thus, the homopolymers used were of the general structure $(dX)_{\overline{100}}$; the hook polymers, $(dX)_{\overline{100}}$-$(dY)_{\overline{25}}$; and the heteropolymers (all combinations of tripolymer and the single tetrapolymer) had a random base composition and ranged in average lengths from 47- to 77-nucleotides. The results of these experiments are presented in Figures 4-7 and Table 1, and they can be summarized as follows:

(1) The affinity of DNA polymerase-α for single-stranded polydeoxynucleotide molecules is dependent on their base composition. Thus, the polymerase exhibits a complex hierarchy of affinities both for deoxyhomopolymers (Fig. 4) (affinity for poly(dT)≥ poly(dG)>poly(dC)≫poly(dA)) and for four random heterotripolymers (Fig. 5) (affinity for poly(dA, dG, dT)>poly(dA, dC, dT)>poly(dG, dC, dT)>poly(dA, dG, dC)). Although the results with the heterotripolymers indicate that A,T-rich polymers are preferred to those containing G and C; that A,T-rich polymers contain G as the third

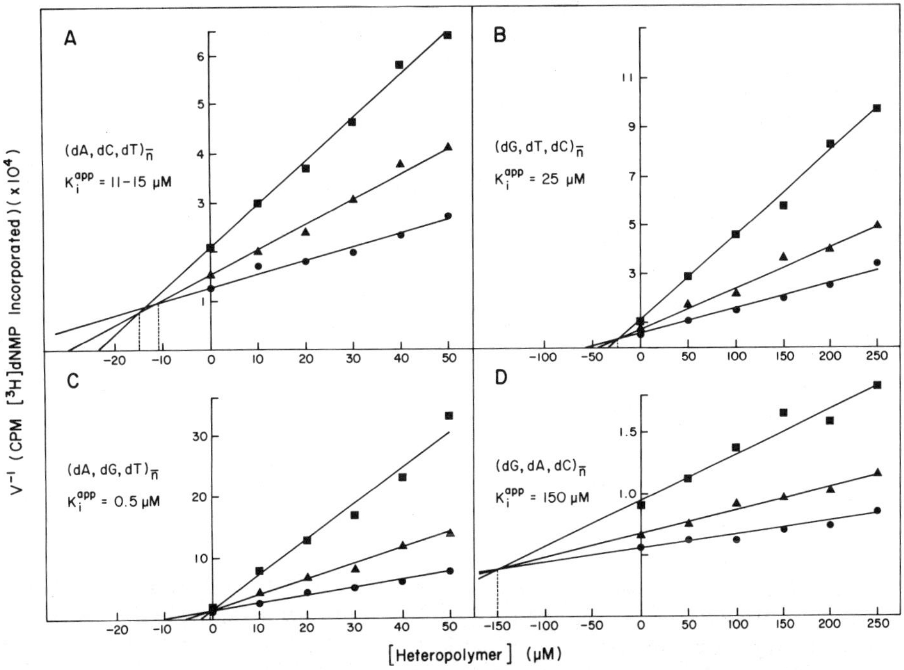

Fig. 5. Effect of synthetic heterotripolymers on the reaction of DNA polymerase-α with activated DNA[11]. The data presented as Dixon plots; substrate DNA concentrations in each panel were as 1X, 2X, 4X. The mean lengths (n) of the several synthetic polymers ranged from 47- to 77-nucleotides.

base are preferred to those containing C; and that G, C-rich polymers containing T are more avidly bound than those containing A; overall, the affinities for the heterotripolymers do not reflect a simple summation of homopolymer affinities. At the extremes, values of apparent K_i (expressed in terms of nucleotide concentration) range over 2 to 3 orders of magnitude; e.g., poly(dT), $K_i = 4\mu M$; poly(dA), $K_i \geq 750\mu M$; poly(dA, dG, dT), $K_i \leq 1\mu M$; poly(dA, dG, dC), $K_i = 150\mu M$. The validity of the interpretation

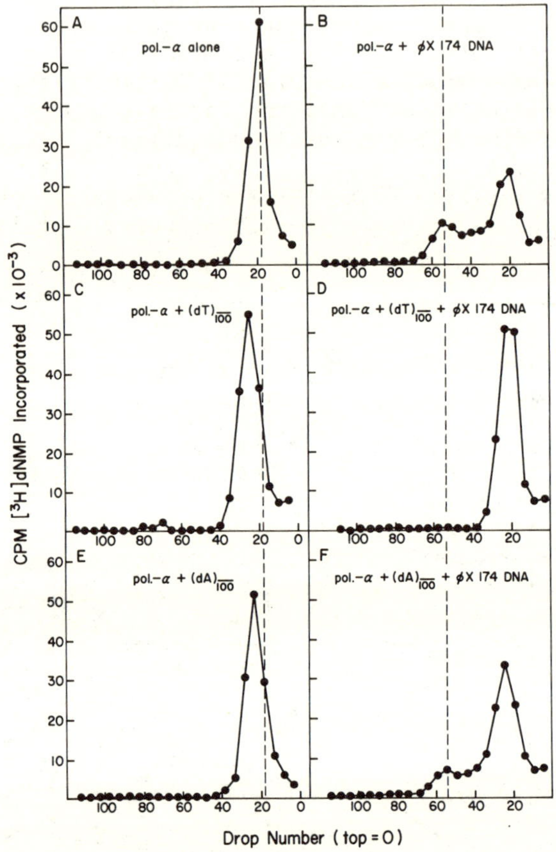

Fig. 6. Glycerol gradient sedimentation analysis of the specificity of the binding interaction between DNA polymerase-α and polydeoxynucleotides[11]. Velocity sedimentation was performed in linear 20-40%(v/v) glycerol gradients, as described[11]. The broken vertical line in panels A, C and E marks the position of DNA polymerase-α sedimented alone (∼7S); the similar line in panels B, D and F marks the position of φX 174 DNA alone (∼17S). For details of load composition and polymerase recovery, see Table 1.

Table 1. Sedimentation Analysis of the Specificity of the Binding Interaction between DNA Polymerase-α and Nucleic Acids[11].

Nucleic Acids in Loaded Sample	Recovery of Polymerase Activity (% initial load)	Position of Recovered Polymerase Activity (% total recovered)	
		0 - 14S	15 - 29S
None	36	100	0
$(dT)_{100}$ (500μM)	50	98	2
$(dA)_{100}$ (500μM)	38	100	0
$(dA, dG, dC)_n$ (40μM)	46	100	0
$(dA, dG, dT)_n$ (40μM)	37	100	0
ϕX 174 DNA (500μM)	39	70	30
ϕX 174 DNA (500μM) + $(dT)_{100}$ (500μM)	46	99	1
ϕX 174 DNA (500μM) + $(dA)_{100}$ (500μM)	38	75	25
ϕX 174 DNA (500μM) + $(dA, dG, dC)_n$ (40μM)	38	68	32
ϕX 174 DNA (500μM) + $(dA, dG, dT)_n$ (40μM)	38	87	13

Velocity gradient sedimentation was performed as in Figure 6. The position of recovered enzyme activity is assigned either to the region of the gradient (0-14S) that envelopes the polymerase activity when sedimented alone, or to the region (15-29S) that includes the profile of ϕX 174 DNA when sedimented alone.

Because of the small size of the synthetic polymers, their binding to the enzyme does not distort the polymerase activity peak and therefore can not be detected directly by this method. The assay is a competition binding assay which measures the effect of the several synthetic polymers on the binding reaction between polymerase-α and ϕX 174 DNA.

that these kinetic measurements do in fact reflect specificity of polymerase-nucleic acid binding interactions has been established unambiguously by semiquantitative "competition-binding" assays, as illustraded in Figure 6 and summarized in Table 1. The principle of these "competition-binding" assays is to compare the ability of the various synthetic polymers to compete for the binding of polymerase-α with ϕX 174 DNA, present at concentrations (nucleotide) ranging from equimolar to 12.5-fold that of the competing polymer. In each instance, the results of the direct binding assays have been entirely consistent with those of the kinetics experiments.

(2) The affinity of polymerase-α for any given DNA template is variable and dependent on the presence or absence of specific competing templates of differing base composition. For example, while poly(dA) has no demonstrable effect on the replication of activated DNA ($K_i \geq 750 \mu M$) (Fig. 4), it behaves as an effective, linearly competitive inhibitor with the substrate $(dA)_{\overline{100}}(dT_{\overline{25}}$ ($K_i \approx 50 \mu M$)[11]; conversely, single-stranded, circular ϕX 174 DNA (Fig. 7), a moderately potent inhibitor of the polymerization reaction with activated DNA ($K_i = 30 \mu M$) has little effect on the replication of $(dT)_{\overline{100}}(dA)_{\overline{25}}$ ($K_i \geq 400 \mu M$) but is an extremely effective competitive inhibitor of the reaction with $(dA)_{\overline{100}}$-$(dT)_{\overline{25}}$($K_i \leq 1 \mu M$). Since the measured binding affinity (K_i) of an enzyme for any ligand competing simply with substrate for a single site or multiple independent binding sites on the protein is a constant, independent of the affinity of enzyme for substrate, and thus of the particular substrate used to assess K_i, such dramatic modulation of the affinity of polymerase-α for one single-stranded polydeoxynucleotide by a second of different base composition requires that each enzyme molecule possess a minimum of two interactive template binding sites. This conclusion is in agreement with our interpretation of the cooperativity of inhibition of polymerase-α by single-stranded, natural heteropolymeric DNA molecules, and it further specifies that base composition (or sequence) must be a primary factor regulating the polymerase-nucleic acid interaction.

On the basis of these several kinetics studies, we proposed the following tentative model of the interaction of KB cell DNA polymerase-α with nucleic acids[11]: The initial binding of the polymerase to DNA is absolutely dependent on single-strand (template) and leads to the allosteric activation of a second template binding site and of the primer binding site. Since we have not observed evidence of cooperativity in the polymerization reaction itself, we have postulated that both template binding sites can not be catalytically active simultaneously (half-of-the-sites reactivity); thus the polymerization reaction exhibits simple saturation kinetics.

In contrast, with respect to inhibition of enzyme activity, since either binding site may be active, two molecules of inhibitor must be bound on each enzyme molecule to block both sites and thereby prevent polymerization. The interaction between the two template binding sites is such that binding of a specific polydeoxynucleotide at one site acts to enhance, either relatively or absolutely, the affinity of the second site for a polymer of similar base composition. When inhibitor (unprimed) template and substrate (primed) template have the same or very similar mean base composition, simple competitive inhibition kinetics are observed. In contrast, when inhibitor and substrate polydeoxynucleotides are of different base composition, there are two possible outcomes. Thus, with two different heterotetrapolymeric molecules (e.g., M13 and KB DNAs), the mean about which the various forms of the polymerase are distributed must change in response to the changing mean base composition of the nucleic acid pool. Because these distributions of base composition are broad, the changes in the polymerase occur gradually over the entire nucleic

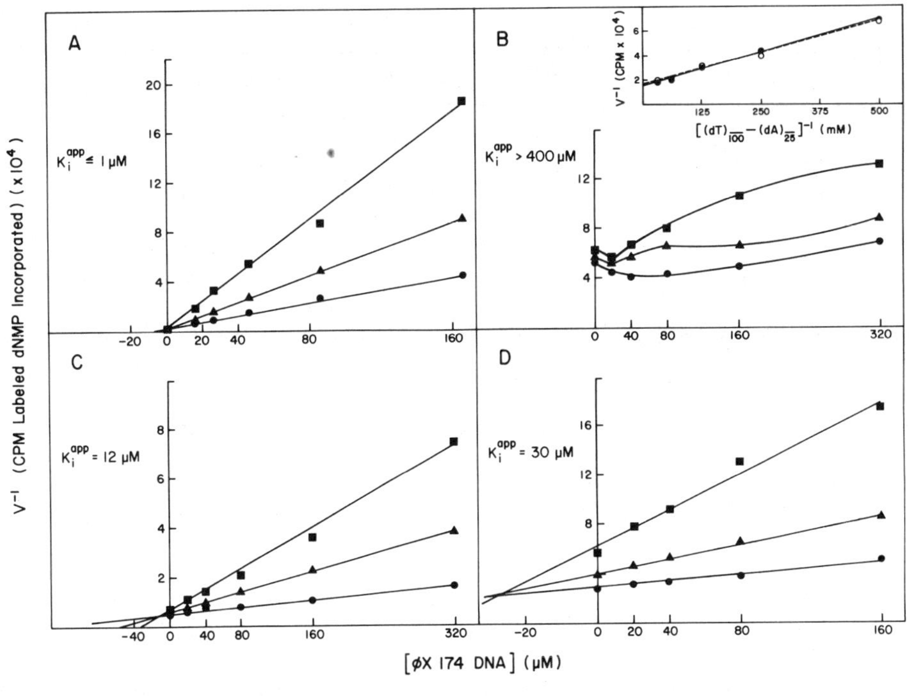

Fig. 7. Effect of primer-template on the apparent kinetic affinity of DNA polymerase-α for inhibitor ϕX 174 DNA[11]. The data are presented as Dixon plots; the concentration of primer-template in each panel was as 1X, 2X, 4X. The primer-templates used were: Panel A, (dA)$_{\overline{100}}$-(dT$_{\overline{25}}$; Panel B, (dT)$_{\overline{100}}$-(dA)$_{\overline{25}}$; the inset in panel B is a Lineweaver-Burk plot of data obtained in reactions that contained ϕX 174 DNA at 0μM (●) or 20μM (○). Panel C, (dC)$_{\overline{100}}$-(dG)$_{\overline{25}}$: Panel D, poly(dA, dG, dT, dC), a random heterotetrapolymer.

acid concentration range and are reflected in nonlinear kinetics of inhibition that yield Hill plot slopes that approach a value of 2. On the other hand, with competing single-stranded polymers of vastly different base composition, linear inhibition kinetics are observed, but values of K_i (for specific dead-end inhibitors) are found to vary over several orders of magnitude. While polymerase transitions must occur to produce these dramatic changes in K_i, the transitions are apparently so sharp as to prevent their resolution in these experiments.

II. The Ordered Sequential Mechanism of Substrate Recognition and Binding by KB Cell DNA Polymerase-α

To pursue some of the implications of the results summarized in the preceeding Section concerning primer-template recognition by DNA polymerase-α , and to explore the possible relationship of those nucleic acid binding events to the important mechanistic problem of appropriate dNTP recognition and insertion, we have carried out a formal steady-state kinetics analysis[12,13] of the interaction of polymerase-α with its several substrates[14].

Although the data reviewed thus far clearly indicated that template and primer behave as mechanistically distinct substrates for KB cell polymerase-α, and thus that we were dealing with a terreactant system (i.e., template, primer, dNTP), we initially examined the kinetic pattern of substrate addition to the enzyme by using unimolecular primer-templates, which could be treated as a single substrate and varied together at constant fixed ratio for comparison with the third substrate, dNTP. These analyses generated[14] linear converging patterns, and linear slope and intercept replots, that were characteristic of a classical sequential reaction mechanism and excluded such alternative possibilities as a ping-pong mechanism or an equilibrium-ordered mechanism. These results did not in themselves, however, permit distinction between random and ordered sequential mechanisms of substrate addition to the polymerase.

To distinguish between these two possibilities, we exploited the chance observation[14] that 2',3'-dideoxyterminated synthetic hook polymers (e.g., $(dX)_n$-$(dY)_m$-$(ddY)_1$) appeared to be more potent inhibitors of DNA polymerase-α than their 3'-hydroxylterminated homologues (e.g., $(dX)_n$-$(dY)_m$). We demonstrated first that the inhibitory potency of both classes of hook polymers is strongly dependent on the presence of the dNTP that is complementary to the template base, but in completely opposite directions (Fig. 8). Thus, for example, in the presence of substrates $(dA)_{\overline{100}}$-$(dT)_{\overline{25}}$ plus dTTP, the inhibitory potency of the hook polymer $(dC)_{\overline{100}}$-$(dG)_{\overline{25}}$ is significantly reduced in the presence specifically of dGTP (Fig. 8A); while the inhibitory potency of the dideoxy-blocked homologue $(dC)_{\overline{100}}$-$(dG)_{\overline{25}}$-$(ddG)_1$ is dramatically increased specifically in response to dGTP (Fig. 8B).

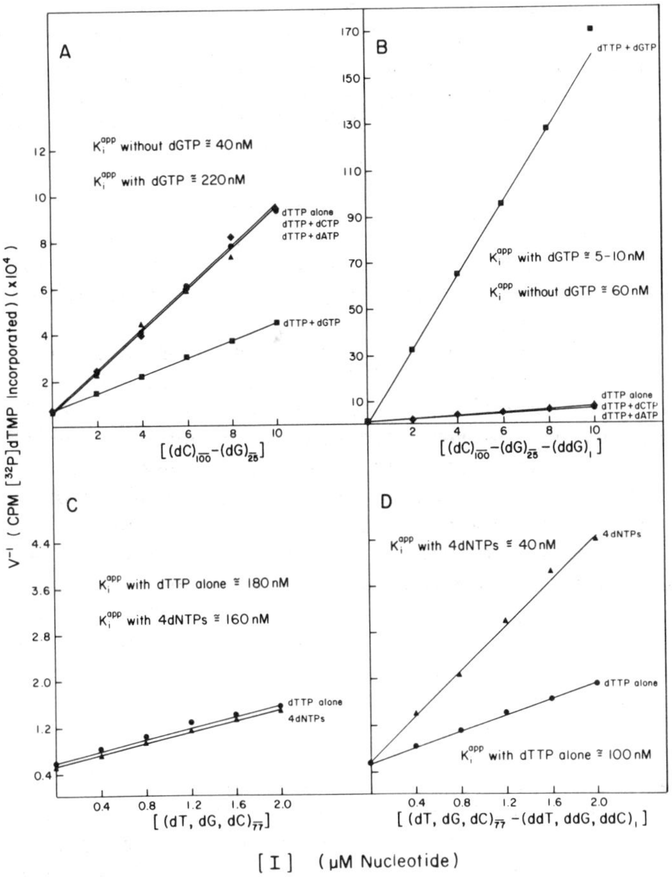

Fig. 8. Effect of complementary and noncomplementary dNTPs on the inhibitory potency of 3'-hydroxyl-terminated and 3'-dideoxy-terminated synthetic polymers. Reactions were carried out as described[14]. The data are displayed as Dixon plots; the concentrations of inhibitor polymers are indicated on the abscissae. Panels A and B: The substrates were $(dA)_{\overline{100}}$-$(dT)_{\overline{25}}$ and dTTP. Additional dNTPs were added as follows: (●) none; (■),dGTP; (▲), dATP; (◆) dCTP. Note the difference in the ordinate scales in panels A and B. Panels C and D: The substrates were $(dA)_{\overline{100}}$-$(dT)_{\overline{25}}$ and dTTP. Additional dNTPs were added as follows: (●), none; (▲) dATP, dGTP and dCTP. Reprinted from Fisher and Korn[14] with permission of American Chemical Society.

Qualitatively identical results were obtained with other pairs of 3'-hydroxyl- and 3'dideoxy-terminated homo- and heteropolymers (e.g., Fig. 8C, D). We had previously noted[11] that in the absence of an inhibitor polymer, the replication of any given hook polymer by DNA polymerase-α is unaffected by the presence in the reaction of the noncomplementary dNTPs; and we could show[14] that the inhibitory potency of single-stranded polydeoxynucleotides alone, i.e., without a base-paired primer moiety, was similarly completely independent of which dNTPs were present in the incubations.

The results presented in Figure 8 are compatible with the following interpretation. The 3'-hydroxyl-terminated inhibitor hook polymers (I) are capable of serving as alternate substrates for polymerase-α, and thus as alternate product inhibitors, if and only if the dNTP complementary to the inhibitor template is present in the reaction. If, under those conditions, the forward pathway to alternate product is rapid relative to the backward dissociation of the E·I complex formed in the absence of (complementary) dNTP, then the availability of the required dNTP will open an additional fast pathway to free enzyme and result in a decreased inhibitory potency of I. In contrast, the 3'dideoxy-terminated inhibitor hook polymer (ddI) can not serve as a polymerase substrate, whether or not the inhibitor-complementary dNTP is present. However, if the sequential reaction sequence is ordered, i.e., if the inhibitor polymer must bind to enzyme prior to the addition of dNTP, and if the E·ddI complex is capable of binding complementary dNTP and thus proceeding one step further down the reaction pathway to E·ddI·dNTP, the result will be a dead-end complex from which the backward dissociation to generate free enzyme will now be that much more difficult.

This scheme explains both the decreased inhibitory potency of I and the increased inhibitory potency of ddI, each demonstrable exclusively in the presence of the inhibitor-complementary dNTP; it also predicts the phenomenon of induced substrate inhibition with ddI, which is "highly diagnostic of ordered mechanisms"[12], and which we could readily demonstrate (Fig. 9) and show, as required by the ordered sequential mechanism of substrate binding[12], to be classically competitive with DNA primer-template (and not, for example, with dNTP) (Fig. 10). With respect to the experiments in Figure 9, it is most important to note that when unprimed single-stranded polydeoxynucleotide inhibitors ($(dA)_{\overline{100}}$, $(dC)_{\overline{100}}$, or φX 174 DNA) are mixed with the corresponding primer-templates, there is no substrate induced inhibition at increasing levels of dNTP; and the pattern of inhibition produced by the unprimed polymers is linearly noncompetitive with respect to dNTP, as expected[14]. Finally, we have also demonstrated that induced substrate inhibition can be driven to completion by saturating levels of inhibitor dNTP[14], thus documenting that the ordered sequence of substrate addition is quantitatively the only available reaction pathway(i.e., a "rigidly ordered" mechanism) and not simply a relatively preferred route with less favored alternatives also likely (i.e., a random mechanism with a preferred binding order).

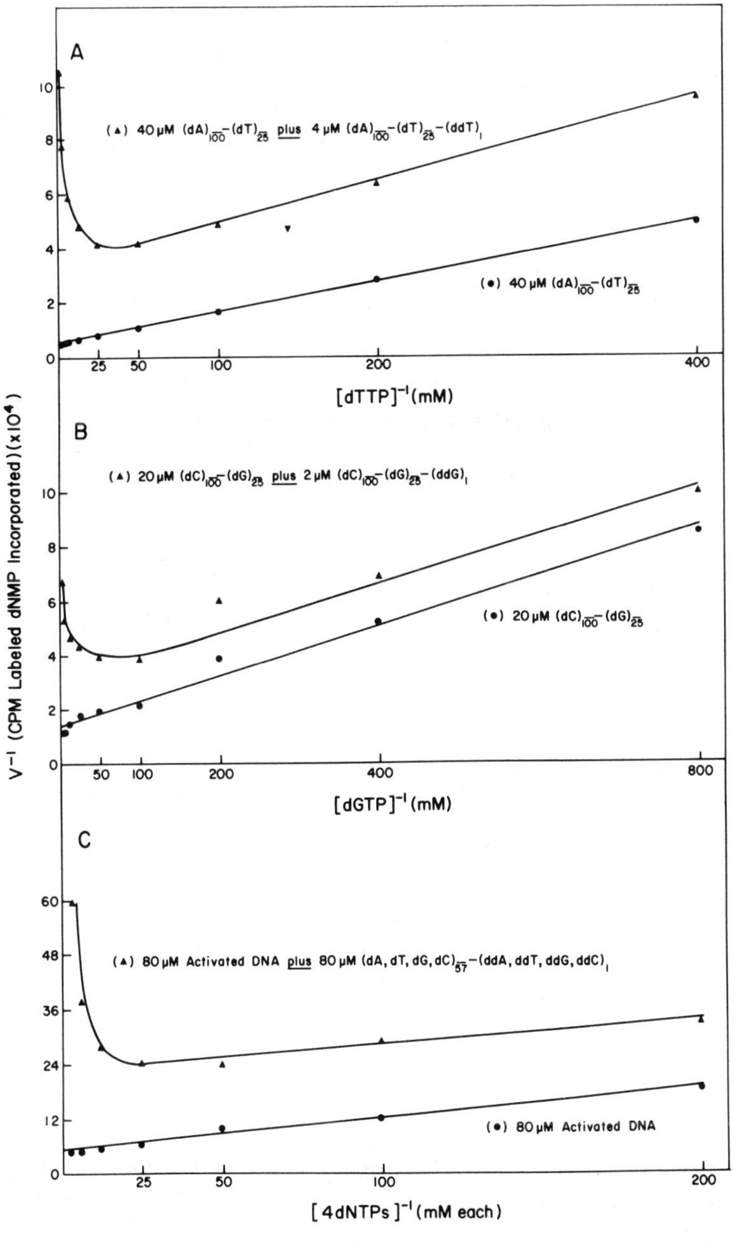

Fig. 9. Dideoxyprimer-induced substrate (dNTP) inhibition of DNA polymerase-α activity. The data are presented as Lineweaver-Burk plots. Reprinted from Fisher and Korn[14] with permission of American Chemical Society.

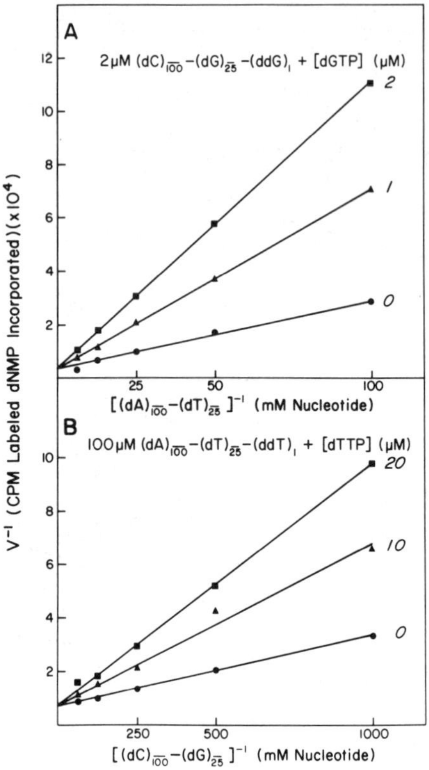

Fig. 10. Dideoxyprimer-induced substrate (dNTP) inhibition of DNA polymerase-α is classically competitive with DNA primer-template. The data are presented as Lineweaver-Burk plots. Panel A: The substrates were $(dA)_{\overline{100}}$-$(dT)_{\overline{25}}$ plus dTTP. Panel B: The substrates were $(dC)_{\overline{100}}$-$(dG)_{\overline{25}}$ plus dGTP. Reprinted from Fisher and Korn[14] with permission of American Chemical Society.

As in the previous kinetics studies, we have employed direct sedimentation binding assays to corroborate the principal kinetic interpretations. An example of one such confirmatory study is presented in Figure 11, in which "competition-binding" assays were used to examine the effects of specific complementary and noncomplementary dNTPs on the ability of two homologous hook polymers, $(dC)_{\overline{100}}$-$(dG)_{\overline{25}}$ and $(dC)_{\overline{100}}$-$(dG)_{\overline{25}}$-$(ddG)_1$, to compete with ϕX 174 DNA as ligands for polymerase-α. The experiment documents several very important points. Panels A-C show that the binding of the polymerase to unprimed template (ϕX 174 DNA) alone is unaffected by the presence of dNTPs, thus confirming the fact that primer is required for any dNTP effect on the enzyme-nucleic acid interaction. (The profiles in panels A-C also illustrate the extremely good reproducibility of the assay).

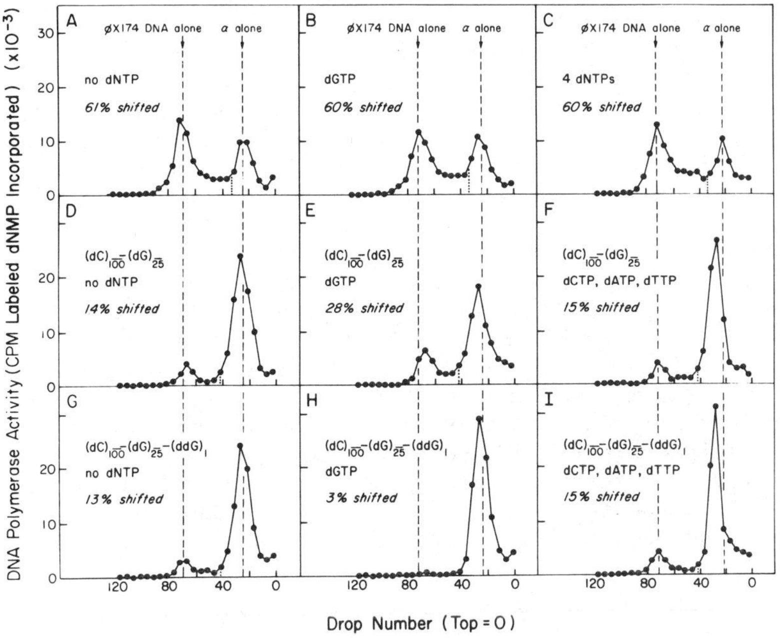

Fig. 11. Competitive sedimentation binding analysis of the interaction of DNA polymerase-α with φX 174 DNA versus 3'-hydroxyl and 2', 3'-dideoxy hook polymers. All loaded samples contained 1mM (nucleotide) φX 174 DNA and hook polymer competitors as indicated on the panels. Both loads and gradients contained the specified dNTPs, each at a final concentration of 100μM. "% shifted" refers to the % of total recovered DNA polymerase activity that had shifted down the gradient from the position of polymerase alone (7S). Reprinted from Fisher and Korn[14] with permission of American Chemical Society.

Panels D-F show that in the absence of dNTPs, or in the presence of three noncomplementary dNTPs, the polymer $(dC)_{\overline{100}}-(dG)_{\overline{25}}$ (3-4S), present in the load at 12.5μM (nucleotide), competes very effectively with φX 174 DNA (at 1mM of nucleotide) such that only about 25% of the polymerase that originally sedimented with the circular DNA molecules continues to do so. In contrast, in the presence specifically of dGTP, the ability of the hook polymer to compete with φX 174 DNA is substantially reduced, leading to a 2-fold increase in the amount of polymerase that sediments with the circular DNA. Panels G-I demonstrate that in the absence of dNTPs, or in the presence of three noncomplementary dNTPs, the blocked polymer, $(dC)_{\overline{100}}-(dG)_{\overline{25}}-(ddG)_1$, is essentially identical to its unblocked homologue in its capacity to compete with φX 174 DNA for the binding of polymerase-α. In the presence specifically of dGTP, however, the binding of the dideoxy polymer by DNA polymerase-α is dramatically enhanced, as evidenced by the almost complete elimination of binding to φX 174 DNA.

By these and other sedimentation binding assays[14], we have thus established that stable, enzyme·dideoxyprimer·template complexes, that are absolutely and specifically dependent on complementary dNTPs, can be identified and quantitated by direct physical means; and in every instance, the results of these binding assays have been in precise agreement with the conclusions drawn from the corresponding kinetics experiments (e.g., Fig. 11 and Figs. 8A, B).

From the studies reviewed in this Section, we have been able to conclude[14] that the interaction of KB cell DNA polymerase-α with its substrates obeys the rigidly ordered sequential terreactant mechanism that is outlined in Figure 12. This scheme prescribes that the polymerase must first bind single-stranded polydeoxynucleotide (template), then primer stem, and finally dNTP in a stricly ordered sequence; only the correct triphosphate that is dictated by the template sequence can participate in the dNTP-binding step. Moreover, these data provide for the first time direct proof, both kinetic and physicochemical, of the template-base specification and the primer dependence of dNTP binding. It should also be noted that it is an important

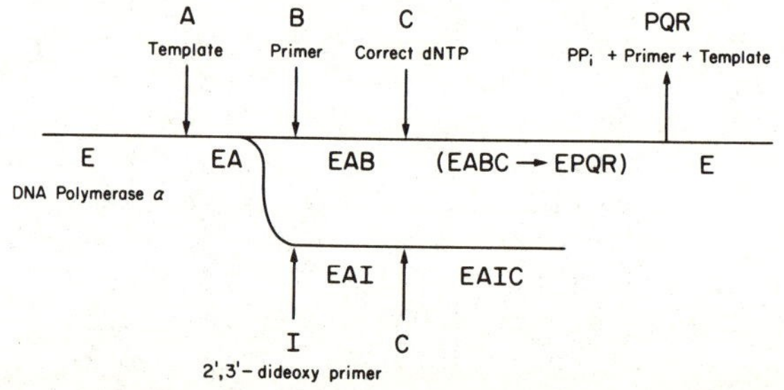

$$E \underset{k_2}{\overset{k_1 A}{\rightleftharpoons}} E\cdot Template \underset{k_4}{\overset{k_3 B}{\rightleftharpoons}} E\cdot Template\cdot Primer \underset{k_6}{\overset{k_5 C}{\rightleftharpoons}} E\cdot Template\cdot Primer\cdot dNTP \xrightarrow{k_7} E + Products$$

Fig. 12. Partial steady-state kinetic model of KB cell DNA polymerase-α. The diagram has been constructed according to the convention of Cleland[15]. Products are grouped together due to lack of information regarding the order of product release and incomplete understanding of polymerase processivity. In the kinetic formulation beneath the diagram, K_7 represents a combination of the several rate constants that characterize individual steps of product generation and release. Note that for the second step, the concentration of primer, B, refers explicitly to that within the partial specific volume of the template to which it is annealed. Reprinted from Fisher and Korn[14] with permission of American Chemical Society.

implication of the ordered mechanism of reactant addition that the binding of each substrate to the enzyme creates, possibly by induction of conformational alterations, the kinetically significant binding site for the substrate that follows. These results thus provide substantial support for our earlier proposal[11] that KB cell DNA polymerase-α is a conformationally active protein that can recognize and respond to template base sequence through its interactions with single-stranded polydeoxynucleotides in its template binding site(s).

III. Properties of the Primer Binding Site and the Role of Mg^{+2} in Primer-Template Recognition by DNA Polymerase-α

In this Section we briefly review a series of studies[15] that have provided important insights into the mechanism of primer recognition by DNA polymerase-α and have clarified the role of Mg^{+2} cation in both the template-binding and the primer-binding steps of the polymerase-nucleic acid interaction. We have earlier noted (Section I) that polymerase-α is unable to bind 3'-PO_4 primer termini, and we have subsequently described further experiments[15] which indicate first, that the enzyme interacts in a kinetically indistinguishable manner with 2'-deoxyribosyl- and ribosyl-terminated primer stems (as well as with 2', 3'-dideoxy termini, as reviewed in Section II); second, that the primer-binding interaction appears to be unaffected by the presence of H, OH or PO_4 in the 2'-terminal position; and third, that efficient priming with a bimolecular primer·template requires an octadeoxynucleotide, and this length requirement is independent of base composition (i.e., A-T versus G-C) and temperature (i.e., 23°C versus 37°C).

In spite of this apparent permissivity of the enzyme with respect to the structure of the 3'-terminal primer sugar residue, we have been able to demonstrate that primer binding requires that the 3'-terminal primer nucleotide must be properly base-paired with template; i.e., a single terminally mismatched deoxynucleotide will prevent both correct primer binding as well as the subsequent step of complementary dNTP addition to the polymerase. The experiment, which is presented in Figure 13, is based on our previous description (Section II) of the inhibitory effects of 3'OH- and 3'-dideoxy-terminated hook polymers on the reactivity of polymerase-α, and examines the kinetic behavior of inhibitor hook polymers that contain a single, 3'-terminal mismatched dideoxynucleotide. We compared the hook polymers $(dC)_{\overline{100}}$-$(dG)_{\overline{25}}$, $(dC)_{\overline{100}}$-$(dG)_{\overline{25}}$-$(ddG)_1$, and $(dC)_{\overline{100}}$-$(dG)_{\overline{25}}$-$(ddA, ddT, ddC)_1$ with respect to their ability to inhibit the polymerization reaction with $(dA)_{\overline{100}}$-$(dT)_{25}$ plus dTTP, and for their ability to induce substrate(dNTP) inhibition by inhibitor-complementary dGTP. In the absence of all 3 non-substrate dNTPs (Fig. 13A), or in the presence of the inhibitor noncomplementary dNTPs, dATP (Fig. 13C) or dCTP (Fig. 13D), the inhibitory potency of the mismatched polymer was clearly less than that of its two homologues;

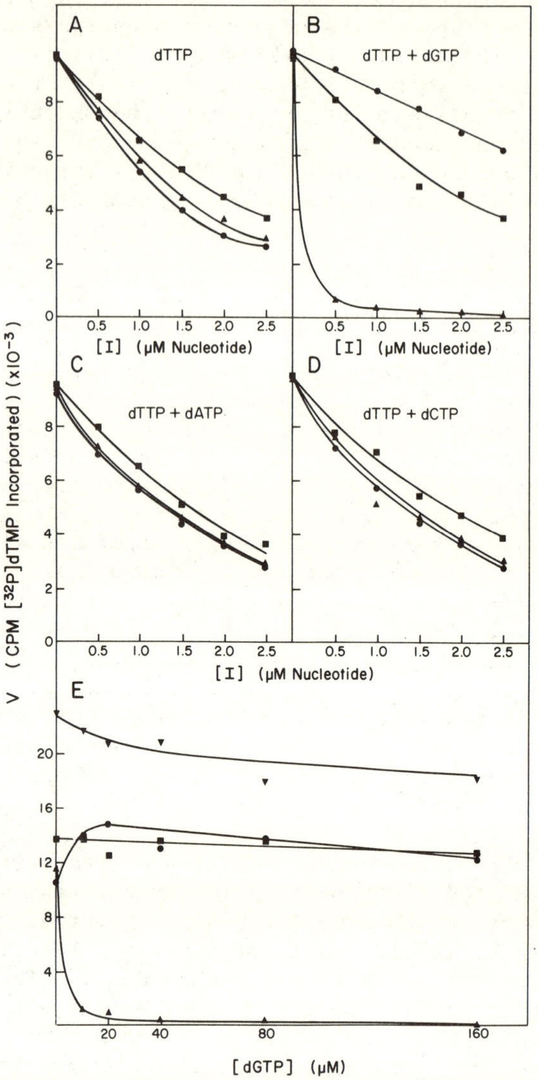

Fig. 13. Effect of complementary and noncomplementary dNTPs on inhibition of DNA polymerase-α by $(dC)_{\overline{100}}$-$(dG)_{\overline{25}}$ (●), $(dC)_{\overline{100}}$-$(dG)_{\overline{25}}$-$(ddG)_1$ (▲), and $(dC)_{\overline{100}}$-$(dG)_{\overline{25}}$-(ddA, ddT, ddC)$_1$ (■). The data in panels A-D are presented as plots of initial velocity (v) versus hook polymer inhibitor concentration (m); in all incubations, the substrates were $(dA)_{\overline{100}}$-$(dT)_{\overline{25}}$ plus dTTP. Panel (E): Activity (v) versus dGTP concentration. (▼), no nucleic acid inhibitor; the other symbols are as defined above. Reprinted from Fisher and Korn[15] with permission of American Chemical Society.

it was in fact comparable to that of $(dC)_{\overline{100}}$ alone, suggesting minimal if any contribution of the terminally mispaired primer moiety to the kinetically measured binding affinity. The addition specifically of dGTP (Fig. 13B) led to the expected decrease (5-fold) in inhibitory potency of $(dC)_{\overline{100}}$-$(dG)_{25}$ and increase (10-fold) in inhibitory potency of $(dC)_{\overline{100}}$-$(dG)_{\overline{25}}$-$(ddG)_1$ but was without significant effect on the inhibitory capacity of the mismatched polymer. Finally, in Figure 13E we show that the mismatched polymer is unable to induce substrate inhibition by dGTP, even at dGTP concentrations 8-fold greater than those required for complete inhibition of the polymerase in the presence of $(dC)_{\overline{100}}$-$(dG)_{\overline{25}}$-$(ddG)_1$; indeed, the mismatched polymer behaves in this experiment identically to the unprimed template, $(dC)_{\overline{100}}$.

In Section II we presented compelling evidence that the binding of complementary dNTP to polymerase-α is template-sequence directed and primer-requiring. The results depicted in Figure 13 further specify that the primer-binding site itself must exhibit remarkable discrimination for correct pairing of (at least) the terminal base, and they identify an additional step in the polymerase reaction mechanism that must certainly contribute to the fidelity of replication.

We turn our attention next to the role of Mg^{+2} in the polymerase-α-primer-template interaction. By measuring values of apparent K_i for unprimed single-stranded polydeoxynucleotide inhibitors at varying concentrations of Mg^{+2}, we observed[15] that increasing concentrations of the cation over a narrow range (2-10mM) were without effect on the apparent kinetic affinity of the enzyme for purine polydeoxynucleotides but led to dramatically enhanced kinetic affinity for pyrimidine homodeoxypolymers and to moderately enhanced affinity for natural heteropolymeric single-stranded DNA. Demonstration of this effect by direct binding assay (and another illustration of the excelent correlation that can be obtained between the kinetics and sedimentation assays) is shown in Figure 14; note particularly (Fig. 14D) the very close agreement between changes in bound polymerase activity measured directly by sedimentation and changes in kinetically estimated binding affinities (apparent K_i), both as a function of Mg^{+2} concentration. From these and similar experiments[15], we have concluded that exogenous Mg^{+2} enhances the binding of polymerase-α to pyrimidine homodeoxypolymers and single-stranded DNA, although it does not appear to be required for the binding of either.

These effects of Mg^{+2} on template binding by DNA polymerase-α led to the reasonable predictions that as a consequence of enhanced nonproductive binding of enzyme at template sites distant from primer termini, increasing Mg^{+2} concentrations (2-10mM) would (1), be inhibitory to the replication of heteropolymeric and pyrimidine polydeoxynucleotide templates, but not of exclusively purine containing templates; (2), that the patterns of inhibition would be essentially noncompetitive with respect to primer-template, and (3), that the

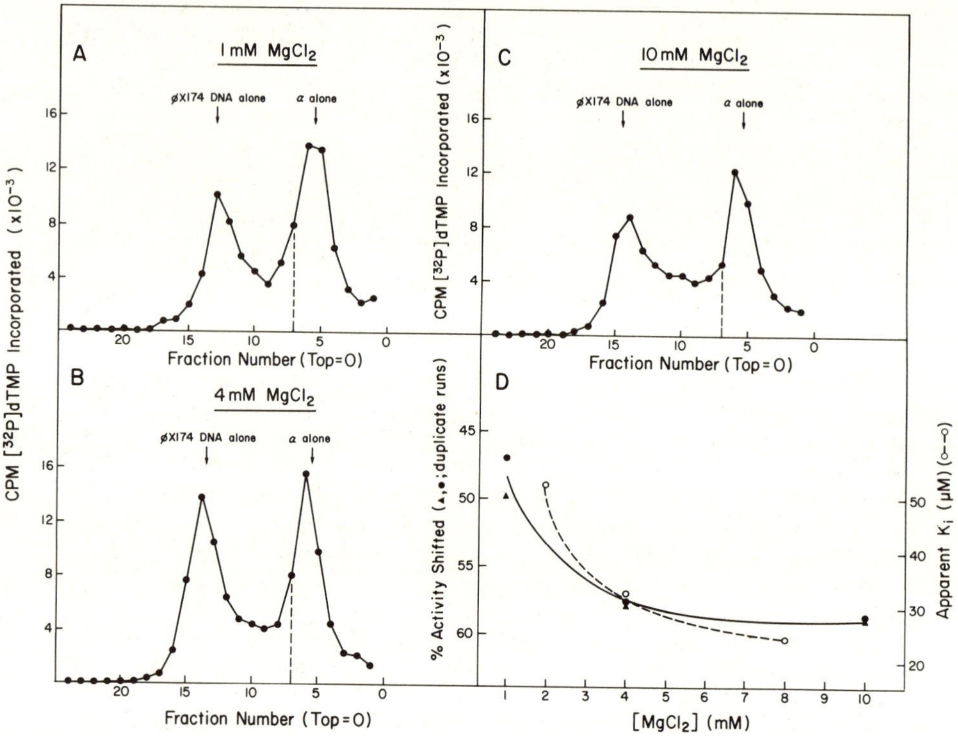

Fig. 14. Sedimentation binding analysis of the effect of $MgCl_2$ on the interaction of DNA polymerase-α with ϕX 174 DNA. Panels A-C: Gradients were formulated and run as described[15]; they contained 1mM (nucleotide) of ϕX 174 DNA and the concentration of $MgCl_2$ that is indicated. The broken lines represent the lower boundary of sedimentation of free polymerase-α in the absence of added DNA, but under otherwise identical centrifugation conditions. Panel D: Plot of % polymerase activity shifted (bound) versus $MgCl_2$ concentration (duplicate determinations), with superimposed kinetically determined values of apparent K_i. Note that the left-hand ordinate scale is decreasing toward the top. Reprinted from Fisher and Korn[15] with permission of American Chemical Society.

enhanced affinity of the enzyme for pyrimidine containing templates would be manifested by corresponding decreases in values of apparent K_m for the primed homologous templates. Direct test of these predictions led to the unexpected observations[15] that variation of Mg^{+2} over the concentration range of interest was in fact without systematic effect on values of apparent K_m measured with activated DNA, $(dA, dG, dC, dT)_{\overline{57}}$, $(dT)_{\overline{100}}$-$(dA)_{\overline{25}}$ and $(dC)_{\overline{100}}$-$(dG)_{\overline{25}}$, but did lead to a dramatic decrease in the apparent affinity of the enzyme for the substrate $(dA)_{\overline{100}}$-$(dT)_{\overline{25}}$. These results thus seemed to require that the divalent cation, in addition to its effects on template binding, must have an opposite and partially compensatory effect of decreasing the affinity of polymerase-α for primer.

A detailed kinetic analysis of the inhibitory effects of $MgCl_2$ on the replication of $(dA)_{\overline{100}}$-$(dT)_{\overline{25}}$ is presented in Figure 15. By Lineweaver-Burk analysis (Fig. 15A, B), it was evident that Mg^{+2} inhibition appeared to be purely competitive with respect to primer-template and noncompetitive with respect to dTTP. When similar experiments were assessed by Dixon plot analysis (Fig. 15C), we observed dramatic concavity consistent with multi-site binding, and Hill plot analysis (Fig. 15D) yielded a slope (Hill coefficient) of 3.9. These results, together with the complete absence of detectable effect of Mg^{+2} on the binding of polymerase-α to $(dA)_{\overline{100}}$ (template) alone, suggested that the divalent cation might compete directly with the primer moiety for the primer-binding site on the enzyme; but they did not exclude the alternative possibilities that (1), the effect of Mg^{+2} on primer binding was modulated indirectly through an effector site on the enzyme, or (2), by virtue of the ordered sequential reaction mechanism, the Mg^{+2} effect was exerted somewhere downstream in the reaction sequence. We were able to exclude the latter possibility completely and show the former to be extremely unlikely by performing sedimentation binding studies[15] and the specific kinetics experiments displayed in Figure 16. In the study shown in Figure 16A, we demonstrate that the effect of Mg^{+2} on the affinity of polymerase-α (apparent K_m) for substrate $(dA)_{\overline{100}}$-$(dT)_{\overline{25}}$ is essentially identical to that on the affinity (apparent K_i) for the dead-end inhibitor $(dA)_{\overline{100}}$-$(dT)_{\overline{25}}$-$(ddT)_1$; thus the possibility that Mg^{+2} acted downstream of dNTP addition to affect primer binding could be eliminated. The study shown in Figure 16B addresses the question of whether Mg^{+2} competes directly with primer or acts indirectly through an effector site, by asking whether the binding of Mg^{+2} and of a second competitive inhibitor, the blocked polymer $(dA)_{\overline{100}}$-$(dT)_{\overline{25}}$-$(ddT)_1$, are mutually exclusive[16]. The pattern of parallel lines obtained supports the interpretation of mutually exclusive binding and the conclusion that the two inhibitors compete with one another for a common binding site on DNA polymerase-α.

On the basis of these results, we have proposed[15] that the normal binding of primer stem by polymerase-α occurs through a Mg^{+2}-primer complex that involves the coordination of 4 Mg^{+2} ions with 7 or 8 phosphate residues in phosphodiester linkage, and thus that the primer-binding site contains 4 Mg^{+2}-primer-binding subsites at which free Mg^{+2} is able effectively to compete. Given the apparent requirement for an octanucleotide primer stem with only 7 internal phosphate groups in the primer backbone[15], we have also suggested that one of the 4 Mg^{+2}-binding subsites might be available to participate in polymerase translocation subsequent to the insertion of the next dNTP.

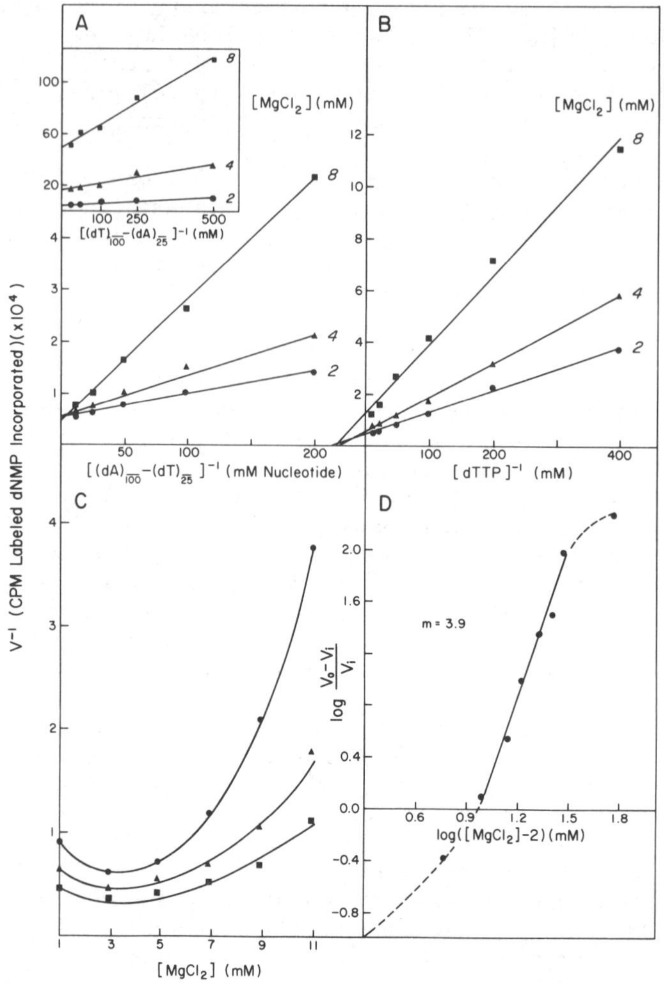

Fig. 15. Effect of $MgCl_2$ on substrate kinetics of DNA polymerase-α with $(dA)_{\overline{100}}$-$(dT)_{\overline{25}}$. Panel A: Lineweaver-Burk analysis, V^{-1} versus $[(dA)_{\overline{100}}\text{-}(dT)_{\overline{25}}]^{-1}$. Panel B: Lineweaver-Burk analysis, V^{-1} versus $[dTTP]^{-1}$. Panel C: Dixon plot analysis, V^{-1} versus $MgCl_2$, at various concentrations of substrate, $(dA)_{\overline{100}}$-$(dT)_{\overline{25}}$, that were present as 1X, 2X, 4X. Panel D: Hill plot analysis of inhibition of polymerase-α by $MgCl_2$. The slope of the Hill plot (m) indicated on the panel was derived from the linear portion of the graph (solid line). The insert in panel A shows a Lineweaver-Burk analysis of the effect of $MgCl_2$ on the replication of $(dT)_{\overline{100}}$-$(dA)_{\overline{25}}$; the pattern of inhibition is noncompetitive with respect to primer-template (as it is for all other primer-templates tested except for $(dA)_{\overline{100}}$-$(dT)_{\overline{25}}$; see text). Reprinted from Fisher and Korn[15] with permission of American Chemical Society.

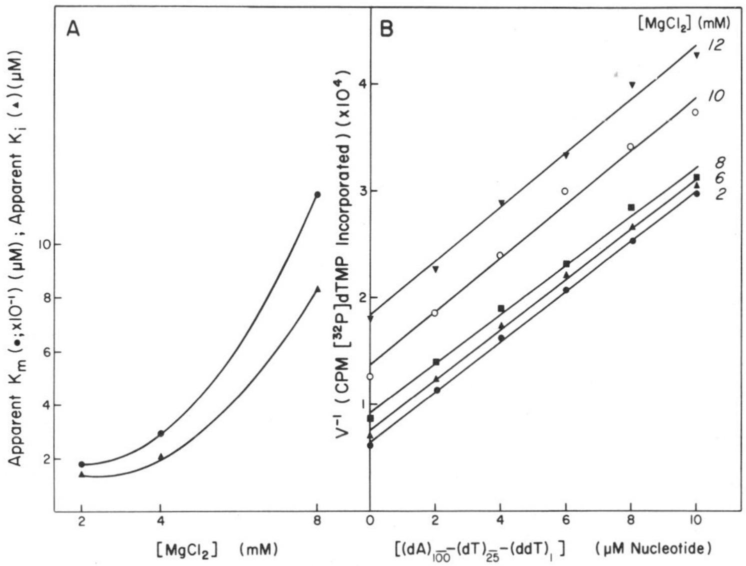

Fig. 16. Effect of $MgCl_2$ on the interaction between DNA polymerase-α and (dA)$_{\overline{100}}$-(dT)$_{\overline{25}}$-(ddT)$_1$. Panel A: Values of apparent K_m for (dA)$_{\overline{100}}$-(dT)$_{\overline{25}}$-(ddT)$_1$ versus $MgCl_2$ concentration. Panel B: Dixon plot analysis of the effect of $MgCl_2$ on polymerase inhibition produced by (dA)$_{\overline{100}}$-(dT)$_{\overline{25}}$-(ddT)$_1$[16]. (The experiments in panel B were carried out at 2.5x levels of dTTP to take advantage of the ordered mechanism of substrate addition and insure that the inhibition produced by the dideoxy-blocked polymer was >95% primer-dependent (see Section II)). Reprinted from Fisher and Korn[15] with permission of American Chemical Society.

The model also supports the hypothesis that Mg^{+2} may be brought to the polymerase·template·primer complex by dNTPs (which have an extremely high affinity for divalent cation) and may then be cycled through the polymerization reaction, univectorially and processively, to each of the 4 Mg^{+2}-primer-binding subsites, thereby facilitating not only phosphodiester bond formation (catalysis) but polymerase translocation as well.

Finally, we note that there is an abundance of published information regarding the deleterious effects of a number of different divalent metals, some of them known mutagens or carcinogens, on DNA polymerase fidelity. With most of all of these metals, the misinsertion frequency of noncomplementary dNMPs tends to increase progressively under conditions of divalent cation concentration that are increasingly inhibitory to the polymerization reaction per se. In light of our evidence that (1), divalent cation (Mg^{+2}) is a powerful effector of the primer-binding reaction, and (2), the primer-binding step is of critical importance to the subsequent step of complementary dNTP recognition and binding by the polymerase, it is attractive to suggest[15] that divalent cation-induced polymerase infidelity may be largely due to perturbation of the polymerase-primer interaction.

IV. Primer-Template Recognition by DNA Polymerase-β

In previous studies we have documented a number of substantial differences in the enzymological properties of human DNA polymerase-α and β. These have included not only the standard parameters routinely examined in optimizing in vitro assays (e.g., pH, monovalent and divalent cation concentrations[2,4]) as well as efficiency of utilization of different synthetic homo- or heteropolymeric polynucleotide primer-templates[4,17], but also dramatic differences between the two polymerases with respect to their ability to replicate a variety of defined, natural DNA substrates. Thus, although both enzymes are highly reactive with gapped ("activated") duplex DNA, polymerase-β is most active on substrates containing short gaps of 10- 15-nucleotides in mean length and is capable of filling those gaps completely; while polymerase-α prefers gaps of 30- to 60-nucleotides in mean length and is incapable of filling them completely, leaving an unfilled stretch of about 20-nucleotides[4,5,7]. DNA polymerase-β, but not polymerase-α, is capable of elongating the natural primer fragment in D-loop mitochondrial DNA molecules[7,18] and polymerase-β, but not polymerase-α, is able to carry out a limited strand-displacement synthesis on nicked duplex DNA substrates to an extent of incorporation of about 15 nucleotides at each nick[5].

In our initial assessment of the latter reaction, we observed that the apparent activation energy (Ea) of the polymerase-β reactions with nicked and gapped DNA was identical, indicating that the frequency of productive interactions of the polymerase with 3'-OH termini at nicks and gaps was indistinguishable and suggesting that thermally-induced, localized destabilization of the 5'-terminated DNA strand at the nick site did not contribute significantly to the rate-determining step(s) of the polymerization reaction[5]. We also noted from a preliminary kinetic assessment of the substrate capacity of duplex DNA molecules that differed only in their mean content of nicks per molecule that polymerase-β appeared to demonstrate measurable affinity only for primer termini, but none for the duplex DNA backbone[5] (Table 2). To explore the implications of these results in greater depth, and to obtain a more comprehensive understanding of the features of the polymerase-β nucleic acid interaction, we have carried out the detailed kinetics experiments[18] that are summarized in this Section.

We first performed a steady-state kinetics study of the mechanism of substrate addition to polymerase-β, and, under conditions in which DNA primer-template was treated as a single reactant, obtained

Table 2. Kinetic Parameters of Reaction of DNA Polymerase-β with Nicked DNA

Mean number of nicks per DNA molecule	Apparent K_m for DNA		V_{max} (fmol of dTMP per minute)
	nucleotide (μM)	3'-OH termini (nM)	
1.3	58	4.3	256
10.0	9	5.1	266
21.4	3.7	4.5	220

Reactions were carried out as described[5] using Mn^{+2} as divalent cation and circular, duplex PM2 DNA substrates that contained the indicated mean number of nicks per molecule. Kinetic parameters were computed by least-squares analysis from Lineweaver-Burk plots. Reprinted from[5] with permission of American Chemical Society. A nick is defined as a single phosphodiester bond scission in one of the two strands of a duplex DNA molecule.

results consistent with the interpretation of a rigidly ordered mechanism of substrate addition to the enzyme, with DNA first, followed by dNTP. Although this mechanism is formally analogous to that demonstrated for DNA polymerase-α (see Section II and[14]), the experimental protocol did not provide information about the individual steps of enzyme binding to template and to primer. Accordingly, we examined the interactions of polymerase-β with several classes of DNA molecules of defined structure to determine whether we could resolve kinetically those putatively separate components of the nucleic acid binding reaction.

The effect of intact duplex DNA molecules on the polymerase-β reaction was tested with a duplex circular relaxed (Form IV) plasmid DNA, which contained no termini; and with a restriction digest of PM2 DNA that contained a mixture of 15, blunt-ended linear duplex fragments ranging from about 100- to 1900-nucleotides in length. Neither population had a detectable effect on the polymerization reaction with gapped DNA substrate, whether tested in the presence of Mg^{+2} or Mn^{+2} as divalent cation (Fig. 17A, B). These results confirmed the apparent indifference of polymerase-β to intact duplex DNA, as earlier inferred, and indicated that the enzyme was capable of discriminating internal(nick) from flush-ended 3'-OH termini, a capacity that contrasts, for example, with the behaviour of *E.coli* DNA polymerase I[19] or of a KB nuclear DNA-dependent ATPase[20].

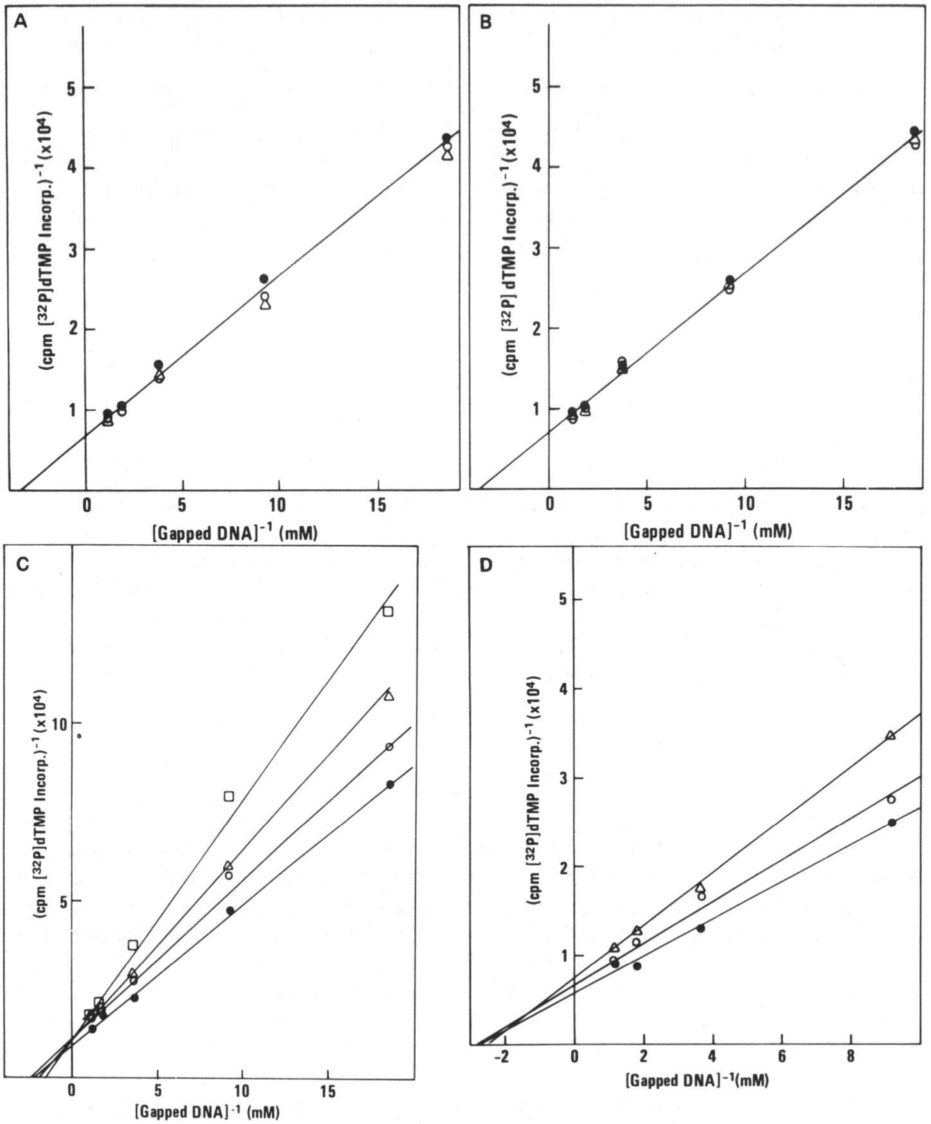

Fig. 17. Effects of intact duplex DNA molecules (A and B) and single-stranded DNA molecules (C and D) on the reactivity of polymerase-β with gapped DNA substrate in the presence of Mg^{+2} as divalent cation[18]. The data are presented as Lineweaver-Burk plots. Panel A: The effect of Form IV plasmid pACYC DNA. Panel B: The effect of blunt-ended duplex restriction fragments of PM2 DNA. In each panel, inhibitor DNA was added as 0 (control), 1X, 2X. Panel C: The effect of circular single-stranded ϕX 174 DNA, added as 0 (control), 1X, 2X, 4X. Panel D: The effect of $(dA)_{\overline{100}}$, added as 0 (control), 1X, 2X.

The kinetics of the interaction of polymerase-β with nicked duplex DNA is illustrated in Figure 18. Duplex linear DNA molecules containing an average of one 3'-OH terminated nick per 1700 nucleotides were tested both as substrates (Figs. 18A, B) and as alternate product inhibitors (Figs. 18C,D) of the polymerization reaction, in the presence of Mg^{+2} and Mn^{+2} as divalent cations[3,5]. As substrates, these nicked molecules exhibited simple saturation kinetics and yielded apparent K_m values, in Mg^{+2}, of 100μM (nucleotide) and 60nM (3'-OH termini); and in Mn^{+2}, of 7μM (nucleotide) and 4nM (3'-OH termini). As alternate product inhibitors, the nicked molecules were linearly competitive with respect to the gapped DNA substrate, and both values of apparent K_i, as well as their response to divalent cation, were essentially identical to those of apparent K_m. When duplex DNA molecules containing an average of one $3'\text{-}PO_4$ terminated nick per 2870 nucleotides were tested as dead-end inhibitors in similar reactions, the results (Figs. 18E, F) were qualitatively identical to those described above; values of apparent K_i were again responsive to divalent cation and measured, in Mg^{+2}, 100nM ($3'\text{-}PO_4$ termini), and in Mn^{+2}, 10nM ($3'\text{-}PO_4$ termini). These results demonstrate, in sharp contrast to the observations with KB cell polymerase-α (see Section I[9,10]), both the very high affinity of polymerase-β for nicked duplex DNA and the roughly comparable avidity of the enzyme for 3'-OH and $3'\text{-}PO_4$ nick termini.

We next examined the interaction of polymerase-β with a variety of natural and synthetic, single-stranded polydeoxynucleotides, including circular φX 174 DNA (no termini)(Fig. 17C), linear denatured fragments of KB DNA, $(dA)_{\overline{100}}$ (Fig. 17D), $(dT)_{\overline{100}}$, $(dC)_{\overline{100}}$, $(dA, dG, dC)_{\overline{51}}$ and its dideoxyterminated homologue. Once again in dramatic contrast to the results with DNA polymerase-α, all of the single-stranded molecules tested produced qualitatively identical patterns of inhibition that were fully and linearly noncompetitive with the gapped DNA substrate (as well as with dNTPs), as evaluated by Lineweaver-Burk and Dixon plot analyses[18]. Slope and intercept replots of the primary data yielded values of apparent K_{is} and K_{ii}, respectively, which were not identical ($K_{is} < K_{ii}$), thus suggesting that these single-stranded polydeoxynucleotides had a somewhat greater affinity for the free polymerase (E) than for the polymerase·DNA (E·S) complex. The performance of these experiments under reaction conditions optimized for replication of the gapped DNA substrates in Mg^{+2} demonstrated that the apparent affinity of polymerase-β for these several single-stranded polymers was extremely low, with apparent K_i values of the order of 500 (heteropolymer) to 3000 (homopolymer) μM (nucleotide). In contrast, when measured under polymerization conditions[3,19] optimized with Mn^{+2}, the apparent affinity of the enzyme for the single-stranded inhibitor species was markedly enhanced, with apparent K_i values in the range of 10 (heteropolymer) to 1000 (homopolymer) μM (nucleotide). These substantial changes in kinetic affinity for single stranded polydeoxynucleotides that are noncompetitive either with DNA or dNTP substrates suggest that the dramatic enhance-

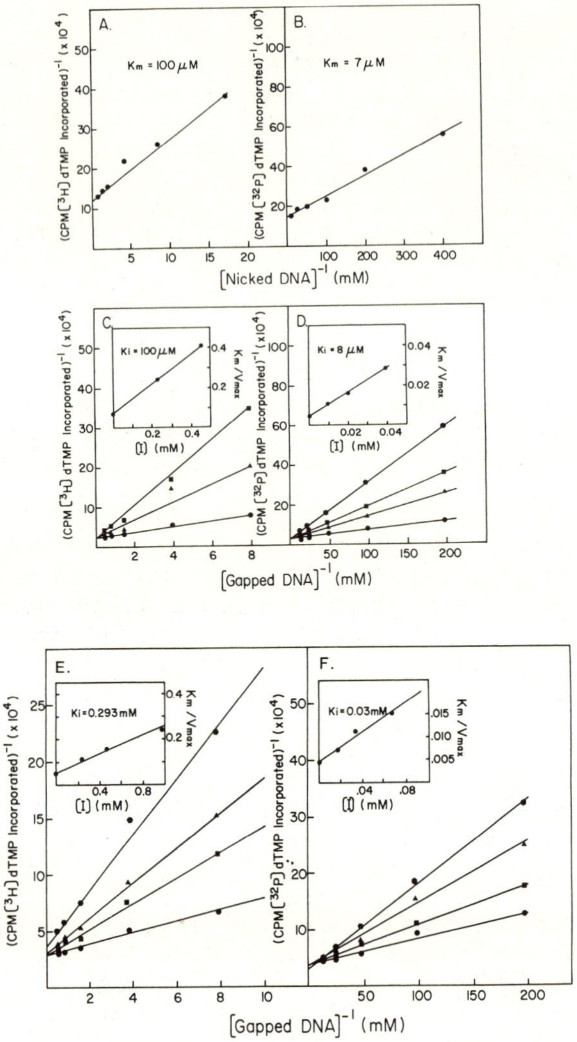

Fig. 18. Interaction of DNA polymerase-β with nicked duplex DNA molecules[18]. The data are displayed as Lineweaver-Burk plots. Panes A and B: Kinetics of dTMP incorporation on 3'-OH nicked duplex DNA in the presence of Mg^{+2} (A) and Mn^{+2} (B). Panels C and D: Inhibition of the polymerization reaction on gapped DNA by 3'-OH nicked duplex DNA in the presence of Mg^{+2} (C) and Mn^{+2}(D). The inserts in panels C-F show slope replots of the primary data, from which values of apparent K_i were computed. Panels E and F: Inhibition of the polymerization reaction on gapped DNA by 3'-PO_4 nicked duplex DNA in the presence of Mg^{+2} (E) and Mn^{+2} (F). The amounts of inhibitor DNA added in the several panels were as 0 (control) 1X, 2X (C) and 0 (control), 1X, 2X, 4X (D-F).

ment of the apparent affinity of polymerase-β for primer-templates[3] in the presence of Mn^{+2} may reflect relatively nonspecific changes in the enzyme (and/or nucleic acid) rather than specific alterations in the active center of the polymerase.

V. Interaction of DNA Polymerase-β with Staggered-End DNA[18]

The data summarized in Sections I and IV demonstrate remarkable differences between human DNA polymerases α and β with respect to their kinetic patterns of interaction with nicked duplex and single-stranded DNA molecules. Although the results in Section IV suggested that a base-paired primer moiety was a primary determinant of catalytically productive binding of polymerase-β to DNA, the data provided no clues that might have helped to resolve partial reaction steps of template binding and primer binding; and yet the capacity of the polymerase to discriminate internal nicks from blunt-ended termini indicated that some other structural elements in the nucleic acid substrate, in addition to the primer moiety, were required. To pursue this problem, we examined the interaction of polymerase-β with a Hind III restriction digest of PM2 DNA that contained 7 duplex fragments, ranging from about 100- to 5000-base pairs in length; each fragment has 5'-PO_4 terminated, 4-nucleotide staggered ends.

In the presence of Mg^{+2} as divalent cation, these fragments were inert as primer-templates, and they had no detectable inhibitory effect on the polymerization reaction with gapped DNA (Fig. 19A); i.e., these molecules were kinetically indistinguishable from blunt-ended duplex fragments. The Hind III-restricted fragments were then subjected to very limited resection with E. coli exonuclease III (exo III) to produce populations of staggered-end molecules with 5'-terminal single-stranded template lengths that ranged from the original 4-nucleotides to an average of 5-, 6- and 7-nucleotides, respectively. These minimally resected fragments proved to be surprisingly good substrates that exhibited simple saturation kinetics (Table 3); although their affinity for polymerase-β was relatively poor, they yielded V_{max} values>50% of that measured with optimally gapped DNA. In contrast to these observations, the original restriction digest was found to provide an effective primer-template population for DNA polymerase-β when reactions were performed in the presence of Mn^{+2} (Fig. 19B, Table 3). These results suggested that with Mn^{+2} as divalent cation, polymerase-β might be capable of filling in staggered-end DNA substrates completely; they also suggested that the minimum length of single-stranded template required for catalytically productive binding of the enzyme to DNA might be strongly modulated by choice of divalent cation.

To resolve these questions conclusively, we first carried out the experiment[18] shown in Figure 19C, in which the Hind III restriction fragments were used as primer-templates in four separate incubations, each containing a different labeled dNTP. The rationale

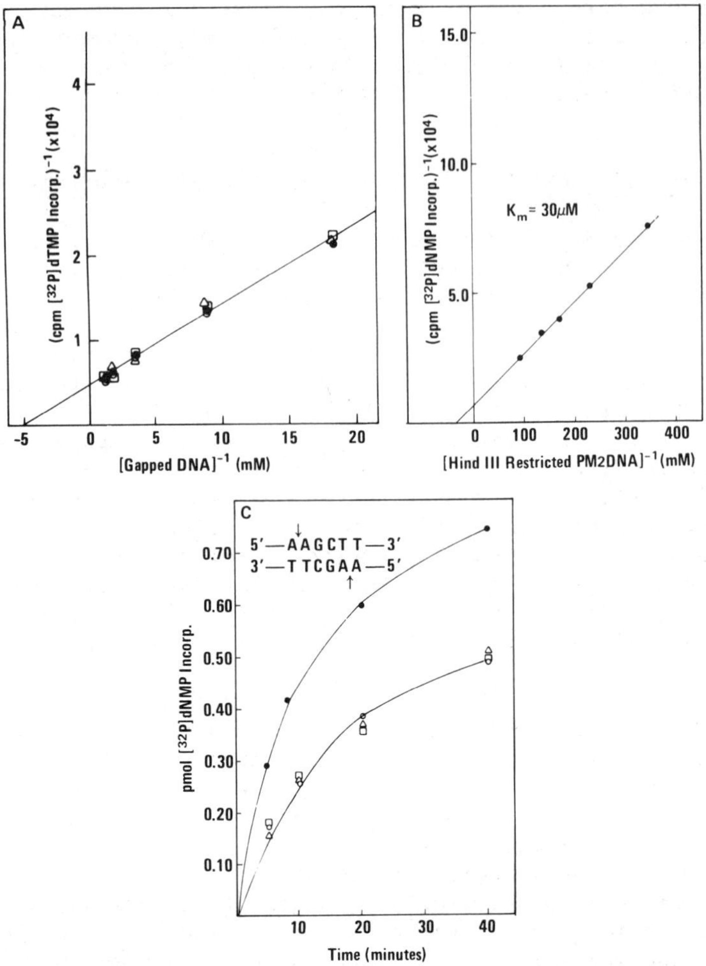

Fig. 19. Interaction of DNA polymerase-β with duplex DNA molecules that contain 4-nucleotide staggered ends. Panel A: Effect of Hind III-restricted PM2 DNA on the polymerization reaction with gapped DNA substrate in Mg^{+2}. The data are presented as a Lineweaver-Burk plot. Inhibitor DNA was added as 0 (control), 1X, 2X, 4X. Panel B: Utilization of Hind III-restriced PM2 DNA as primer-template by polymerase-β in the presence of Mn^{+2}. The data are presented as a Lineweaver-Burk plot. Panel C: Estimation of minimal effective template length that can be copied by polymerase-β in Mn^{+2}. The Hind III staggered-end sequence (template) is shown in the upper left corner of the panel. Incorporation of dAMP (●); dGMP (□), dCMP (△); dTMP (○). Reprinted from Wang and Korn[18] with permission of American Chemical Society.

was based on the known sequence of the 4-nucleotide stagger (template), viz., T-C-G-A, and the results clearly demonstrated that all 4 dNMPs could in fact be incorporated, including the last product nucleotide, dTMP. We then performed the final set of experiments that are schematized in Figure 20. The principal of these experiments was to measure the extent of deoxynucleotide incorporation by DNA polymerase-β on a population of defined, staggered-end primer-templates that could be specifically recovered by restriction of a sequenced DNA molecule, pBR 322, and then analyzed on high resolution polyacrylamide gels. As shown in Figure 20, pBR 322 DNA molecules were first restricted at a single site with Hind III and then resected with exonuclease III to generate a population of staggered-end DNA molecules containing a Poisson distribution of (staggered-end) template lengths. This DNA preparation was used as primer-template in reactions with polymerase-β that were carried to extent, either with Mg^{+2} or with Mn^{+2} as divalent cation. The polymerization products at extent were then separated from the pBR 322 molecules by restriction with Eco RI, again at a single site, and were analyzed by gel electrophoresis. The separation between the two unique restriction sites in the original population of pBR 322 DNA molecules was such that the maximum length of product that could be achieved, i.e., by complete filling in of the staggered-end templates, was 33-nucleotides.

The results of these experiments[18] (Fig. 20) demonstrated that the extent products synthesized by polymerase-β in Mg^{+2} consisted of two predominant chains, a 27-mer and a 29-mer, with trace amounts of 26-mer, 28-mer and 30-mer also detected. However, no fragments longer than 30-nucleotides were present. In contrast, the extent products synthesized in the presence of Mn^{+2} contained only two chains of 32- and 33-nucleotides in length, and the longer fragment was the more abundant product. These data thus provide unambiguous documentation of the fact that in the presence of Mn^{+2}, DNA polymerase-β is capable of filling in a staggered-end template completely. The results of the gel analysis are in excellent agreement with those presented in Figures 19B, C, and they support the conclusion that with Mn^{+2} as cation, the minimum template length required by polymerase-β is of the order of a single nucleotide. On the other hand, with Mg^{+2} as cation, polymerase-β is unable to copy the terminal 3 or 4 nucleotides of a staggered-end template. Once again, the results of the gel analysis are entirely corroborative of the kinetics data presented in Figure 19A and Table 3, and they permit the conclusion that in the presence of Mg^{+2}, polymerase-β requires a minimum template length of at least 5 nucleotides for catalytically productive binding to DNA.

We have also demonstrated that the choice of divalent cation affects the processivity of polymerization of DNA polymerase-β[18]. Thus, in the presence of Mg^{+2}, on either nicked or gapped DNA substrates, polymerization is essentially distributive, with an average

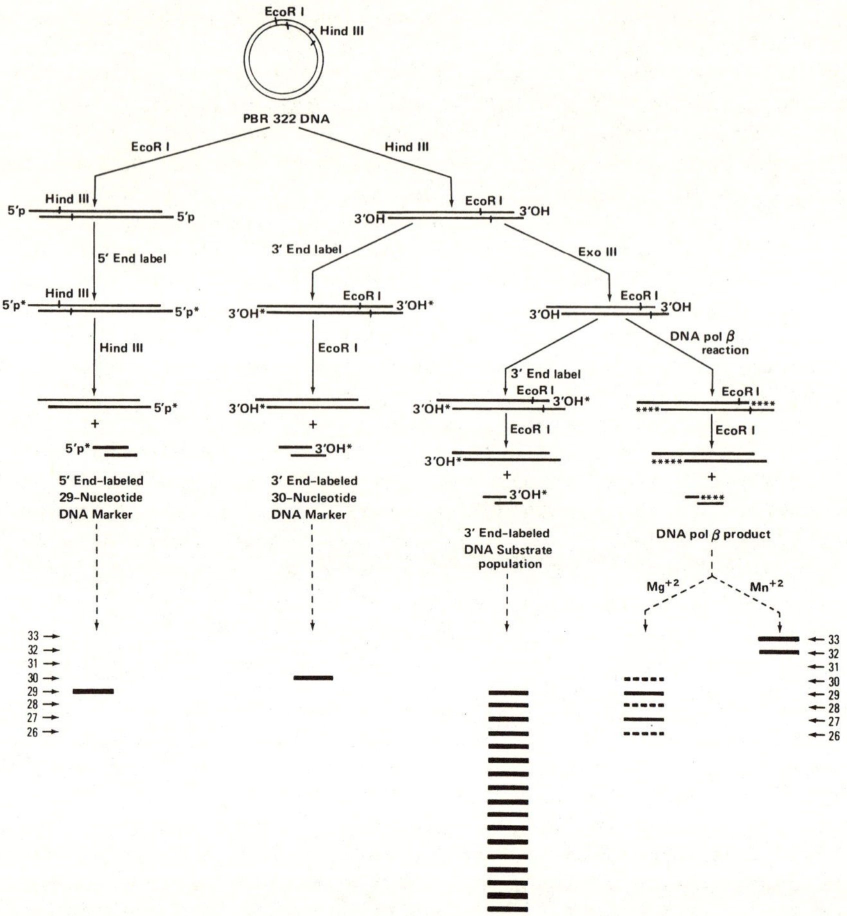

Fig. 20. Schematic representation of experimental protocol designed to demonstrate difference in minimum template length required for catalytically productive binding by human DNA polymerase-β in the presence of Mg^{+2} and Mn^{+2}, respectively. The upper portion of the figure shows the generation and recovery for analysis of unique populations of staggered-end primer-templates prepared from pBR 322 DNA by specific restriction endonuclease cleavages and exonuclease III resection. The lower portion of the figure is a schematic that demonstrates the patterns of fragment lengths experimentally observed following high resolution polyacrylamide gel electrophoresis. The 2 lanes at the right side of the gel diagram show the product fragments recovered after polymerization to extent in the presence of Mg^{+2} and Mn^{+2}, respectively[18]. (See text for details).

Table 3. Comparison of Primer-Template Capacity of Specific Staggered-End DNA Molecules

Primer-Template	Mean Number of Nucleotides per Staggered End	Mg^{+2} App. K_m (μM)	Mg^{+2} V_{max} (fmol dTMP/min)	Mn^{+2} App. K_m (μM)	Mn^{+2} V_{max} (fmol dTMP/min)
Optimally gapped DNA	-	300	440	7	59
Nicked DNA	-	100	46	7	16
Hind III-restricted, PM2 DNA	4.0	--- Not used ---		30	35
Exonuclease III-restricted, Hind III-restricted PM2 DNA					
Substrate 1	5	930	135	-	-
Substrate 2	6	830	138	-	-
Substrate 3	7	730	138	-	-

Reactions were carried out for 10 min. at 35°C as described[18]. "Optimally gapped" (activated) DNA contained gaps of 14-nucleotides mean length[18]. "Nicked" DNA is native calf thymus DNA that contained about 1 nick per 1700 nucleotides. The restricted and resected PM2 DNA substrates are described in the text. Kinetic parameters were calculated from Lineweaver-Burk plots by the method of least-squares.

of 1 or 2 nucleotides incorporated per binding cycle. In the presence of Mn^{+2}, however, the mechanism of polymerization becomes modestly processive, on both gapped and nicked DNA substrates, with an average incorporation of 4 to 6 nucleotides per cycle.

Table 4. Effect of Divalent Cations on Catalytic Properties of Human DNA Polymerase-β

Property	Primer-Template	Mg^{+2}	Mn^{+2}
App K_m[a] (DNA)	Gapped DNA		
	(nucleotides)	330μM	7μ
	(3'-termini)	195nM	4nM
	Nicked DNA		
	(nucleotides)	100μM	7μM
	(3'-termini)	60nM	4nM
V_{max}[a] (fmol dTMP/min)	Gapped DNA	440	59
	Nicked DNA	46	16
Ea (Kcal/mol)	Gapped DNA	17.5	11.5
	Nicked DNA		11.4
Q_{10}	Gapped DNA	2.7	1.9
	Nicked DNA		1.9
Processivity	Gapped DNA	1.9	5.6
	Nicked DNA	1.5	3.4
Minimum Effective Template Length	Staggered-end DNA	≥ 5	≥ 5

[a]Values of apparent K_m and V_{max} are averages from 17 separate experiments.

The results discussed in this section thus identify two additional catalytic parameters[3,5,8] of the polymerase-β reaction, minimum effective template length and processivity, that are significantly altered by substitution of Mn^{+2} for Mg^{+2} as divalent metal activator. Table 4 presents a summary of all to the divalent cation effects on catalysis that we have identified to date in the reaction of DNA polymerase-β with DNA primer-templates.

VI. Concluding Comments

We may summarize our present understanding of the catalytic mechanism of KB cell DNA plymerase-α as follows:

(1) The interaction of the polymerase with its substrates obeys a rigidly ordered sequential terreactant mechanism, with template (single-stranded polydeoxynucleotide) as the first substrate, followed by primer as the second substrate and dNTP as the third.
(2) Although kinetically significant dNTP binding is absolutely dependent on antecedent primer binding, specification of which dNTP can add to the enzyme is strictly determined by template sequence.
(3) Catalytically significant primer binding requires a minimum length of primer stem of 8-nucleotides, of which it appears that at least the terminal 3 to 5 nucleotides must be template-complementary.
(4) A single mispaired terminal primer nucleotide is sufficient to prevent correct primer binding and also the subsequent step of complementary dNTP addition to the polymerase.
(5) The enzyme can properly bind a primer stem bearing a 3'-terminal OH or H residue, but the presence of a 3'-terminal PO_4 substituent blocks primer recognition.
(6) Primer binding appears to occur through the coordinated participation of 4 Mg^{+2}-primer-binding subsites which may serve as a "Mg^{+2}-shuttle": thus, Mg^{+2} may be brought to the polymerase·template·primer complex by dNTPs and may then be cycled univectorially and processively to each of the 4 postulated primer-binding subsites, thereby facilitating not only phosphodiester bond formation (catalysis) but polymerase translocation as well.
(7) Each catalytically active polymerase-α molecule appears to possess at least 2 strongly interactive (positively cooperative) single-stranded polydeoxynucleotide binding sites (and possibly 2 complete active centers[14]).

On the basis of these data, we have suggested that the properties of KB cell DNA polymerase-α are compatible with those of a conformationally active protein which is capable of catalytically significant response to signals that are generated by template sequence and transduced via template binding site(s) interactions.

In contrast to results obtained with polymerase-α, under reaction conditions optimized for gapped DNA substrate with Mg^{+2} or Mn^{+2} as divalent cation, human DNA polymerase-β exhibits only a relatively low apparent affinity for single-stranded, natural or synthetic polydeoxynucleotides. That affinity appears not to be modulated by base composition and is not enhanced by the presence of potentially base-pairable 3'-OH termini (although it does vary with choice of divalent cation), and the pattern of inhibition produced by these polymers under all conditions tested is fully and linearly noncompetitive with substrateDNA (or with dNTPs). In each of these respects, the kinetics of the interaction between polymerase-β and the single-stranded deoxypolymers is dramatically different from that exhibited by KB cell polymerase-α.

Although human polymerase-β, like polymerase-α, has no kinetically detectable affinity for intact duplex DNA molecules, whether covalently-closed circles of flush-ended linear fragments, polymerase-β has an extremely high affinity for nicked duplex DNA; and unlike polymerase-α, polymerase-β demonstrates a comparable binding avidity for nick sites (primer termini) bearing either 3'-OH or 3'-PO_4 residues. Also in marked contrast to polymerase-α, polymerase-β exhibits brisk reactivity with primers that contain 1 to 3 terminally mispaired nucleotide residues. These several results suggest that the requirements for primer recognition and binding by DNA polymerase-β may be substantially less stringent than those of polymerase-α with respect (at least) to the 3' terminal primer domain.

Although the mechanism of substrate addition to DNA polymerase-β is formally analogous to that of polymerase-α, in that it can be described as a rigidly ordered sequential pathway, with DNA first, followed by dNTP, our studies of the interaction of polymerase-β with DNA molecules of defined structure have not resolved kinetically distinguishable steps of primer binding and template binding. Thus we have tentatively concluded[18] that both of these steps in the nucleic acid binding reaction may take place by a concerted mechanism.

From our studies of the patterns of interaction of human polymerase-β with staggered-end DNA primer-templates of known sequence, we have obtained strong evidence that a primary signal for catalytically productive nucleic acid binding by this enzyme is an (at least partially) base-paired primer moiety that must be adjacent to a short length of (potentially) single-stranded template. Both the minimum length of required template, as well as the processivity of polymerization, are affected by the choice of divalent metal activator. Thus, in the presence of Mg^{+2}, the minimum template length is $\geqslant 5$ nucleotides, and the polymerization mechanism is essentially distributive, with an average of 1 to 2 nucleotides incorporated per binding cycle. In contrast, with Mn^{+2} as divalent cation, the minimum required template length appears to be but a single nucleotide, and the reaction mechanism becomes modestly processive, with an average of 4 to 6 nucleotides inserted per cycle.

Although it is prudent to be cautious in attemting to extrapolate results obtained with highly purified, near-homogeneous enzymes from in vitro assays to in vivo processes, particularly for enzymes that likely function in DNA replication and repair in coordination with many other replication factors, the substantial body of data reviewed in this paper provides an extraordinarily rich and detailed enzymological framework that should prove invaluable to future investigations of the mechanisms and the regulation of mammalian DNA replication and repair, e.g., in efforts to reconstitute complex replication and repair "systems". In addition to their important implications for the mechanisms of human DNA polymerase catalysis, these data also identify a number of intriguing features that distinguish DNA polymerases α and β with respect to the nucleic acid signals that modulate their catalytically significant binging interactions with primer-templates. At least some of those signals may prove to be of physiological significance.

REFERENCES

1. P. A. Fisher and D. Korn, J. Biol. Chem. 252:6528-6535 (1977).
2. T. S.-F. Wang, W. D. Sedwick and D. Korn, J. Biol. Chem. 249:841-850 (1974).
3. T. S.-F. Wang, D. C. Eichler and D. Korn, Biochemisty 16:4927-4934 (1977).
4. P. A. Fisher, T. S.-F. Wang and D. Korn, J. Biol. Chem. 254:6128-6135 (1979).
5. T. S.-F. Wang and D. Korn, Biochemistry 19:1782-1790 (1980).
6. D. C. Eichler, T. S.-F. Wang, D. A. Clayton and D. Korn, J. Biol. Chem. 252:7888-7893 (1977).
7. D. Korn, P. A. Fisher, J. Battey and T. S.-F. Wang, Cold Spring Harbor Symp. Quant. Biol. 43:613-624 (1978).
8. D. Korn, P. A. Fisher and T. S.-F. Wang, Prog. Nuc. Acid Res. & Molec. Biol. 26:63-81 (1981).
9. P. A. Fisher and D. Korn, J. Biol. Chem. 254:11033-11039 (1979).
10. P. A. Fisher and D. Korn, J. Biol. Chem. 254:11040-11046 (1979).
11. P. A. Fisher, J. T. Chen and D. Korn, J. Biol. Chem. 256:133-141 (1981).
12. W. W. Cleland, Methods Enzymol. 63A:500-513 (1979).
13. F. B. Rudolph, and H. J. Fromm, Methods Enzymol. 63A:138-159 (1979).
14. P. A. Fisher and D. Korn, Biochemistry. 20:4560-4569 (1981).
15. P. A. Fisher and D. Korn, Biochemistry. 20:4570-4578 (1981).
16. W. W. Cleland, "The Enzymes," P. D. Boyer, ed., 3rd Edition, Vol.II, pp. 1-65, Academic Press, New York (1970).
17. T. S.-F. Wang, W. D. Sedwick and D. Korn, J. Biol. Chem. 250: 7040-7044 (1975).
18. T. S.-F. Wang and D. Korn, submitted for publication.
19. P. T. Englund, R. B. Kelly and A. Kornberg, J. Biol. Chem. 244: 3045-3052 (1969).
20. L. M. Boxer and D. Korn, Biochemistry 19:2623-2633 (1980).

STRUCTURAL ANALYSIS OF EUKARYOTIC DNA POLYMERASE-α

Marcel Mechali and Anne-Marie de Recondo

Unite d'Enzymologie
Institut de Recherches Scientifiques sur le Cancer
BP n° 8, 94802 Villejuif Cedex, France

INTRODUCTION

Three major species of DNA polymerases have been described in eukaryotic cells. The first one to be identified in 1958 by Bollum was DNA polymerase-α[1]. DNA polymerase-β was described in 1971 by Weissbach et al.[2], Baril et al.[3], and Chang and Bollum[4]. DNA polymerase-γ was the last DNA polymerase identified in 1973 by Fridlender et al.[5]. Byrnes et al. claimed, in 1976, the existence of a fourth class of DNA polymerase, named DNA polymerase-δ[6]. Unfortunately, it is not clear at present whether DNA polymerase-δ is a distinct DNA polymerase species, or rather the association of DNA polymerase-α with a 3'-5' exonuclease activity[6,7]. Over the last few years, the description of the properties of these enzymes and their functional roles in eukaryotic DNA replication have been well documented[8-11]. Circumstantial evidence for involvement of DNA polymerase-α in the DNA replication process has been found in a great variety of experimental systems. They include cells stimulated to divide[12-14], regenerating rat liver[15,16], cells infected by DNA viruses[17-20], perinatal development of neurons[21], and cardiac muscle differentiation[21,22].

Although DNA polymerase-α was described twenty-three years ago[1], progress in defining its molecular structure has been restricted by the difficulty in obtaining appreciable quantities of this protein. Moreover, the enzyme soon appeared to be very heterogeneous[23]. Different factors, such as association with other proteins of the replication complex, proteolysis and aspecific aggregation, have been mentioned to account for the unusual difficulty in achieving a complete purification of this enzyme[23-25]. However, during the last few years, studies dealing with the structure of DNA polymerase-α from human KB cells[26], regenerating rat liver[27], _Drosophila melanogaster_ embryos[28], mouse myeloma cells[7], and calf thymus[24,29,30], have been reported. In all cases, DNA polymerase-α appeared as a multisubunit protein, although the results were somewhat conflicting concerning the characterization of the catalytic core component. We report here our data which indicate that DNA polymerase-α can exist in two structural forms, at least partly interconvertible : the DNA polymerase-holoenzyme and the catalytic core component. This result could explain the reported heterogeneity of this enzyme, which may be related to the existence of these two structural forms in the eukaryotic cell[27].

PURIFICATION OF DNA POLYMERASE-α

Numerous difficulties were encountered in most laboratories, including ours, in working out a reproducible purification procedure that yields homogenous DNA polymerase-α.

The purification protocol used to obtain DNA polymerase-α from regenerating rat liver is presented in Table 1. Details of the purification procedure have been published elsewhere[27]. The methodology we used is classical, but some important points should be noted.

Table 1. Purification of DNA polymerase-α

Fraction	Total protein (mg)	Total activity (units)	Total activity (units/mg)	Yield %
I. Cytoplasmic extract	5,400	81,000	15	100
II. Ammonium sulfate	327	63,000	192	77
III. DEAE - cellulose	75	44,500	594	54
IV. Phosphocellulose	4.0	24,000	6,000	29
V. Hydroxylapatite	0.40	16,000	40,100	20
VI. DNA - cellulose	0.059	2,673	45,300	3

DNA polymerase activity was assayed using activated calf thymus DNA[27]. One unit corresponds to 1 nmol of total nucleotide incorporated in 1 hour at 37°C, under initial velocity conditions. DNA polymerase activity eluting during the DNA-cellulose chromatography was fractionated in different pools (see text and Fig. 2). Fraction VI represents the holoenzyme fraction eluting between 0.09 and 0.11M KCl. Other fractions were analyzed in Fig. 2.

For best results, the entire procedure should be performed within a minimum time, particularly after step III. DNA polymerase-α activity was pooled by taking into account not only the peak of total activity but also the peak of specific activity. This was done in order to minimize contaminating proteins at each step of the purification. During our adjustments of the purification procedure, we found that it was important to eliminate contaminating material very early in purification. If this was not the case, the difficulty in obtaining a homogeneous α-polymerase preparation by using different chromatographic procedures was greatly increased. Polyethylene glycol precipitation of the α-polymerase from crude extract in the presence of 2M NaCl allowed us to rapidly enhance the specific activity of the α-polymerase preparation[31]. However, since, under these conditions DNA was also precipitated, we looked for other methods. We selected an ammonium sulfate fractionation carried out by successive backwashes of a first precipitate made by the addition of 0.27g/ml ammonium sulfate. This could also be done by a first addition of only 0.22g/ml ammonium sulfate. As shown in Figure 1, DNA polymerase-α was eluted from the precipitate between 12 and 17% (w/v) ammonium sulfate in the eluting buffer. The same procedure could be applied to separate DNA polymerase-β, which was eluted between 32 and

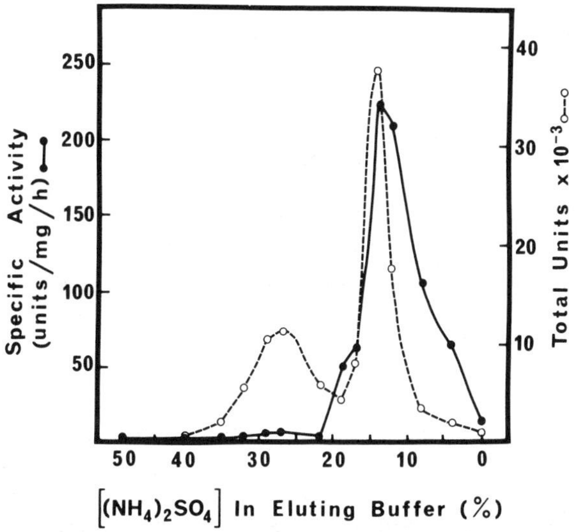

Fig. 1. Ammonium sulfate fractionation of DNA polymerase activities from a crude extract. The fractionation was performed by a salting-out process, as indicated in the text. DNA polymerase activities and proteins from a 0.5g/ml ammonium sulfate precipitate, were eluted with buffer containing decreasing ammonium sulfate concentrations. DNA polymerase activity was expressed as total units (o---o) and with respect to the protein concentration (•--•).

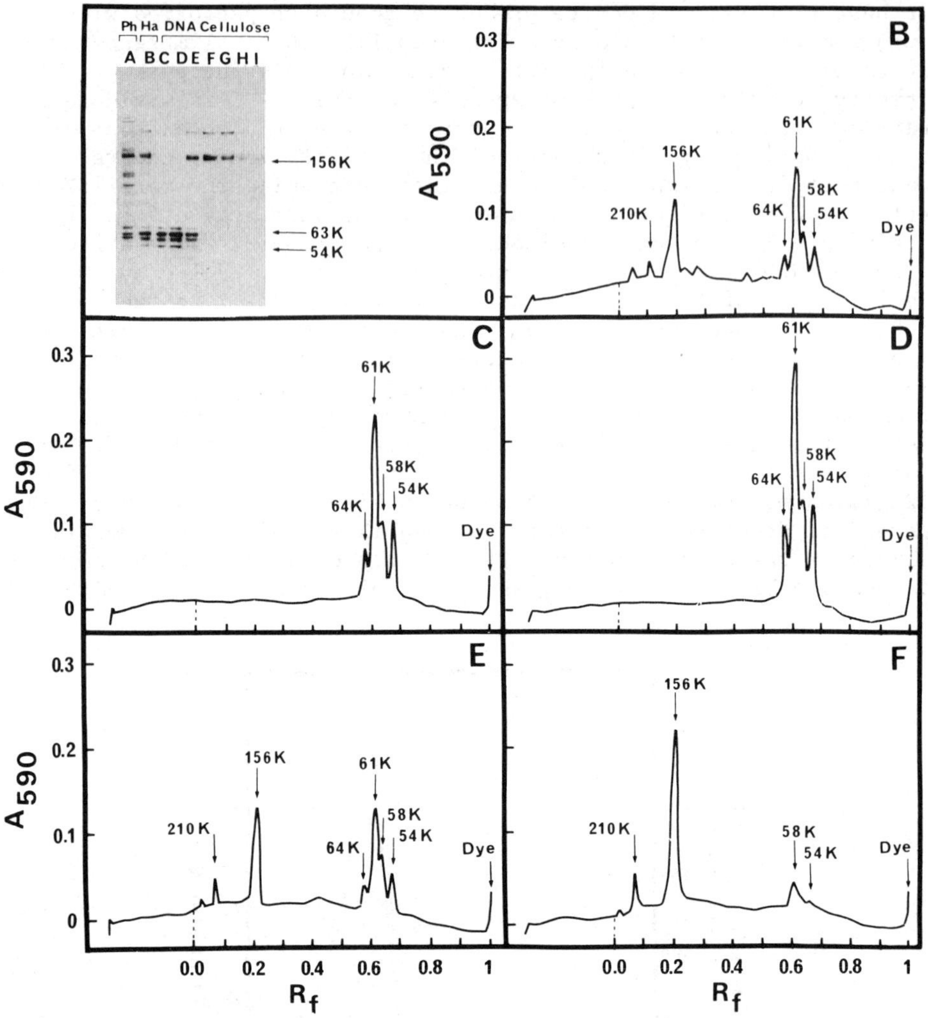

Fig. 2. Sodium docecyl sulfate polyacrylamide gel analysis of the last steps of purification. Enzyme fractions obtained from the phosphocellulose chromatography (Fraction IV), hydroxylapatite chromatography (Fraction V), and DNA-cellulose chromatography (Fraction VI) were analyzed on an SDS-8% polyacrylamide gel overlaid with a 4% stacking gel. Lanes A to I and densitometer scans B to F refer to: Fraction IV (A), Fraction V (B), DNA cellulose fractions eluting at 40-60mM KCl (C), 60-90mM KCl (D), 90-110mM KCl (Fraction VI enzyme, E), 110-125mM KCl (F), 125-150mM KCl (G), 150-170mM KCl (H), 170-200mM KCl (I). Total DNA polymerase activity eluting during the DNA-cellulose chromatography was distributed in fractions (C), undetectable, (D), 4%;(E), 52%; (F), 21%; (G), 10%; (H), 8%; (I), 5%. The molecular weight

24% (w/v) ammonium sulfate in the eluting buffer. This backwash procedure allowed us to reproducibly eliminate between 92 and 96% of contaminating proteins from a crude extract. It also removed most of the nucleic acid which did not elute from the precipitates under these conditions. The A_{280}/A_{260} ratio of Fraction II was between 1.4 and 1.5, compared to 1.0 for the crude extract. The order of the steps which followed was chosen so that dialysis of the fractions could be omitted. Only two dialysis steps were necessary during the purification, after Fraction II and V. Hydroxylapatite chromatography was performed in the presence of 0.5M NaCl, so that aggregation effects were reduced. DNA cellulose chromatography allowed us to separate the core enzyme from the holoenzyme. To do so, a shallow KCl gradient from 0.03 to 0.15M KCl was applied to a ds DNA cellulose column. The holoenzyme (Fig. 2E) eluted between 0.09 and 0.11M KCl and the core enzyme eluted over 0.125M KCl (Fig. 2F to I). Fraction VI enzyme was relatively stable when stored at -80°C in the presence of 50% glycerol. Stability was greatly reduced at low protein concentration (< 15μg/ml).

It has been shown that bovine serum albumin[7] (and unpublished results), ampholine[30] or D. glucose (G. Villani, personal communication) prevent the inactivation of the α-polymerase.

Other methods of purification can also be used. For example, oligo dT-cellulose[7], gel filtration[27,29], glycerol gradient sedimentation[28], Blue dextran-Sepharose[28,30], heparin-Sepharose[30] and gel electrophoresis[7,26], have been found to be efficient in achieving purification of the α-enzyme.

Proteolysis appears to be responsible, at least in part, for the heterogeneity of the α-enzyme[30,32]. It is thus very important to prevent this during the purification of DNA polymerase-α. Addition of the protease inhibitors, phenylmethylsulfonyl fluoride (PMSF) and sodium metabisulfite, in buffers used during the chromatographic procedures[26-28], as well as leupeptin and pestatin[33], made possible the isolation of a homogeneous enzyme.

The final α-enzyme preparation must be free of endo- and exonuclease, RNA polymerase and ATPase activities. Endonuclease activity can be tested using a highly sensitive test of conversion of supercoiled DNA in either open circular or linear molecules[34].

---

standards used were Escherichia-coli RNA polymerase subunits (165K, 155K, 87K and 39K), β-galactosidase (130K), phosphorylase a (97K), bovine serum albumin (67K), catalase (57.5K), glutamate dehydrogenase (53K), ovalbumin (43K), soybean trypsin inhibitor (21.5K).

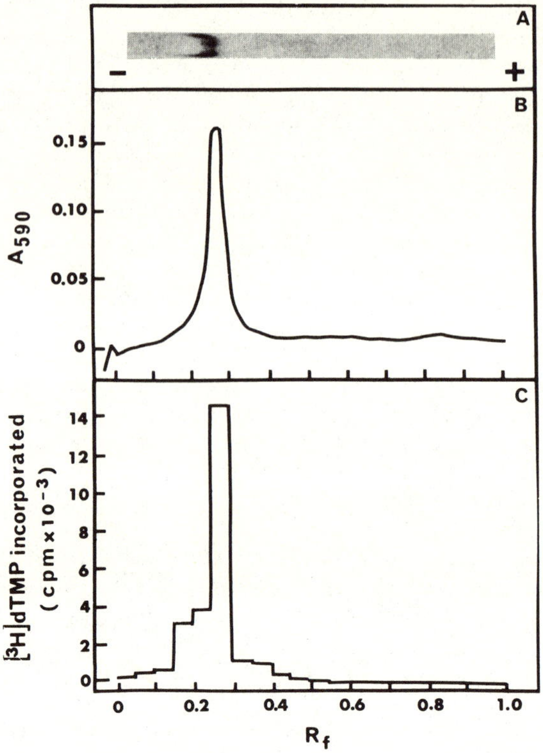

Fig. 3. Nondenaturing polyacrylamide gel electrophoresis of DNA polymerase-α, Fraction VI. Fraction VI enzyme (1.5μg, 68 units) was applied to two lanes of a 4.5% polyacrylamide gel. One lane was stained with Coomassie blue (A) and scanned densitometrically (B). The second lane was cut in slices which were individually eluted. Aliquots were assayed for DNA polymerase activity (C). Recovery of loaded enzyme activity was 26%.

The homogeneous α-enzyme gives a single stained band on native polyacrylamide gel (Fig. 3) and DNA polymerase-α activity coincides with this band.

STRUCTURAL ANALYSIS

Sedimentation Coefficient

In 5 to 20% sucrose gradient sedimentation performed at 5 or 350mM KCl, DNA polymerase-α activity sedimented at 9.7 and 7.4S, respectively (Fig. 4). This difference between the sedimentation coefficient determined in the presence or absence of salt was also reported for the α-enzyme from Drosophila melanogaster embryos[28], mouse myeloma[35], and Hela cells[36]. On the other hand, the purified α-enzyme from KB cells has the same sedimentation coefficient of 7.1 to 7.2S at high or low ionic strength[26].

Size and Molecular Weight under Native Conditions

When the rat liver enzyme was analyzed by Sephadex G-200[34] or Sephacryl S-200 filtration[28], the polymerase activity eluted with the void volume. Thus, no correlation could be found between the molecular weight of the enzyme determined by sucrose gradient (155K at high ionic strength) and its behavior during gel filtration. Similar results have been obtained for the enzyme extracted from other sources[8-10] and these data can be interpreted to mean that DNA polymerase-α has a non spherical shape[8-10,27,28]. A different result was obtained for the KB cell α-polymerase, which has an apparent molecular weight of 140,000 when filtered on Sephacryl S-200; this data is in agreement with the sedimentation coefficient of this enzyme[26].

The molecular size of the rat liver DNA polymerase-α was also measured by using pore gradient gel electrophoresis in which sample proteins pass through a gradient of continuously increasing polyacrylamide concentration until their "pore limit" is reached[37].

As shown in Fig. 5, DNA polymerase-α migrated as a single protein band of unexpectedly large molecular radius (72Å). Assuming DNA polymerase-α to be a globular protein, its migration would correspond to a molecular weight of about 1.3×10^6. It seems more likely that the size of the molecule obtained is due to its asymmetry. A large diameter of the molecule in comparison with its height could explain the results obtained by gel filtration and pore gradient gel electrophoresis. Using different polyacrylamide gels with varying degrees of cross-linking, the molecular weight obtained for the α-polymerase from Drosophila melanogaster was 550,000[28] and 280,000 for the α-polymerase from KB cell[26]. In view of these overall results, it appears very difficult to assess the true molecular weight or size of α-polymerase under native conditions.

Subunit Composition of the α-polymerase from Rat Liver

The results obtained in different laboratories concerning the polypeptide composition of the α-polymerase are somewhat conflicting, even concerning the characterization of the catalytic core component. DNA polymerase-α from human KB cells has two different polypeptides of 76 and 66K[26]. A similar structure is observed for the α-enzyme from mouse myeloma[7]. On the other hand, DNA polymerase-α from calf

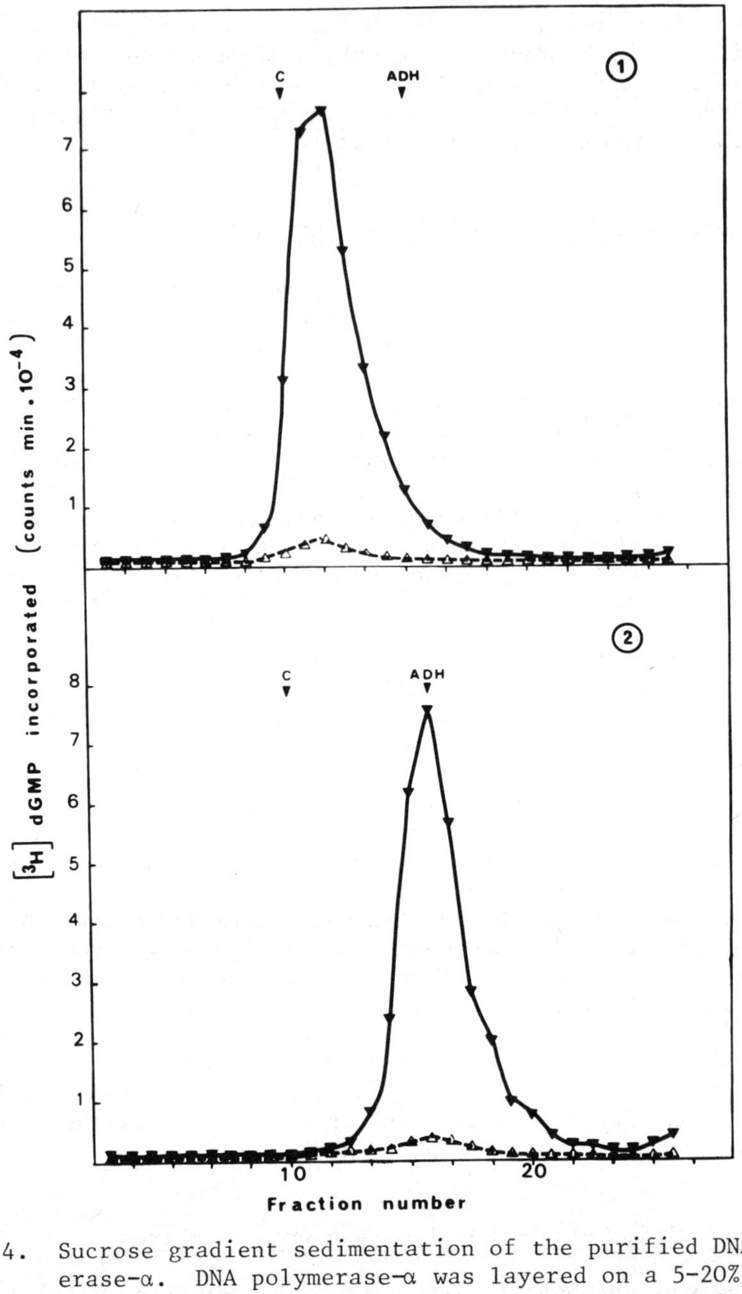

Fig. 4. Sucrose gradient sedimentation of the purified DNA polymerase-α. DNA polymerase-α was layered on a 5-20% sucrose gradient made in 50mM Tris-HCl pH 7.5, 1mM MgCl2, 10mM 2-mercaptoethanol and containing 10mM KCl (1) or 350mM KCl (2). Sedimentation was carried out at 4°C in a spinco SW 50.1 rotor for 15 hours at 37,000r.p.m. Marker proteins used were yeast alcohol dehydrogenase (7.4S) and bovine liver

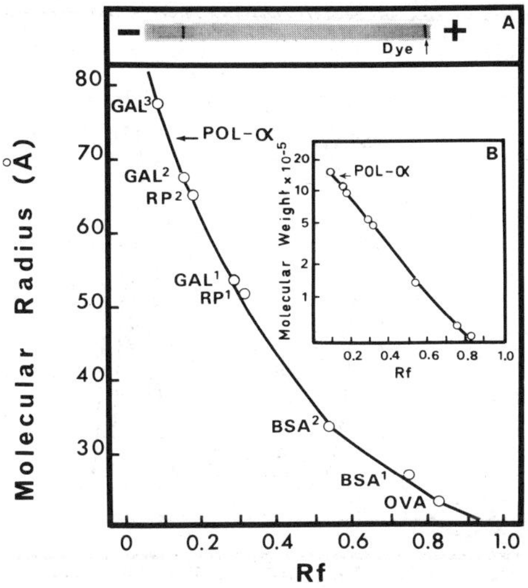

Fig. 5. Determination of the molecular radius of DNA polymerase-α by pore gradient gel electrophoresis under nondenaturing conditions. DNA polymerase-α, Fraction VI (5μg) previously dialysed against 40mM Tris-HCl (pH 7.5), 5mM KCl, 2mM 2-mercaptoethanol, 12% glycerol was electrophoresed on a non-denaturing 3.2 to 20% polyacrylamide gradient gel. The molecular radii of the marker proteins used were computed from published values of Mr an V and aligned on the calibration curve. OVA : ovalbumin; BSA^1 and BSA^2 : bovine serum albumin monomer and dimer; RP^1 and RP^2 : E. Coli RNA polymerase monomer and dimer; GAL^1, and GAL^2 and GAL^3 : β-galactosidase monomer, dimer and trimer.

thymus contains a 155,000 dalton polypeptide associated with smaller subunits[24]. The Drosophila embryo DNA polymerase also contains a high molecular weight polypeptide of 148K, associated with three smaller subunits of 58, 46 and 43K[28]. A similar result was obtained with the enzyme from chick embryo (Matsukage et al., this issue).

When the α-polymerase from regenerating rat liver (Fraction VI) was submitted to SDS polyacrylamide gel electrophoresis, five polypeptides of 156,000, 64,000, 61,000, 58,000 and 54,000 daltons were observed (Fig. 2E and Fig. 6). An additional higher molecular weight polypeptide could also be detected in some fractions, but in its absence (Fig. 2F), DNA polymerase-α activity was equally recovered.

catalase (11.3S). DNA polymerase activity was tested with poly (dC)·$(dG)_{12-18}$ in the presence of 6mM KCl (▼---▼) or 100mM KCl (▽---▽).

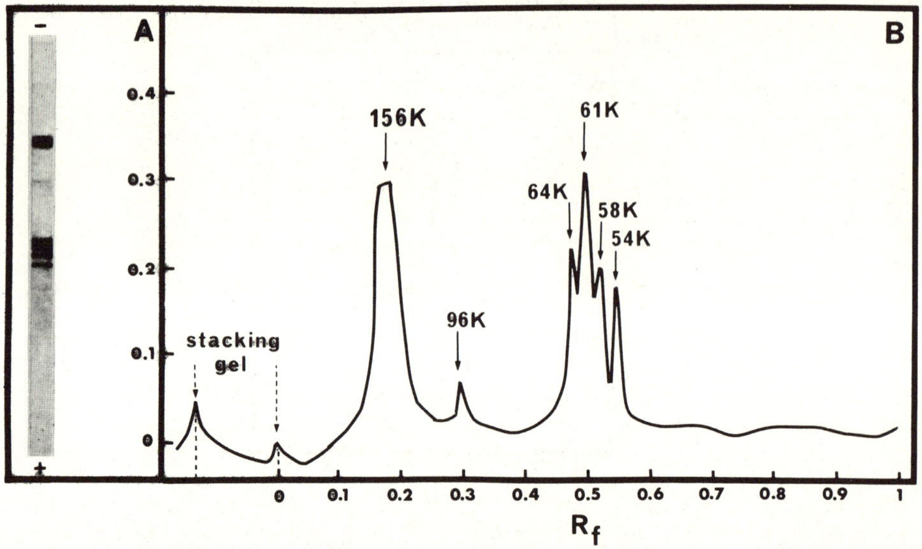

Fig. 6. Sodium dodecyl sulfate polyacrylamide gel analysis of DNA polymerase-α. Fraction VI enzyme (4.7μg) was analyzed on an SDS-polyacrylamide gel. An SDS-6 to 20% gradient polyacrylamide running gel was used to increase the resolution of protein bands. Samples were first treated by 3.7M urea for 1 hour before adjustment to the Laemmli final sample buffer. The molecular weight standards used were E. Coli RNA polymerase core enzyme subunits (165K, 155K and 39K), phosphorylase a (97K), bovine serum albumin (67K), ovalbumin (43K) and chymotrypsinogen (25.5K). (A), stained gel, (B), densitometer scan.

The same subunit composition was obtained by transferring the stained protein band from a native polyacrylamide gel electrophoresis (Fig. 3) onto a sodium dodecyl sulfate gel. The same result was also obtained after different treatments of the α-polymerase fraction before its loading onto the SDS gel.

The holoenzyme was separated from the core-enzyme by chromatography on DNA-cellulose[27], as shown in Figure 2. The catalytic activity correlated with the 156K polypeptide (Fig. 2F), whereas no activity was detected with the isolated 54-64K polypeptides which first eluted (Fig. 2C and D). The fraction possessing the best specific activity always contained 1 mol of the core subunit (156K polypeptide) associated with 0.7 to 1.2 mol of each of the 54K-64K polypeptides (Fig. 2E). However, when the 156K polypeptide was separated from its associated 54-64K polypeptides, its specific activity could be estimated up to sixfold lower[27].

Cross linking experiments have shown that the 54-64K polypeptides can constitute a heteropolymeric subunit which can also interact with the catalytic core polypeptide[27].

Table 2. Stimulation of the Core-Enzyme by the 54-64K Associated Subunit

Reconstitution	% DNA polymerase activity with Activated DNA	poly(dA).$(dT)_{12-18}$	poly(dT).$(A)_{10-12}$
core enzyme	100	100	100
54-64K associated subunit	ε	ε	ε
core enzyme associated subunit (% w/w)			
5/75	106		
15/85	143		
50/50		108	
38/62		123	
17/83		139	
38/62			140
24/76			136
24/76 (+0.2mM ATP)			164
14/86			218

The isolated core subunit and 54-64K subunit were mixed using different ratio w/w. After 20 min incubation at 4°C in the presence of 500μg/ml BSA, the DNA polymerase-α activity was tested with different template-primers. The activity of the core subunit and 54-64K subunit isolated were tested in each experimental case. The results are expressed as percentage of at least two determinations.

Electron microscopic studies of the holoenzyme, the core enzyme and the associated 54-64K subunit confirmed the biochemical results[27]. The four 54-64K polypeptides interacted to constitute a highly regular structure of 158 ± 10Å mean diameter, whereas the isolated core unit appeared as a smaller and less characteristic structure[27]. No structure of larger size appeared in the holoenzyme fraction; thus, the entire structure of the holoenzyme would seem to be imposed by the structural arrangement of the associated 54-64K subunit.

In vitro reconstitution experiments show that the 54-64K subunit can stimulate the catalytic activity of the core enzyme (Table 2). It is interesting to note that the best stimulation was registered with RNA-primed template.

Moreover, the stimulation became apparent only when the ratio (w/w) of the core subunit over the associated subunit reached the ratio obtained in the holoenzyme fraction. Experimental conditions however remain to be improved in order to fully reconstitute the holoenzyme.

Recently, it has been shown that the high molecular weight polypeptide from Drosophila embryo DNA polymerase-α was also the catalytic component of the enzyme, and that the smaller subunits increased its activity[33].

It is clear however, that no unified conclusion can be made at present concerning the subunit structure of the α-enzyme. Thus, in the same kind of species and tissue, different results were obtained in different laboratories. The calf thymus α-enzyme as isolated by Holmes et al. was composed of a 155,000 dalton catalytic core component which could be associated with a 50-70K subunit[24]. The calf thymus α-enzyme isolated by Grummt et al. was composed of seven polypeptides of 52-64K[29]. The calf thymus α-enzyme as isolated by Grosse and Krauss was composed of two polypeptides of 123 and 134K[30]. Proteolysis could be responsible for such a heterogeneity in the experimental data. In the case of rat liver, we have shown by tryptic fingerprinting that the different polypeptides of the enzyme are structurally distinct (to be published). A similar result has been obtained for the α-polymerase from Drosophila melanogaster embryos[33]. In the two cases, the presence or absence of protease inhibitors does not alter the polypeptide pattern in the purified α-polymerase[27,33]. However, in some cases, proteolysis could account for the presence of low molecular weight polypeptides in the α-polymerase preparation. Thus, the α-polymerase from calf thymus described by Grosse and Krauss seems to be a proteolysis product derived from a larger DNA polymerase[30]. Endogenous proteolysis has also been shown to produce different subspecies of α-polymerase from Drosophila melanogaster embryos[32].

CATALYTIC PROPERTIES OF RAT LIVER DNA POLYMERASE-α

Reaction Requirements

Maximum activity required the four deoxyribonucleoside triphosphates, DNA, and Mg^{2+}. The activity was inhibited at high ionic strength[38,39]. Optimim pH was around 8.2, although the enzyme was rather active at pH 9.1. The activity was strongly inhibited by N-ethylmaleimide, phosphonoacetic acid and aphidicolin (Table 3).

Table 3. Effects of Various Inhibitors on Regenerating Rat Liver DNA Polymerase-α Activity

Inhibitor	% activity remaining
100mM KCl	35
150mM KCl	7
2mM N-ethylmaleimide	5
0.05mM Phosphonoacetic acid	38
0.1mM Phosphonoacetic acid	20
0.1μg/ml Aphidicolin	55
10μg/ml Aphidicolin	25

DNA polymerase-α was assayed in the presence of the indicated concentrations of various inhibitors. The results are expressed as percent of activity detected in the absence of an inhibitor.

Table 4. Template-primer Specificity of DNA Polymerase-α from Regenerating Rat Liver

Template-primer	% activity
Activated calf thymus DNA	100
Poly(dC).$(dG)_{12-18}$	115
Poly(dT).$(A)_{10-20}$	35
Poly(dT).poly(A)	15
Supercoiled SV40 DNA I	<1
Denatured calf thymus DNA	<1

DNA polymerase-α activity was tested as described in[27]. The results are expressed as per cent of the activity detected with activated DNA under standard experimental conditions[27].

DNA activated by treatment with pancreatic DNase was the best effective primer-template for DNA polymerase-α. As shown in Figure 7, the optimum activity was reached when about 6% of the DNA had been converted to acid soluble products. Denatured DNA and supercoiled DNA were very bad templates for the α-enzyme. The α-polymerase could use an RNA-primer template with relatively good efficiency, as shown in Table 4.

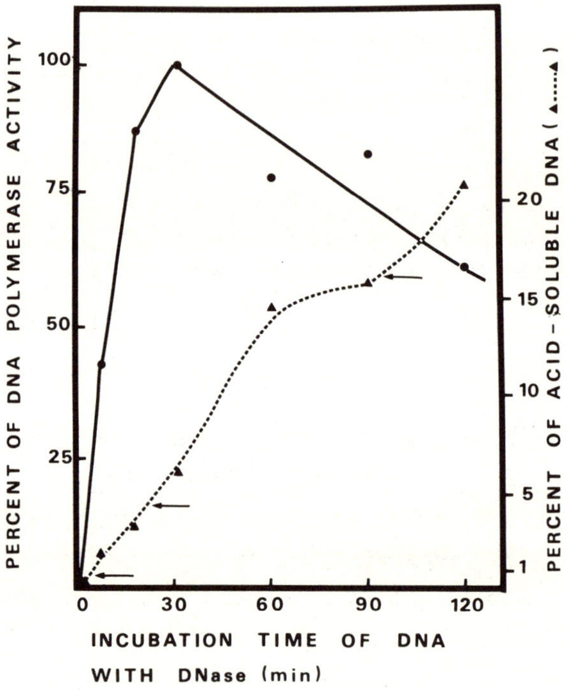

Fig. 7. Activation of calf thymus DNA by pancreatic DNase. Calf thymus DNA was incubated in 50mM Tris-HCl pH 7.5, 2.5mM $MgCl_2$, 0.05mM $CaCl_2$. The reaction was started by the addition of DNase 5μg/ml. After 20 min and 95 min at 35°C, the DNase concentration was raised to 7μg/ml and 9μg/ml (arrows). At different times, one aliquot was used to determine the acid-soluble DNA, 5mM EDTA and 0.1mM EGTA was added to a second aliquot which was then heated for 30 min at 65°C, in the presence of 0.5M NaCl. DNA was precipitated by two volumes of ethanol, dialyzed, and then tested for its template capacity in the presence of 0.4 units DNA polymerase-α. (Δ---Δ) per cent acid-soluble DNA, (o——o) per cent of DNA polymerase activity.

Thermal Sensitivity of the α-polymerase

The DNA polymerase-α from regenerating rat liver was found to be heat-sensitive[40]. As shown in Figure 8, an initial inactivation corresponding to 50% less of enzyme activity occurred in the first 5 min, and was then followed by inactivation at a slower rate. The same result was obtained when the residual activity was tested with different kinds of template-primers[40]. Thus, the ability of α-polymerase to use a DNA template of different base sequences in the presence either of a DNA or RNA 3'-hydroxyl primer to initiate DNA synthesis was equally sensitive to the heat treatment. This result indicates that a major conformational change occurs in the molecular structure of the enzyme during the first step of thermal denaturation. The first initial step of inactivation of the α-polymerase could be due to the dissociation of the oligomeric structure of the holoenzyme, whereas the second step of inactivation could be due to the inactivation of the catalytic core itself.

As shown in Figure 9, the presence of the DNA template during the heat treatment protects the α-polymerase against thermal denaturation, probably by stabilizing the structure of the holoenzyme.

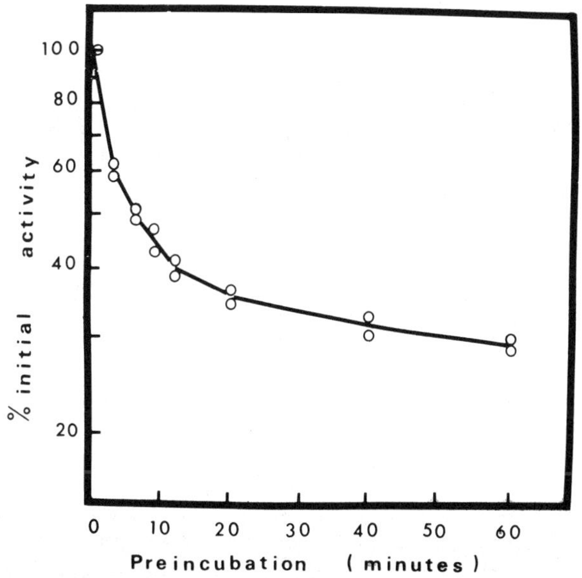

Fig. 8. Heat inactivation of DNA polymerase-α. 6.8 units DNA polymerase-α were incubated at 50°C in 90μl 50mM Tris-HCl (pH 7.6), 10mM KCl, 5mM 2-mercaptoethanol, 0.2mM EDTA, 45% glycerol and 1mg/ml bovine serum albumin. Samples (5μl) were withdrawn at the indicated times and immediately assayed using activated calf thymus DNA.

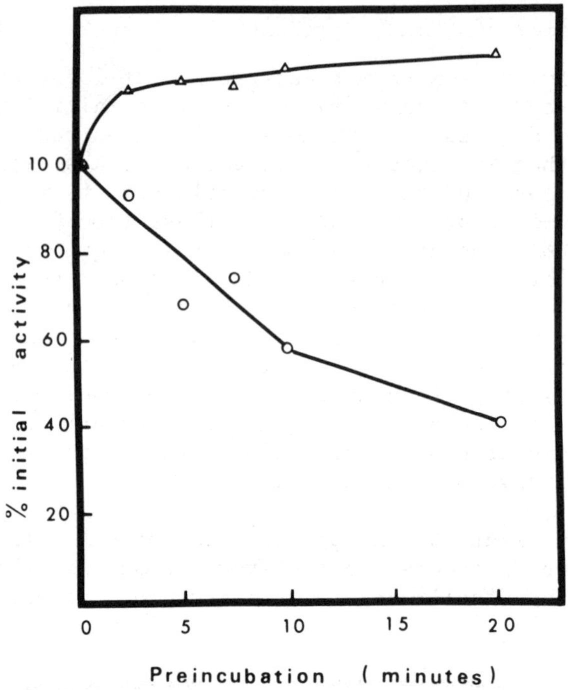

Fig. 9. Protection of DNA polymerase-α against heat inactivation by activated calf thymus DNA. 35 units DNA polymerase-α in the same buffer as in Figure 8, were separated in two equal fractions. Preincubation was then carried out at 37°C with (Δ---Δ), or without (○——○) activated calf thymus DNA (375μg/ml). Samples were withdrawn at the indicated times and assayed using activated calf thymus DNA. The activated DNA level in the polymerase assay was adjusted when relevant. 100% initial activity corresponded to 244 and 251 pmol incorporated in samples taken at zero time from inactive mixture with and without added activated DNA, respectively.

COMPARISON BETWEEN RAT LIVER AND DROSOPHILA ENZYMES

Banks et al.[28] and Villani et al.[33], recently described the subunit structure of DNA polymerase-α from Drosophila melanogaster embryos. This enzyme is composed of four polypeptides with molecular weights 148,000, 58,000, 46,000 and 43,000. Its subunit composition is very similar to that obtained for the enzyme from rat liver. This similarity was also observed when the two enzymes were compared on the same SDS-polyacrylamide gel. In our hands, the two high molecular weight polypeptides had the same mobility, corresponding to 155-160K, whereas slight differences were observable with the molecular weights of the smaller polypeptides. In both cases, the high molecular weight subunit was required for DNA polymerase activity[27,33]. This led us to make a direct comparison of the structure of the catalytic subunits of the α-polymerases from rat liver and Drosophila embryos, by means of the tryptic peptide mapping method.

The two catalytic subunits isolated from an SDS-polyacrylamide gel were [^{125}I] labeled, oxidized with performic acid, and digested with trypsin. Fingerprints were prepared at different pH in the first dimension, followed by chromatography in the second (to be published). Under these conditions, 25 to 30 [^{125}I] tryptic peptides were resolved. The position of 8 to 10 peptides was similar in the fingerprints of the rat liver and Drosophila embryo catalytic subunits (Fig. 10). The difference in the number of polypeptides found in common reflects only the capacity to discriminate between each peptide under the different fingerprint conditions. These results strongly suggest that the catalytic unit of the rat liver and Drosophila embryo α-polymerase are not structurally identical, but share some common tryptic peptides. Thus, a partial structural relationship appears between the DNA polymerase-α from a mammal (rat) and an insect (_Drosphila melanogaster_). This result adds further support to the numerous enzymatic and structural similarities between these two enzymes; this relationship is also of interest from a phylogenic point of view.

DISCUSSION

The DNA polymerase-α from regenerating rat liver can be purified to near homogeneity in two structural forms, at least partly interconvertible. The core-enzyme is related to a 156,000 dalton polypeptide which can associate with a subunit composed of four polypeptides of 64,000, 61,000, 58,000 and 54,000 daltons, to constitute the putative holoenzyme. The associated subunits stimulate the activity of the core subunit. Our results concerning the subunit structure of the enzyme agree relatively well with those obtained with calf thymus enzyme by Holmes et al.[24], and with the _Drosophila melanogaster_ embryo enzyme[33]. On the other hand, the α-polymerase from human KB cells[26] and from mouse myeloma cells[7] appears to be constituted of only low molecular weight polypeptides. The results obtained with the calf thymus enzyme, when purified in different laboratories[24,29,30], indicate that proteolysis could partially account for the disparity in the data (see above). However, we would like to emphasize that the subunit structure of the α-polymerase, as purified in different laboratories, is still confusing. The heterogeneity of DNA polymerase-α may be due to two essential factors. Proteolysis of a catalytic high molecular weight polypeptide can lead to the appearance of small, still active polypeptides. In this case, primary sequence homology would be clearly detected in the smaller polypeptides. Whether or not this proteolysis is artifactual, or

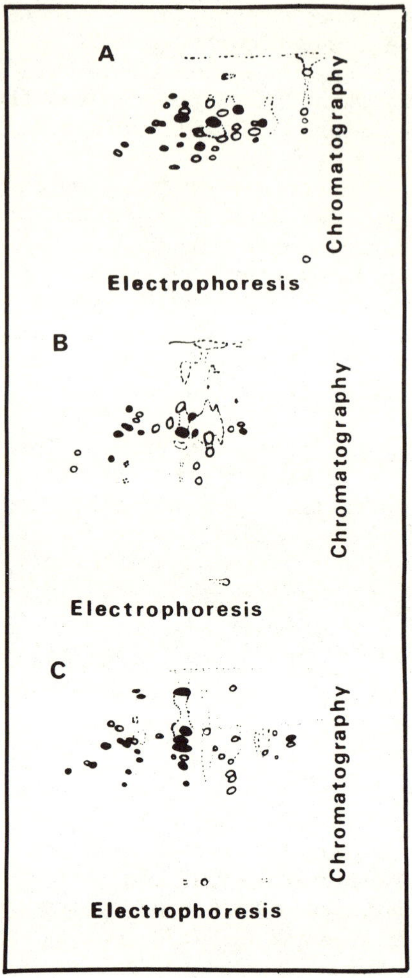

Fig. 10. Two-dimensional tryptic fingerprints of mixed digests of the catalytic core of regenerating rat liver and drosophila embryo DNA polymerase-α. Equal amounts of the tryptic digests of the [^{125}I] labeled catalytic core of regenerating rat liver and Drosophila embryo DNA polymerase-α were mixed, Electrophoresis was at pH 1.6 in the first dimension in (A), pH 4.7 in (B), and pH 6.5 in (C). Diagrams of the autoradiograms are shown. Shaded spots indicate the peptides found in common in the fingerprints of the two catalytic units analyzed separately (to be published). Dotted lines indicate unidentified or ambiguous areas.

whether it may be partly related to the normal biology of the enzyme in cells under different growing conditions, remains to be demonstrated. We cannot exclude the possibility that post-translational events modulate the activity of the α-enzyme during the cell cycle. The presence of the very high level of α-polymerase activity in oocytes[41,42] in the absence of DNA synthesis is of interest in terms of the metabolism of this enzyme.

On the other hand, the results obtained with the rat liver[27] and Drosophila DNA polymerase-α[28,33], as well as those obtained with the calf thymus enzyme by Holmes et al.[24], strongly suggest that α-polymerase may exist in two different structural forms : the catalytic core component, and the holoenzyme as the association of the core component with additional smaller subunit(s) which stimulate its activity[27]. We suggest, as a working hypothesis, that the associated subunit acts as an initiating factor which enhances the stabilization of the α-polymerase to the template-primer.

We propose a model which can explain our previously published results concerning the enzymatic properties and the binding of the α-polymerase to the DNA (Fig. 11). Template competition experiments performed in our laboratory indicated that the binding of regenerating rat liver α-polymerase to its template could be distinguished from the polymerization itself[39]. The holoenzyme could be the structural form involved in the binding to the DNA template-primer. In this case, the associated subunit stabilizes the binding of the core subunit to the 3'-hydroxyl primer. Once the initiation of synthesis begins, the complex between the core subunit and the 3'-hydroxyl primer allows the elongation to proceed. In DNA competition experiments, the stabilizing subunit could dissociate during the elongation, depending on the nature of the initiator-template or the nature of the competitor used[39].

When the holoenzyme is submitted to a heat treatment, the core subunit and the stabilizing unit are dissociated. In this case, the recognition of the template-primer by the core subunit drops rapidly, as shown in Figure 8 and in[40]. The same kind of result seems to be obtained by an urea treatment of the holoenzyme. Thus, Holmes et al.[43] with calf thymus enzyme, and Villani et al.[33] with the Drosophila embryo enzyme, report that the catalytic units of these two α-polymerases can be dissociated from their associated subunits by an urea treatment, with a consequent loss of activity.

We have also shown that the holoenzyme can bind to a supercoiled covalently closed DNA molecule, in the absence of free 3'OH ends which initiate the catalytic process of replication[25]. In this case, we also showed that, after centrifugation of the holoenzyme-DNA complex in sucrose gradients, the catalytic core was dissociated from its stabilizing subunit which remained bound to the DNA. A loss of activity concomitently occurred[25]. Since this kind of DNA does not contain free 3'OH ends, no stabilization of the structure of the holoenzyme could occur. On the other hand, in the presence of a

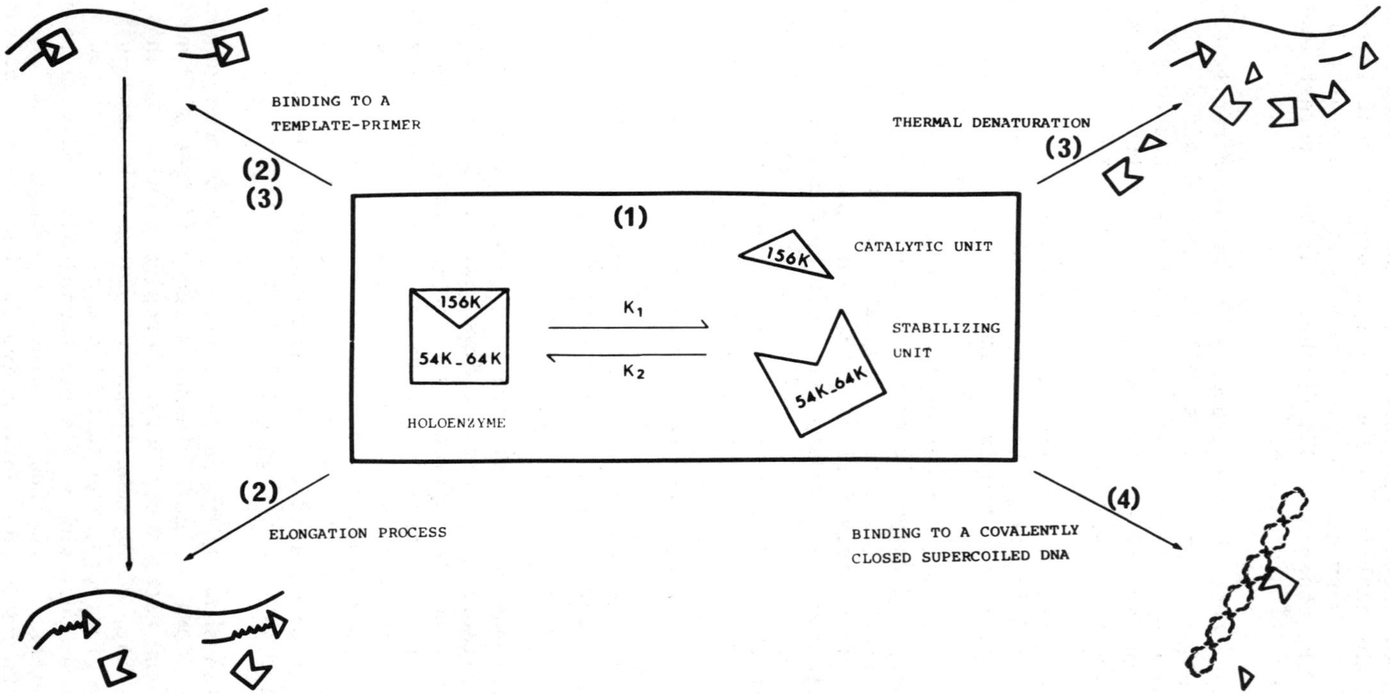

Fig. 11. Eukaryotic DNA polymerase-α : a structural model. Our results obtained concerning the *in vitro* functions of the α-polymerase can be related to the presence of two DNA polymerase-α species in the cell. The holoenzyme is defined as the most active structure implicated in the initiation of DNA synthesis using template-primers. The catalytic unit, or core-enzyme could be mainly involved in the elongation process. (1) to (4) refer to our results previously published (see p. 77).

template-primer, the structure of the holoenzyme was stabilized by its binding, both to the template and to the 3'OH primer (Fig. 9 and Ref.[40]).

Our working hypothesis could be tested in different ways. Thus, one could expect that the holoenzyme would be more protected by the template-primer against heat denaturation than the core enzyme. The holoenzyme would also seem to be more processive than the core-enzyme. This seems to be the case with the α-polymerase from Drosophila embryos (G. Villani, personal communication).

The presence of two α-polymerase species in eukaryotic cell can also be related to the normal biology of this enzyme. Thus, Pedrali Noy and Weissbach reported that blockage of protein synthesis by cycloheximide in growing HeLa cells leads to a 50% drop in activity of DNA polymerase-α, together with the appearance of a new chromatographic form of α-polymerase[44]. It would be interesting to know what is the structural modification of the α-polymerase after a block in protein synthesis, and whether or not the regulation of DNA synthesis is partly related to the regulation of synthesis of the subunits of the holoenzyme. From the same point of view, it would be interesting to study the structural relationship between the α-polymerases from oocytes and from dividing eggs, or the subunit composition of α-polymerase from tumor cells.

From a phylogenic point of view, we found that a partial structural relationship existed between the DNA polymerase-α from a mammal and an insect. It is also interesting to note that the replicative enzyme from E.coli, the holoenzyme DNA polymerase III, also presents some analogies with the rat liver and Drosophila embryo replicative enzyme. E. Coli DNA polymerase III holoenzyme was composed of six polypeptides with molecular weight of 140,000, 52,000, 42,000, 32,000, 25,000 and 10,000, and the high molecular weight polypeptide is the catalytic subunit of the enzyme[45,46]. Since the replication of DNA

(1) M. Mechali, J. Abadiedebat and A. M. de Recondo (1980) Eukaryotic DNA polymerase-α. Structural analysis of the enzyme from regenerating rat liver. J. Biol. Chem. 255:2114-2122. (2) O. Fichot, M. Pascal, M. Mechali and A. M. de Recondo (1979) DNA polymerase-α from regenerating rat liver. Catalytic properties of the highly purified enzyme. Biochem. Biophys. Acta 561:29-41. (3) M. Mechali and A. M. de Recondo (1980) Thermal sensitivity of eukaryotic DNA polymerase-α and protection by its templates. FEBS Letters 109:219-222. (4) M. Mechali and A. M. de Recondo (1978) Detection of a DNA binding factor associated with mammalian DNA polymerase-α. Biochem. Biophys. Res. Commun. 82:255-264.

enzymatically complex, one may suppose that all of these analogies are only a reflection of the constraints imposed by the structure of the substrate DNA itself. This could explain why the same kind of structure for the replicative enzyme was always selected. On the other hand, due to the importance and the required fidelity of replication, linked to its frequency, one may also suppose that some aspects of the DNA replication mechanism were submitted at a very early stage to a high selective pressure. To a certain extent, preservation of the replication mechanism could be associated with some retention of coding sequences for important replicative functions. In this respect, the partial homology between the catalytic unit of DNA polymerase-α from mammals and insects may reflect an essential structure conserved for the enzymatic synthesis of DNA. If such a homology among the DNA polymerase-α from different species is confirmed, it will be another argument for the major role of this enzyme in DNA replication. DNA polymerase-α could be the key-enzyme of a "replication assembly" containing the factors involved in the different steps of DNA replication and its regulation. Thus, characterization of the structure of this enzyme would be a good manner in which to understand the replication function.

ADDENDUM

Recently the correlation of the catalytic activity with the high molecular weight polypeptide from HeLa cells, Drosophila embryo, and various eukaryotic α-polymerases was further confirmed (P. Lamothe, B. Baril, A. Chi, L. Lee and E. Baril. Proc. Natl. Acad. Sci. U.S.A., in press; A. Spanos, S. G. Sedgwick, G. T. Yarranton, U. Hubscher and G. R. Banks (1981) Nucl. Ac. Research, 9:1825-1839). These articles also emphasize the heterogeneity of this enzyme.

REFERENCES

1. F. J. Bollum, J. Ann. Chem. Soc. 80:1766 (1958).
2. A. Weissbach, A. Schlabach, B. Fridlender and A. Bolden, Nature 231:167-1701971).
3. E. F. Baril, O. E. Brown, M. D. Jenkins and J. Laszlo, Biochemistry 10:1981-1992 (1971).
4. L. M. S. Chang and F. J. Bollum, J. Biol. Chem. 246:5835-5837 (1971).
5. B. Fridlender, M. Fry, A. Bolden and A. Weissbach, Proc. Natl. Acad. Sci. U.S.A. 69:452-455 (1972).
6. J. J. Byrnes, K. M. Downey, U. L. Black and A. S. So, Biochemistry 15:2817-2823 (1976).
7. Y. C. Chen, E. W. Bohn, S. R. Planck and S. H. Wilson, J. Biol. Chem. 254:11678-11687 (1979).
8. F. J. Bollum, Progr. Nucl. Ac. Res. 15:109-144 (1975).
9. A. Weissbach, Ann. Rev. Biochem. 46:25-47 (1977).

10. A. Falaschi and S. Spadari, in:"DNA synthesis, Present and Future," I. Molineux and M. Kohyama, eds., NATO Advanced Study Institute Series A, No. 17, pp. 487-515, Plenum Press, New York (1978).
11. M. L. de Pamphilis and P. M. Wassarman, Ann. Rev. Biochem. 49: 627-666 (1980).
12. L. M. S. Chang, M. Brown and F. J. Bollum, J. Mol. Biol. 74:1-8 (1973).
13. S. Spadari and A. Weissbach, J. Mol. Biol. 86:11-20 (1974).
14. U. Bertazzoni, M. Stefanini, G. Pedrali-Noy, G. Giulotto, F. Nuzzo, A. Falaschi and S. Spadari, Proc. Natl. Acad. Sci. U.S.A. 73:785-789 (1976).
15. L. M. S. Chang and F. J. Bollum, J. Biol. Chem. 247:7948-7950 (1972).
16. A. M. de Recondo and J. Abadiedebat, Nucl. Ac. Res. 3:1823-1837 (1976).
17. M. Mechali, M. Girard and A. M. de Recondo, J. Virology 23:117-125 (1977).
18. U. Wintersberger and E. Wintersberger, J. Virology 16:1095-1100 (1975).
19. H. J. Edenberg, S. Anderson and M. L. de Pamphilis, J. Biol. Chem. 9:3273-3280 (1978).
20. H. Krokan, P. Schaffer and M. L. de Pamphilis, Biochemistry 18: 4431-4443 (1979).
21. U. C. Hubscher, C. C. Kuenzle and S. Spadari, Nucl. Ac. Res. 4: 2917-2929 (1977).
22. W. C. Claycomb, 250:3229-3235 (1975).
23. A. M. Holmes, I. P. Hesslewood and I. R. Johnston, Europ. J. Biochem. 43:487-499 (1974).
24. A. M. Holmes, I. P. Hesslewood and I. R. Johnston, Europ. J. Biochem. 62:229-235 (1976).
25. M. Mechali and A. M. de Recondo, Biochem. Biophys. Res. Commun. 82:255-264 (1978).
26. P. A. Fisher and D. Korn, J. Biol. Chem. 252:6528-6535 (1977).
27. M. Mechali, J. Abadiedebat and A. M. de Recondo, J. Biol. Chem. 255:2114-2122 (1980).
28. G. R. Banks, J. A. Boezi and I. R. Lehman, J. Biol. Chem. 254: 9886-9892 (1979).
29. F. Grummt, G. Waltl, H. M. Jantzen, K. Hamprecht, U. Hubscher and C. C. Kuenzle, Proc. Natl. Acad. Sci. U.S.A. 76:6081-6085 (1979).
30. F. Grosse and G. Krauss, Nucl. Ac. Res. 8:5703-5714 (1980).
31. M. Duguet, M. Mechali and J. M. Rossignal, Anal. Biochem. 88:399-405 (1978).
32. C. L. Brackel and A. B. Blumenthal, Biochemistry 16:3137-3143 (1977).
33. G. Villani, B. Sauer and I. R. Lehman, J. Biol. Chem. 255:9479-9483 (1980).
34. M. Mechali and A. M. de Recondo, Eur. J. Biochem. 58:416-466 (1975).

35. A. Matsukage, E. W. Bohn and S. H. Wilson, Biochemistry 14:1006-1020 (1975).
36. S. Spadari, R. Muller and A. Weissbach, J. Mol. Biol. Chem. 249: 2991-2992 (1974).
37. J. Margolis and K. G. Kenrick, Anal. Biochem. 25:347-362 (1968).
38. A. M. de Recondo, J. A. Lepesant, O. Fichot, I. Grasset, J. M. Rossignol and M. Cazillis, J. Biol. Chem. 248:131-137 (1973).
39. O. Fichot, M. Pascal, M. Mechali and A. M. de Recondo, Biochem. Biophys. Acta 561:29-41 (1979).
40. M. Mechali and A. M. de Recondo, FEBS Letters 109:219-222 (1980).
41. R. M. Benbow, R. Q. W. Pestell and C. C. Ford, Developmental Biology 43:159-174 (1975).
42. G. Martini, F. Tato, D. Gandini Attardi and G. P. Tocchini-Valentini, Biochem. Biophys. Res. Commun. 72:875-879 (1976).
43. A. M. Holmes, I. P. Hesslewood and I. R. Johnston, Nature 255:420-422 (1975).
44. G. Pedrali Noy and A. Weissbach, Biochem. Biophys. Acta 477:70-83 (1977).
45. C. Mc Henry and A. Kornberg, J. Biol. Chem. 252:6478-6484 (1977).
46. T. Kornberg and M. L. Gefter, J. Biol. Chem. 247:5369-5375 (1972).

DNA CHAIN ELONGATION MECHANISM OF DNA POLYMERASES α, β AND γ

Akio Matsukage, Masamitsu Yamaguchi, Kazushi Tanabe,
Yukari N. Taguchi, Miwako Nishizawa and Taijo Takahashi

Laboratory of Biochemistry
Aichi Cancer Center Research Institute
Chikusa-ku, Nagoya 464, Japan

1. INTRODUCTION

There are many lines of evidence that indicate the existence of at least two kinds of DNA replication mechanisms in eukaryotic cells.

(1) One is that observed in nuclear DNA replication, where DNA chains (at least the lagging strand) are synthesized in relatively short pieces (3-5*s*) that are later elongated and joined together[1-6]. These short DNA intermediates are also observed in the replication of viral DNA such as polyoma virus[7,8] and simian virus (SV)40[9,10].
(2) The other is that for adenovirus DNA[11-12] and mitochondrial DNA[13]. These DNA's are not replicated via Okazaki pieces as the intermediates, but replicated in a continuous mode.

A number of circumstantial observations[14,15] such as the subcellular localizations or the effects of inhibitors of DNA polymerases suggest that DNA polymerase α, β, γ are ivolved in nuclear DNA replication, DNA repair and mitochondrial DNA replication, respectively. In addition it is suggested that DNA polymerase α and γ are involved in the replication of SV40 DNA[16] and adenovirus DNA[17,18], respectively. Does the difference between the two types of replication mechanism depend on the difference in the reaction properties of DNA polymerases?

The present paper first describes the purification and the structural analysis of chick embryo DNA polymerases, which gives firm bases for the kinetic analysis. This paper deals, then, with the quantitative analysis of DNA chain elongation mechanisms by DNA polymerase α, β and γ.

2. PURIFICATION AND STRUCTURAL ANALYSIS OF CHICK EMBRYO DNA POLYMERASES

DNA polymerase β[19]

It has been well recognized that mammalian DNA polymerase-β consists of a single polypeptide of about 40,000 dalton[20-24]. On the other hand, chick embryo DNA polymerase β has been believed to consist of Mr=27,000 polypeptides[25,26]. We have been trying extensive purification of all three DNA polymerases from chick embryos. The systematic fractionation and purification procedures are summarized in Figure 1.

Table 1. Purification of DNA polymerase β

	Fraction	Protein		Specific activity
		mg[a]	units[a] (%)	units/mg
Crude extract	I	120	51(100)	0.43
First phosphocellulose and first ammonium sulfate fractionation	II	2.7	120(240)	44
DEAE-cellulose	III	2.4	100(200)	42
Second phosphocellulose and second ammonium sulfate fractionation	IV	$1.1x10^{-2}$	200(390)	18,000
Sephadex G-150	V	$3.4x10^{-3}$	200(390)	59,000
Third phosphocellulose P-11	VI	$1.5x10^{-3}$	130(250)	87,000
Blue Agarose	VII	$2.7x10^{-4}$	57(110)	210,000
Single-stranded DNA-cellulose	VIII	$2.9x10^{-5}$	24	830,000
	(VIII-e)	$4.4x10^{-6}$	(47)	1,100,000[b]
	VII	$2.9x10^{-5}$	24	490,000[c]
	(VIII-e)	$6.5x10^{-6}$		750,000[b,c]

[a]Quantities are expressed per g wet weight of chick embryos.
[b]Fractions in the peak of DNA polymerase activity (see Fig. 3).
[c]The protein value was derived by densitometry of a stained gel. All other protein amounts were determined from UV absorbance.

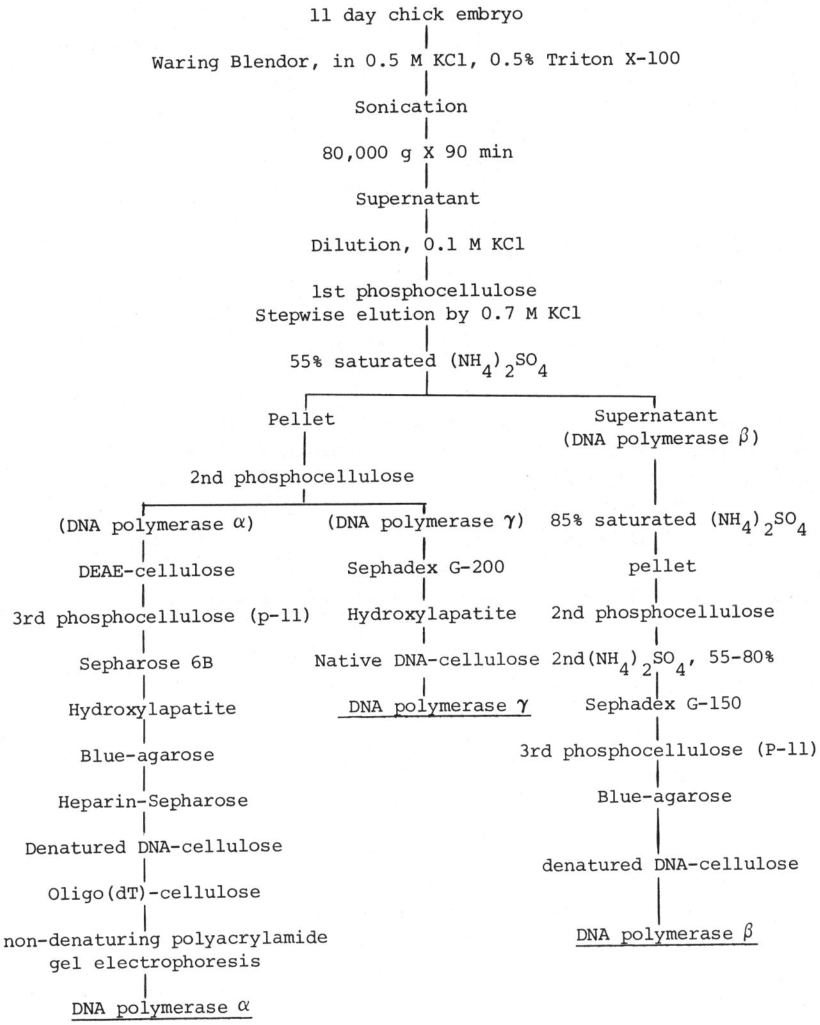

Fig. 1. Systematic fractionation and purification of chick embryo DNA polymerase α, β and γ.

An example of the purification of DNA polymerase-β was shown in Table 1. The extract was made from chick embryos obtained from 11-day fertilized eggs. About 500g of material was used for one operation. After the first phosphocellulose column chromatography (a stepwise elution), DNA polymerase α and γ were precipitated by 55% saturated $(NH_4)_2SO_4$ while DNA polymerase β remained in the supernatant, from which it was precipitated by 80% saturated $(NH_4)_2SO_4$[27]. DNA polymerase β was, then, purified by a series of column chromatographies including affinity chromatographies on a Blue-Agarose column and a denatured DNA-cellulose column. DNA polymerase β activity was eluted from a Sephadex G-150 column as a single peak and its apparent molecular weight was estimated as 40,000 (Fig. 2). No activity larger or smaller than 40,000 was detected.

Figure 3 shows the final step of the purification, a denatured DNA-cellulose column chromatography. The polypeptides in the indicated fractions were analyzed by a sodium dodecyl sulfate (SDS)-polyacrylamide gel electrophoresis (the inset of Fig. 3). A Mr=40,000 polypeptide was the only major one and accounted for more than 95% of total protein. The final preparation had a specific enzyme activity of 1,100,000 units/mg protein with $(rA)_n$-$(dT)_{12-18}$ as a template·primer.

No polypeptide was detected around Mr=27,000. However, when the purification was performed omitting three steps (the second $(NH_4)_2SO_4$ fractionation, the third phosphocellulose column chromatography and the Blue-Agarose column chromatography), the preparation contained a Mr=27,000 polypeptide in addition to the Mr=40,000 one. The structures of Mr=27,000 and Mr=40,000 polypeptides were compared by two-dimensional tryptic peptide mapping[19,28,29]. No common spot of peptide was detected between these polypeptides (Fig. 4). The evidence indicated that the Mr=27,000 polypeptide is not the proteolytically degraded product of Mr=40,000 polypeptide but a contaminant. On the other hand, the Mr=40,000 polypeptide of chick embryo DNA polymerase β has a partial homology with Mr=40,000 polypeptide of DNA polymerase β from rat ascites hepatoma cells[24]. Fourteen out of twenty four spots of tryptic ^{125}I-peptides of the chick enzyme were found in the map of the rat enzyme.

The partial homology in structure of the avian and the mammalian DNA polymerase β's was further confirmed by an immunological method. An antibody against chick embryo DNA polymerase β represented partial cross reaction with mouse DNA polymerase β.

In conclusion, chick embryo DNA polymerase β, like the mammalian enzyme, consists of a single Mr=40,000 polypeptide.

DNA polymerase γ[27]

Although there have been several reports[30-34] dealing with the purification from various organisms, DNA polymerase γ has not been purified to homogeneity sufficient for the structural analysis. This section describes the first success of the structural analysis of DNA polymerase γ.

A purification procedure for the chick embryo DNA polymerase was summarized in Figure 1 and an example was shown in Table 2. The final preparation has a specific activity of 570,000 units/mg protein, which is more than 20 times that reported previously[32,33]. DNA polymerase γ was separated from DNA polymerase α by the 2nd

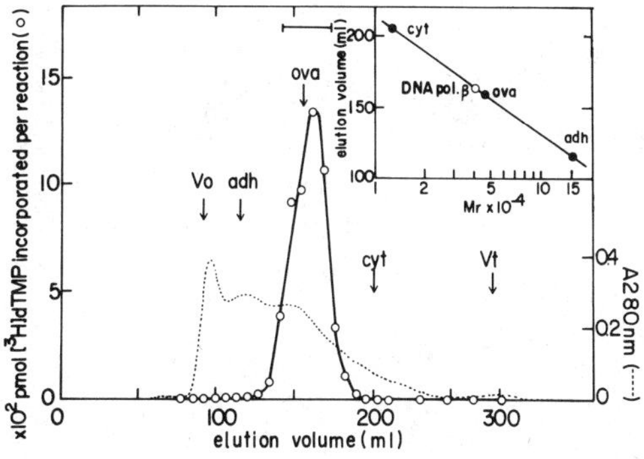

Fig. 2. Gel filtration on a Sephadex G-150 column. Fraction after the second phosphocellulose column chromatography and the following $(NH_4)_2SO_4$ fractionation, DNA polymerase β was chromatographed on a column. The chromatography was developed by a solution containing 0.5M KCl, 50mM Tris-HCl (pH 7.6), 20% glycerol, 0.1mM EDTA, 0.5mM dithiothreitol. The DNA polymerase activity in a 1-μl portion of each fraction was measured. adh, alcohol dehydrogenase (yeast); ova, (ovalbumin); and cyt, cytochrome c (horse).

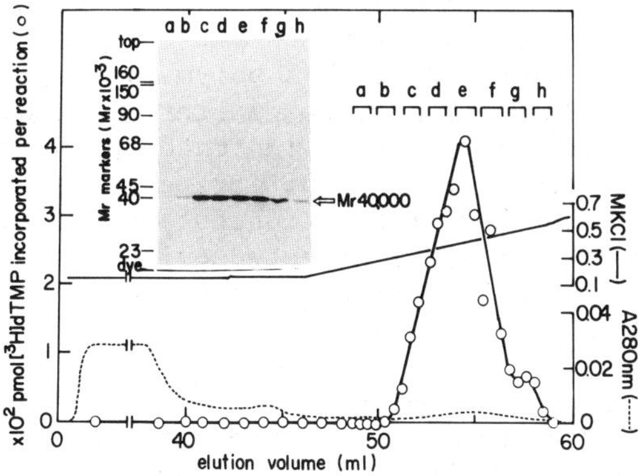

Fig. 3. Single-stranded DNA-cellulose column chromatography and SDS-polyacrylamide gel electrophoresis of the purified DNA polymerase β. After the Blue-Agarose column chromatography, DNA polymerase β was applied to a single-stranded DNA cellulose column. The protein was eluted with a linear gradient of 0.15 - 0.6M KCl. DNA polymerase activity in a 1-μl portion of each fraction was measured (5 min at 37°C). One hundred microliters of each fraction was pooled as indicated, and each sample (300μl) was subjected to SDS-polyacrylamide gel electrophoresis (inset).

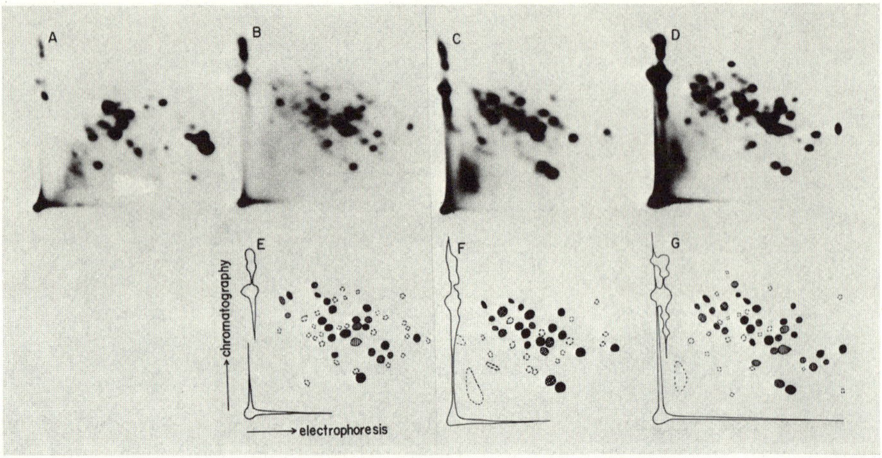

Fig. 4. Tryptic peptide mapping of ^{125}I-iodized polypeptides. Polypeptide bands of Mr=40,000 and 27,000 in the polyacrylamide gel, along with a Mr=40,000 polypeptide band of DNA polymerase β from rat ascites hepatoma AH130 cells[24], were taken and separately ^{125}I-labelled. Tryptic peptides were mapped and autoradiography was carried out. A, Mr=27,000 polypeptide from the chick DNA polymerase β preparation; B, Mr=40,000 polypeptide from the chick preparation; C, Mr=40,000 polypeptide from the rat DNA polymerase β; D, mixture of ^{125}I-labelled tryptic peptides of Mr=40,000 polypeptides from chick and rat DNA polymerase β; E, the traced diagram of B; F, the traced diagram of C; G, the traced diagram of D. Solid spots indicate spots common to both the chicken and rat enzymes; striped spots indicate spots present in chicken preparation but absent in rat preparation; meshed spots are spots detected only in rat preparation; and spots indicated by dotted lines are minor ones.

phosphocellulose column chromatography and the hydroxylapatite column chromatography. Two peaks of the enzyme activity were observed in the gel filtration on a Sephadex G-200 column. The first peak has Mr=280,000 while the second has Mr=180,000. Two activities were separately purified through the final step. Affinity chromatographies on native DNA-cellulose columns were shown in Figure 5 and the peptide contents in the fractions containing the enzyme activity were analyzed. Only one polypeptide of Mr=47,000 was detected in the preparation from the Mr=280,000 active form, and its amount changed in proportion to the amount of the enzyme activity. Although there were several polypeptides in the preparation from Mr=180,000 activity, Mr=47,000 polypeptide was again proportional to the enzyme amount. A minor polypeptide of Mr=135,000 seemed to be proportional to the enzyme, while others were not.

Table 2. Purification of DNA polymerase γ

Step	Fraction	Protein[a]	Activity[a]	Specific activity
		mg	units (%)	units/mg
Crude extract	I	80	30(100)	0.38
First phosphocellulose and ammonium sulfate fractionation	II	4.8	17(57)	3.5
Second phosphocellulose	III	4.1×10^{-1}	23(77)	56
Sephadex G-200	IV-1	1.0×10^{-1}	8.4(28)	84
	IV-2	9.0×10^{-2}	9.0(30)	100
Hydroxylapatite	V-1	8.4×10^{-3}	4.1(14)	490
	V-2	8.1×10^{-3}	5.0(17)	620
Double-stranded DNA-cellulose	VI-1	9.0×10^{-6}	3.6(12)	400,000
	(VI-1-d)[b]	1.3×10^{-7}		570,000
	VI-2	1.3×10^{-5}	4.5(15)	350,000

[a]Quantities are expressed per g wet weight of chick embryos.
[b]Fractions in the peak of DNA polymerase activity (see Fig. 6A).

We concluded that the only Mr=47,000 polypeptide is the component of the purified chick embryo DNA polymerase γ. The reasons are as follows:

(1) The preparation of Mr=280,000 activity contained Mr=47,000 polypeptide almost exclusively and not the Mr=135,000 polypeptide.
(2) Specific activities of DNA polymerase γ in Mr=280,000 and Mr=180,000 forms were equal with respect to Mr=47,000 polypeptide (660,000 units/mg and 650,000 units/mg, respectively). However, the specific activity with respect to Mr=135,000 was estimated extremely high (2,600,000 units/mg).
(3) As shown later, the initial rate of poly(dT) chain elongation by this enzyme was equal to the turnover number which was calculated on the assumption that only Mr=47,000 polypeptide is the component of DNA polymerase γ.

As described earlier, the minimum molecular weight of the activity was about 180,000 which agrees with that estimated from _s_-value (7.5_s_) determined by a glycerol gradient centrifugation in the presence of 1M KCl and 0.5% Triton X-100. Thus, the basic active

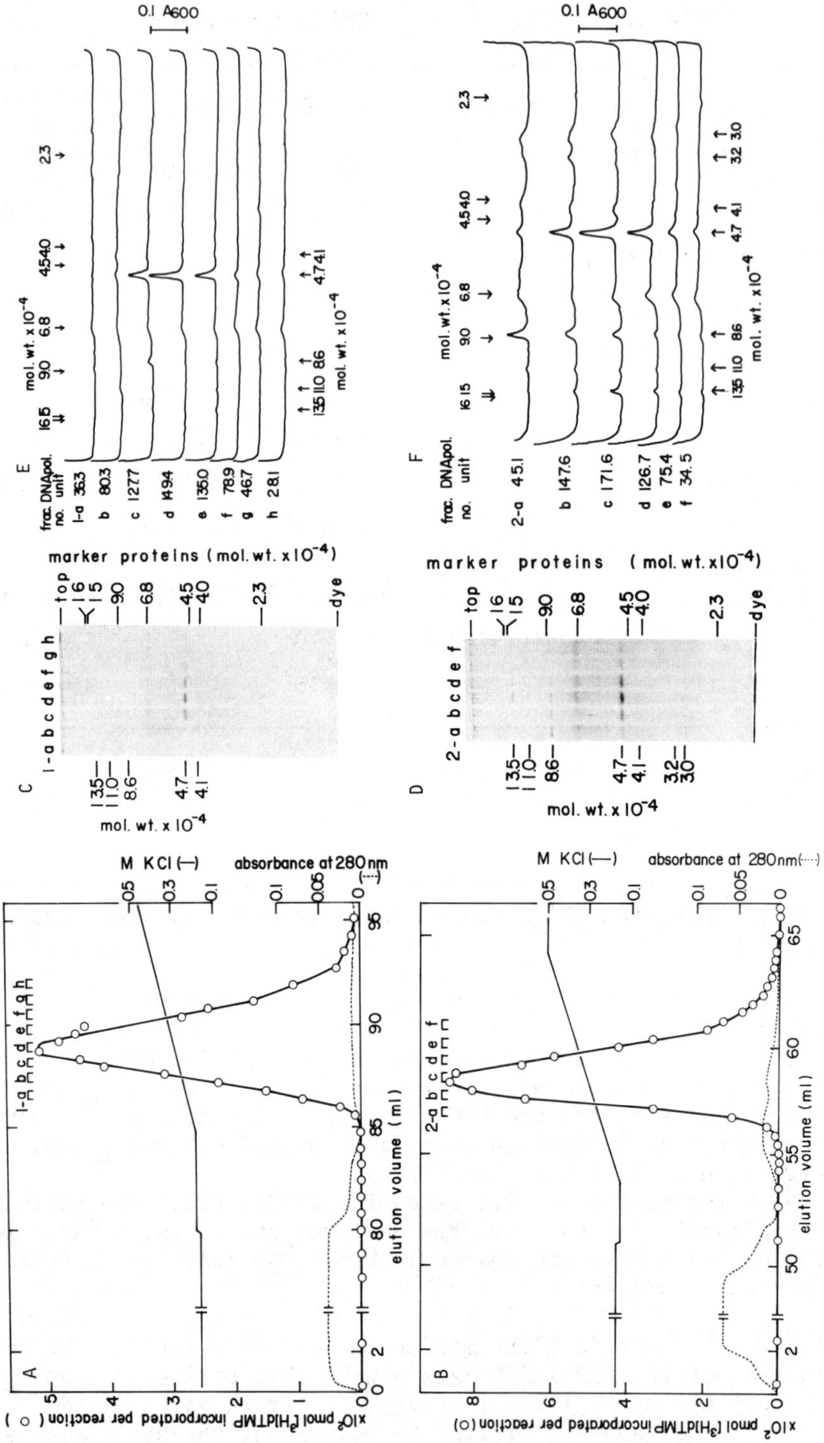
A
B
C
D
E
F
M KCl (—)
absorbance at 280 nm
elution volume (ml)
x10² pmol [³H]dTMP incorporated per reaction (○)
marker proteins (mol. wt. x 10⁻⁴)
mol. wt. x 10⁻⁴
0.1 A600
frac. no.
DNA pol. unit
top
dye

form of the chick embryo DNA polymerase γ might consist of the tetramer of Mr=47,000 polypeptides. It is left to further examination to resolve whether the Mr=280,000 form seen in the earlier step of the purification has a different assembly of the subunit, such as a hexamer, or whether it is associated with the other protein component(s).

Recently, we extensively purified DNA polymerase γ from the mouse myeloma MOPC104E cells. Although the preparation is not as pure as the chick enzyme, the amount of the Mr=47,000 polypeptide was proportional to the enzyme activity through the DNA-cellulose column chromatography. Furthermore, the tryptic peptide mapping of ^{125}I-labelled polypeptides indicates that the Mr=47,000 polypeptides from chick and mouse enzymes share partially the same structure (our unpublished data). A serological cross-reaction between chick and mouse enzymes was also observed. We have already reported that the molecular weight of mouse DNA polymerase γ was between 180,000-280,000[(32)]. Therefore, the basic structure of the mouse enzyme might be a tetramer of Mr=47,000 polypeptides. The specific activity of the mouse enzyme is 490,000 units/mg of Mr=47,000 polypeptide.

DNA polymerase α

Chick embryo DNA polymerase α was eluted as a broad peak or split into several peaks of activity in various column chromatographies as well documented with mammalian enzymes[35-38]. The biological significance of the heterogeneity was discussed by us[39-40] and others[41]. The heterogeneity, the lability of the activity in the purified preparation and the weak binding ability to DNA-cellulose caused the extreme difficulty in obtaining the homogeneous preparation. After twelve purification steps, we obtained an electrophoretically pure preparation. SDS-polyacrylamide gel electrophoresis of the prep-

Fig. 5. Double-stranded DNA-cellulose column chromatography and SDS-polyacrylamide gel electrophoresis of the purified DNA polymerase γ. Fractions from the hydroxylapatite column were each chromatographed on a double-stranded DNA-cellulose column. Fraction from the Mr=280,000 form (A) and fraction from the Mr=180,000 form (B) were applied. Protein was eluted with a linear gradient of 0-15 to 0.5M KCl. The DNA polymerase γ activity in 2μl of each fraction was measured (o-o). Fractions were combined as indicated in A and B (1-a to h and 2-a to f). Then 300μl of 1-a to h (C) and a 200μl portion of 2-a to f (D) were subjected to the electrophoresis in SDS-polyacrylamide gel. C and D, photographs of stained gels; E and F, traces obtained with a scanning densitometer. In E and F, the DNA polymerase γ activity (in units) applied to each slot of the gel is indicated.

aration represented 5 polypeptide bands and their molecular weights were 130,000-150,000 (a broad band), 59,000, 56,000, 54,000, and 51,000. The result is similar to those reported with DNA polymerase-α's from calf thymus[36], regenerating rat liver[42], and Drosophila embryo[43,44]. The tryptic peptide mapping indicates that four polypeptides between 51,000 and 59,000 daltons have very similar structures, but no homology was detected between the 130,000-155,000 group and the 51,000-59,000 group. Results indicate that the small group is not the proteolytic product of the large group and the small group polypeptides might be generated from one polypeptide by some minor modification(s) (unpublished data).

Estimation of the Molecular Number per Cell and the Turnover Numbers of Chick Embryo DNA Polymerase β and γ

Information about the pure enzyme made it possible to estimate the exact number of enzyme molecules from the amount of activity. The calculation was done from the values of the specific activities of the pure enzymes, their molecular weights, enzyme contents per g tissue and cell number per g tissue which was obtained from the measurement of the DNA content (1.4mg/g chick embryo tissue[19]) and DNA content per cell (2.3×10^{-12}g per fibroblastic cell[45]). The enzyme numbers per g cell were estimated as 4,980 for DNA polymerase β and 250-750 for DNA polymerase γ. It should be pointed out that these numbers represent only "extractable and enzymatically active" molecules. Therefore, these are "minimum" numbers.

The turnover numbers (the number of nucleotides polymerized per unit time per enzyme molecule) are 740 nucleotides/min for DNA polymerase β and 2,080 nucleotides/min for DNA polymerase γ. These values give the very accurate bases for the analysis of DNA chain elongation mechanisms of these enzymes as described in the following section.

3. MODE OF DNA CHAIN ELONGATION BY CHICK EMBRYO DNA POLYMERASE β AND γ

$(r)_n \cdot (dT)_{12-18}$ was employed as a template·primer for the analysis of DNA chain elongation mechanisms of DNA polymerase β and γ. Reaction was carried out in the mixture containing the template·primer, the highly purified DNA polymerase β or γ and the other requirements including $[^3H]$dTTP. At various times of incubation, the reaction mixture was mixed with an equal volume of the denaturing solution containing 1M KCl, 0.6N NaOH, 1.5% Sarcosyl and 30mM EDTA. Then, the size of the product was determined by alkaline sucrose gradient centrifugation method[46].

Table 3. Estimation of Molecular Number/Cell and Turnover Number of Chick Embryo DNA Polymerase β and γ

	DNA polymerase β	DNA polymerase γ
1. Observed results		
a) Specific activity of the final preparation (units/mg)	1,100,000	570,000
b) Purity (%)	>95	86
c) Molecular weight (estimated from gel filtration, s-value)	40,000	180,000 and 280,000
d) Polypeptide (molecular weight)	40,000	47,000
e) Enzyme content (unit/g tissue)	200[a]	90[b]
f) Cell number/g tissue	6.1×10^{8}[c]	
2. Expected values		
1) Specific activity of homogeneous enzyme (units/mg)	1,100,000	660,000
2) Polypeptide constituent	$(40{,}000)_1$	$(47{,}000)_4$ (a basic form)
3) Molecular weight	40,000	188,000
3. Calculated values		
1) Enzyme number/cell	4,980	750
2) Turnover number (nucleotides/min/enzyme)	733	2,040

[a]See Table 1. after the 2nd phosphocellulose step.
[b]The highest value ever obtained.
[c]Determined from the measurement of DNA content (see text).

DNA polymerase β

The result of zone sedimentation analysis of the product synthesized by DNA polymerase β was shown in Figure 6. The sizes of 10, 20, 30 and 45min products were 2.9, 3.6, 4.0 and 4.5s, respectively. The length (in nucleotide number) was calculated from s-values. The products were elongated very slowly at the rate of about 4 nucleotides per min. The product number can be calculated by dividing the number of dTMP incorporated by the average nucleotide number per product. The enzyme number per reaction was also calculated from the DNA polymerase activity added as described earlier. The product number per enzyme molecule is constantly about 200 throughout the reaction time. Results indicate that one DNA polymerase β molecule polymerized dTMP on many primers in a highly distributive fashion; the enzyme added one or so nucleotides(s) to one primer, then left from the growing point to work on another primer in a random fashion. Thus, all available primers might be equally elongated. Essentially the same mechanism was observed with the mouse DNA polymerase β[47].

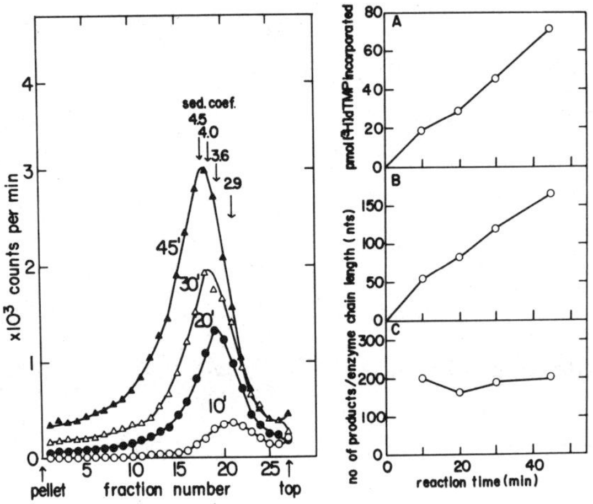

Fig. 6. Alkaline sucrose gradient centrifugation of poly(dT) products synthesized by DNA polymerase β. Reaction mixtures were incubated at 37°C for 10, 20, 30, and 45min in a final volume of 100μl containing the similar component described in the legends of Figure 7 except that 1 : 1 mixture of $(rA)_n$ and $(dT)_{12-18}$ was used. The product was analyzed in alkaline sucrose gradient centrifugation (left panel). The amounts ^3H dTMP incorporated into polymer were measured using 5μl aliquots of reaction mixture (right-A). Molecular weights were calculated from s-values and expressed in nucleotide number (right-B). Product number/enzyme was calculated from the $[^3H]$dTMP incorporated, molecular weight of the product and enzyme number (c.f. text) (right-C).

DNA polymerase γ[46]

The size analysis of products synthesized by chick embryo DNA polymerase γ was shown in Figures 7 and 8. Polymerization of H dTMP continued almost linearly for at least 2h. The sizes of products synthesized in the reaction for 0.4, 0.9, 1.5, 2 and 3min were 7.7s, 10s, 11.9s, 13,0s and 13.9s, respectively. After 3min, the rate of the chain elongation leveled off and the size of product synthesized

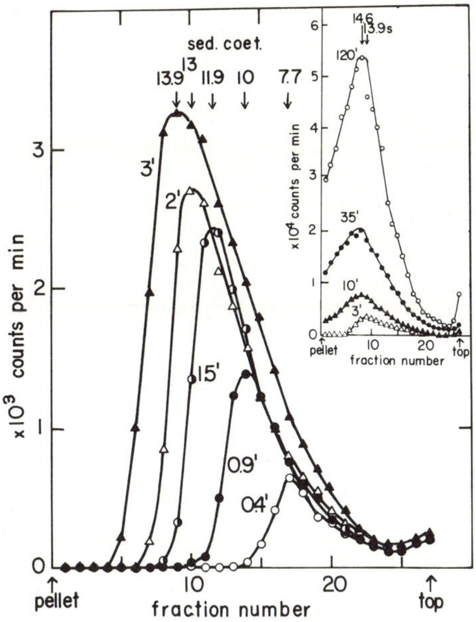

Fig. 7. Alkaline sucrose gradient centrifugation of poly(dT) products synthesized by DNA polymerase γ. Reaction mixtures were incubated at 37°C for 0.4, 0.9, 1.5, 2, 3, 10, 35 and 120min in a final volume of 100μl containing the following components: 50mM Tris-HCl (pH 8.5), 0.5mM $MnCl_2$, 1mM dithiothreitol, 110mM KCl, 20mM potassium phosphate, 400μg per ml bovine serum albumin, 14% glycerol, 80μg per ml $(rA)_n$, 16μg per ml $(dT)_{12-18}$, 12.7μM [³H]dTTP and 0.65 units DNA polymerase γ. After incubation, the reaction was mixed with an equal volume of a 2 X denaturing solution, then incubated at 37°C for 20min. 100μl aliquots of the resulting solutions were layered over 4.8ml 5-20% sucrose gradients made in the solution containing 1M NaCl, 0.3M NaOH and 5mM EDTA, and then centrifugation was carried out for 15h at 38,000r.p.m. at 5°C. The gradient was fractionated into 27 fractions from the bottom and a 100-μl portion was placed on a DEAE-cellulose paper disk and washed. Radioactivity in poly(dT) retained to the disk was measured.

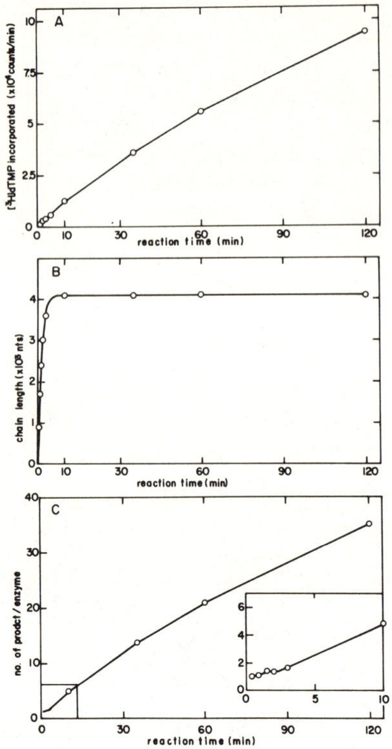

Fig. 8. The time course of incorporation of [^{3}H]dTMP into poly(dT) (A), elongation of poly(dT) chains (B) and the number of products per DNA polymerase γ molecule (C). Incorporation of ^{3}H dTMP was measured as described in Figure 6 legend.

in the reaction lasting 10min or longer was always 14.6*s*. Molecular weights were calculated from *s*-values and represented in nucleotide numbers. The initial chain elongation rate was 1,900 nucleotides/min. The final length (average about 4,100 nucleotides) was reached at about 4min, after which the size did not change but the number of products with the final length increased with reaction time. The initial DNA chain elongation rate coincided with the turnover number (2,080 nucleotides/min) described earlier. Results indicate that a DNA polymerase γ molecule completed the synthesis of one long poly(dT) chain before it started with the next primer. The number of products per enzyme molecule was one for the initial 3min of the reaction, and thereafter increased in proportion to reaction time. It is concluded that the mechanism of DNA polymerase γ is highly processive; if the enzyme started with one primer, it worked continuously on the growing point of the DNA chain until completion. This mechanism is quite different from that of DNA polymerase β. The mechanism is essentially the same with DNA template (unpublished data).

The chain elongation mechanism by mouse myeloma DNA polymerase γ is the same as that of the chick enzyme[47]. The elongation rate was 1,260 nucleotide per min, which is again similar to the turnover number (1,530 nucleotides/min) calculated from the assumption that this enzyme consists of Mr=47,000 polypeptide (unpublished data).

The Effect of the Amounts of Template·Primer on the Chain Elongation Rate and the Product Number

Since the modes of DNA chain elongation are quite different between DNA polymerase β and γ (distributive vs.processive), the effect of the relative ratio of the template.primer to the enzyme would differently affect this enzyme. As seen in Figure 9, the increase in the amount of $(rA)_n \cdot (dT)_{12-18}$ resulted in the dramatic reduction of the elongation rate of each DNA chain and the increase of the product number per DNA polymerase β molecule. On the other hand, the variation in the amount of the template·primer in the reaction containing DNA polymerase γ is virtually non-effective to the chain elongation rate, the product number per enzyme molecule and also the final size of product. These results are reasonable for the difference in mechanisms of chain elongation by DNA polymerase β and γ.

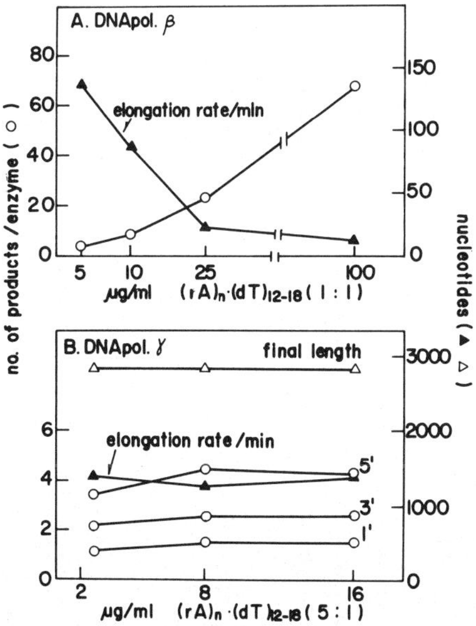

Fig. 9. The effect of the amount of template·primer on the DNA chain elongation rate and the product number per enzyme. Reactions and product analysis were carried out as described in Figures 7, 8 and 9. Reaction mixtures contained equal amounts of DNA polymerase β or γ and indicated amounts of $(rA)_n \cdot (dT)_{12-18}$.

Table 4. Km's of DNA Polymerase β and γ for dNTP and 3'-OH Termini of Primer[a]

Template·primer	[^{3}H] dNTP	unlabelled dNTP	Km for [^{3}H] dNTP (μM) Pol. β	Km for [^{3}H] dNTP (μM) Pol. γ	Km for 3'-OH of primer (μM) Pol. β	Km for 3'-OH of primer (μM) Pol. γ
Activated calf thymus DNA	dTT	dATP, dTTP dGTP	5.2	1.34	400-600 (μg/ml)	13.4 (μg/ml)
"	dATP	dCTP, dTTP dGTP	18.7	0.13		
"	dGTP	dCTP, dTTP dATP	28.6	0.37		
$(rA)_n \cdot (dT)_{12-18}$	dTTP		3.4	0.18	0.64	0.12
$(dA)_n \cdot (dT)_{12-18}$	dTTP		12.9	2.0	1.0	0.07
$(dT)_n \cdot (rA)_{12-18}$	dATP		87.0	1.40	0.37	—
$(dT)_n \cdot (dA)_{12-18}$	dATP		50	0.27	2.8	0.06
$(dC)_n \cdot (dG)_{12-18}$	dGTP		133.0	0.48	0.37	0.008

[a]Rat DNA polymerase β and mouse DNA polymerase γ were used in this work.

Km's for the Deoxynucleoside Triphosphates and the Primers

In Table 4, Km values for dNMP[48,49] and primer[50] of DNA polymerase β and γ are summarized. With various template·primers, Km's of DNA polymerase β are much higher than those of DNA polymerase γ. These differences can be easily understood from the differences in the reaction mechanisms of these enzymes: DNA polymerase β requires the simultaneous binding of the 3'-OH of the primer and dNTP for the reaction to continue while DNA polymerase γ requires only the binding of dNTP, since it might remain on the growing point. The processive property of DNA polymerase γ might be supported by the multimeric structure, i.e., multiple active sites, of this enzyme.

4. SYNTHESIS OF SHORT DNA PIECES BY DNA POLYMERASE α[51]

The present studies were performed to clarify the DNA chain elongation mechanism in vitro of DNA polymerase α from mouse myeloma MOPC104E cells along with DNA polymerase β. The major purpose of this study was to see whether the reaction property of DNA polymerase α is responsible for the size of the Okazaki pieces (3-5s) in eukaryotic cells. We chose the single-stranded calf thymus DNA with the defined starting region of DNA replication as a template in combination with RNA primer. The principle of the experiment is shown in Figure 10. Poly(dT) tails were added to the 3'-termini of denatured DNA using terminal deoxynucleotidyl transferase[52], and the primer $(rA)_{12-20}$ was hybridized with the poly(dT)-tail of the template. This template·primer system has two possible initiation points from which DNA

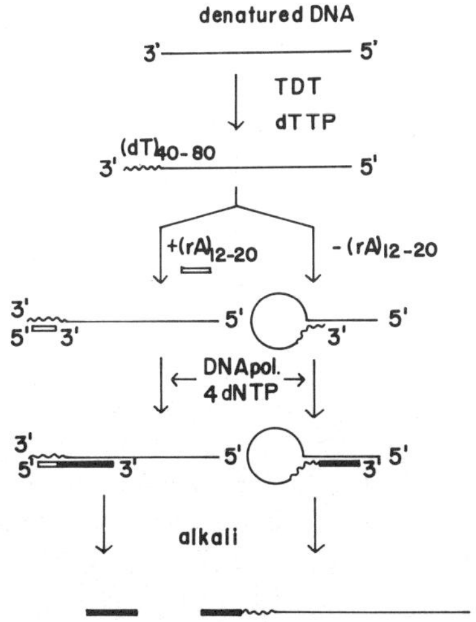

Fig. 10. Principle of the experiment. TDT stands for terminal deoxynucleotidyl transferase.

polymerase can start replication; one is the 3'-OH end of $(rA)_{12-20}$ primer and the other is the 3'-OH end of the template. In the latter case, the template is thought to form a hairpin structure at the 3'-terminal region, and therefore DNA synthesis is not dependent on the addition of the primer. If the DNA chains started with $(rA)_{12-20}$ primers are significantly shorter than the templates, they can be separated from the DNA chains started from the 3'-OH end of the templates. Eventually, the former could be separated from the latter by an alkaline sucrose gradient centrifugation.

Figure 11 shows an example of the analysis of product synthesized by DNA polymerase α. In spite of the long reaction time (120min), the product thought to be started with $(rA)_{12-20}$ sedimented at 3-5s region. The radioactivity was also detected in the region of the size of the template, and was almost identical to the product synthesized without primer. Therefore, the larger product was synthesized depending on the hairpin structure of the template without the $(rA)_{12-20}$ primer. The RNA-linkage to the 5'-termini of short DNA pieces was confirmed as described previously[53].

An argument for the limitation of the product size is that DNA polymerase α copied only poly(dT)-tail and shut off further elongation

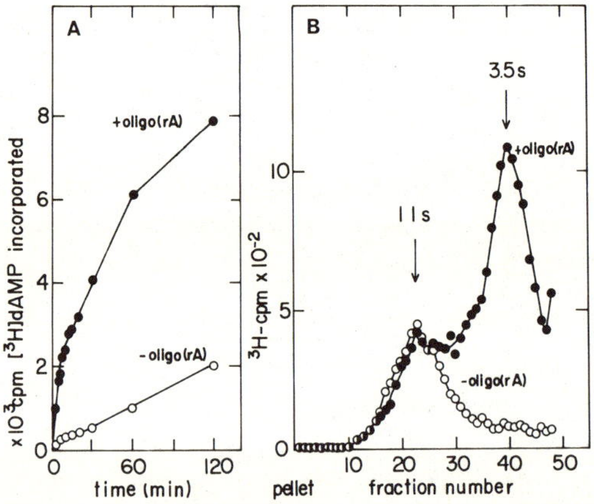

Fig. 11. Alkaline sucrose gradient centrifugation of the product synthesized by DNA polymerase α. The reaction mixture (150μl) contained 0.2 unit of DNA polymerase α, dDNA-3'-$(dT)_{78}$ (average 11s) with or without $(rA)_{12-20}$ and other components. A 5μl aliquot was withdrawn at each indicated time and used to measure the [^{3}H]dAMP incorporated into polymer DNA (A). At 120min, 50μl portions of the reaction mixtures were taken for analysis by alkaline sucrose gradient centrifugation (B). Centrifugation was carried out for 10h at 55,000r.p.m. o, without $(rA)_{12-20}$; •, with $(rA)_{12-20}$.

beyond the boundary of poly(dT) and calf thymus DNA strand. But this is not the case. As seen in Figure 12, the sizes of products were not proportional to the sizes of poly(dT) tails (from 17 to 67 average nucleotides lenght), but were all 3-5s. Therefore, the product sizes are longer than tails. Furthermore, not only [^{3}H]dAMP but also [^{3}H]dTMP was incorporated into products. Since dTMP was incorporated into the DNA strand complementary to the calf thymus DNA strand, the termination of chain elongation occurred in the calf thymus DNA strand.

The sizes of products synthesized in reactions for 15, 30 and 60min with DNA polymerase α were compared to those with DNA polymerase β. The sizes of all product synthesized by DNA polymerase α were constantly 3-5s (Fig. 13). The products synthesized from $(rA)_{12-20}$ by DNA polymerase β were also clearly separated from the products synthesized depending on the hairpin structures of templates. The sizes of product in this case, however, increased in proportion

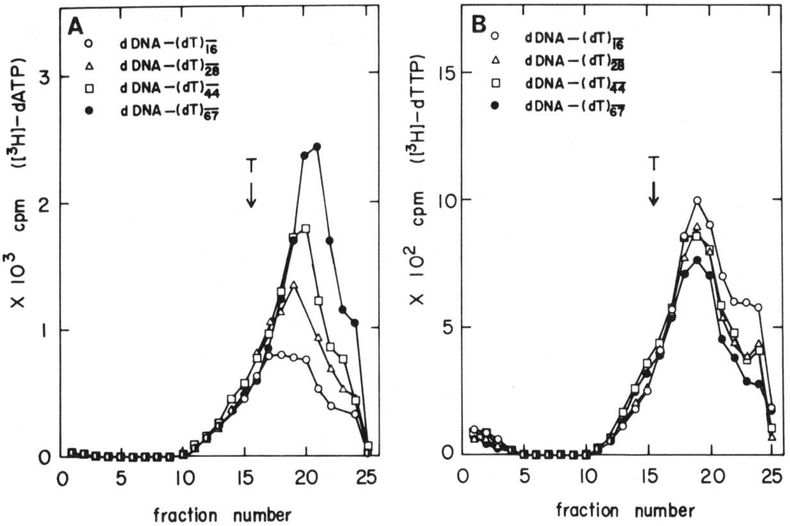

Fig. 12. Effect of the size of $(dT)_n$ tails at the 3'-end of dDNA templates on the product sizes. dDNA-3'-$(dT)_n$ with various sizes of $(dT)_n$ tails (n=average 16 (○), 28 (△), 44 (□), and 67 (●) was synthesized by changing the incubation time with terminal deoxynucleotidyl transferase. The average size of the template thus obtained was about 9s. Products were labelled in the presence of [^{3}H]dATP (A) or [^{3}H]dTTP (B) with other unlabelled dNTP's $(rA)_{12-20}$, and DNA polymerase α. The reaction time was 30min. Products were analyzed by alkaline sucrose gradient centrifugation. The position of the template is indicated by "T". The sedimentation direction was right to left.

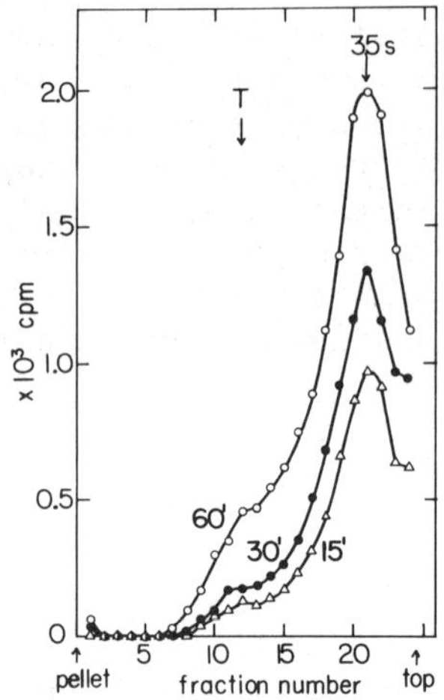

Fig. 13. Effect of incubation time on the size of product synthesized by DNA polymerase α. At the indicated times, aliquots were withdrawn for size analysis by alkaline sucrose gradient centrifugation. Centrifugation was carried out for 10h at 55,000rpm at 5°C. ▲, 15min; ●, 30min, and O, 60min.

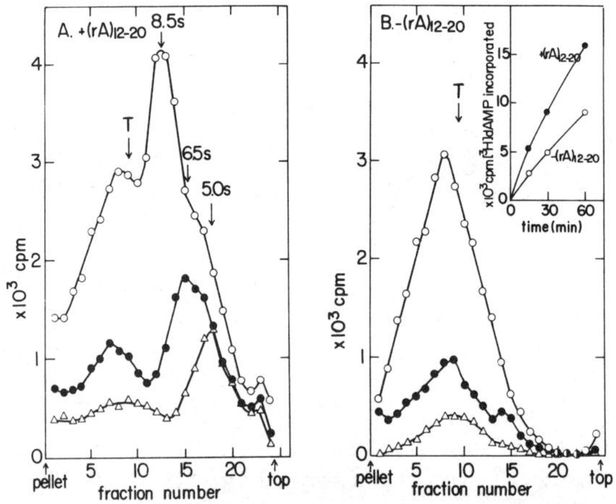

Fig. 14. Effect of incubation time on the size of product synthesized by DNA polymerase β. The reaction mixture contained [^{3}H]dATP, dDNA-3'-$(dT)_{\overline{45}}$ with (A) or without (B) $(rA)_{12-20}$, DNA polymerase β and other components. At 15, 30 and 60min, aliquots were withdrawn and analyzed. △, 15min; ●, 30min, and O, 60min. The time course of incorporation of [^{3}H]dAMP into polymer is shown in the inset panel B.

to the reaction time; products at 15, 30 and 60min sedimented at 5.0<u>s</u>, 6.5<u>s</u> and 8.5<u>s</u>, respectively. Their lengths were calculated to be 238, 475 and 962 nucleotides. Thus, the DNA chain grew at the constant rate of about 16 nucleotides per min (Fig. 14).

The amount of [^{3}H]dAMP incorporated into products started with $(rA)_{12-20}$ primers increased almost linearly in containing either reaction DNA polymerase α or β. Relative numbers of products (expressed as pmol dNMP incorporated per average chain length of products) increased linearly with time in the case of DNA polymerase α, but remained constant for 60min in the case of DNA polymerase β. The reaction mechanism of DNA polymerase β in this system agrees with

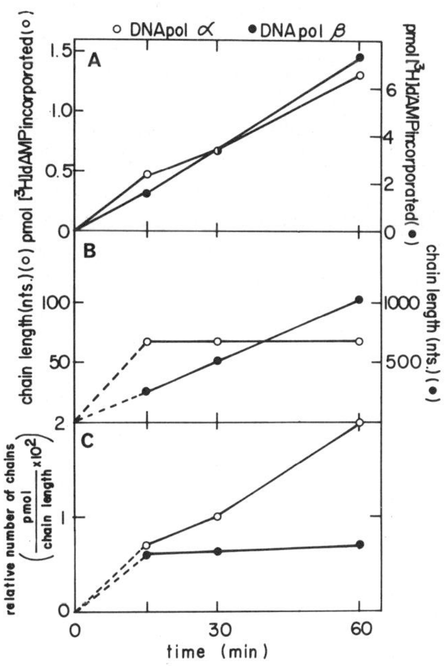

Fig. 15. Time-dependent variation in sizes and amounts of products synthesized by DNA polymerase α and β. The experiments were carried out like those shown in Figures 13 and 14 under identical conditions, except that 80mM KCl was added to the reaction mixture for DNA polymerase β. The products were analyzed by alkaline sucrose density gradient centrifugation. The radioactivity in the region of products smaller than template size, which were thought to start with $(rA)_{12-20}$ primers, was summed up to determine the amount of [^{3}H]dAMP incorporated (A). Molecular weights of products were calculated from the s-values and were expressed in nucleotide numbers (B). The relative amounts of product molecules were calculated by dividing [^{3}H]dAMP incorporated by the molecular weight of the product (C).

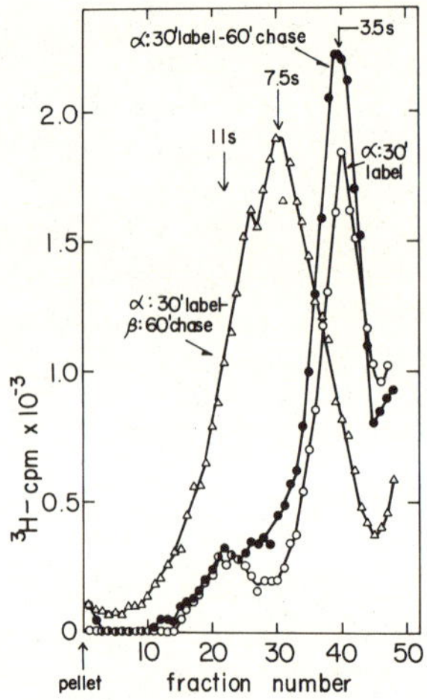

Fig. 16. Label-chase experiments indicating that DNA polymerase β can utilize short DNA pieces which DNA polymerase α synthesized as primers for further elongation. The reaction mixture contained $[^3H]$dATP, dDNA-3'-$(dT_{\overline{78}})$ and DNA polymerase α, and was incubated for 30min at $37^{\circ}C$. The mixture was then divided into three 60μl aliquots, which were treated as follows: i) the reaction was stopped at this point (○), ii) 3μl of 25mM dATP were added (●), and iii) 3μl of 25mM dATP and 6μl DNA polymerase β were added (Δ); ii) and iii) were further incubated for 60min at $37^{\circ}C$. The products were analyzed by alkaline sucrose gradient centrifugation.

that described previously in the system containing $(rA)_n \cdot (dT)_{12-18}$. Distributive mode of the reaction by DNA polymerase β was again confirmed with the DNA template.

A label-chase experiment was carried out to examine whether short DNA pieces synthesized by DNA polymerase α were further elongated by DNA polymerase α or β. In the first step reaction, short pieces were labelled by $[^3H]$dATP in the presence of DNA polymerase α. Then, excess unlabelled dATP was added and the 2nd step reaction was carried out for 60min in the presence or absence of DNA polymerase β (DNA polymerase α added in the first step was still present). As seen in Figure 16, DNA polymerase α did not convert ^{3}H-labelled short DNA pieces into longer molecules, but DNA polymerase β converted all the 3-5_s_ pieces into longer molecules of about 7.5_s_. Results indicate that DNA polymerase β can efficiently utilize short pieces synthesized by DNA polymerase α as primers for further elongation. On the other hand, DNA polymerase α might recognize the size of the short piece and shut off further elongation of such DNA pieces.

5. CONCLUSION

The results described in this paper represent the following conclusions:

(1) The purified chick embryo DNA polymerase β consists of a single Mr=40,000 polypeptide. This conclusion was further confirmed by the tryptic peptide mapping method and the serological methods, which also indicate that the avian and the mammalian DNA polymerase β share a partial homologous structure.

(2) Chick embryo DNA polymerase γ, and probably mouse DNA polymerase γ, also have a tetramer structure of Mr=47,000 polypeptides. This tetramer structure may support the highly processive property of this enzyme. The partial homology in the structure of the avian and the murine enzymes was also confirmed by tryptic peptide mapping and the serological method. It is interesting to point out that antisera prepared against chick DNA polymerase β and γ may be useful for the study on mammalian enzymes in addition to avian enzymes.

(3) The structure of the chick embryo DNA polymerase α is not well enough clarified. But we have obtained results suggesting that the enzyme consists of two groups, large and small, of polypeptides, and there are some minor modifications in the small subunits.

(4) The information about the nearly homogeneous enzyme permits us to estimate the number of enzymes per cell and the exact turnover number of these enzymes. The latter especially, represented the firm bases for the quantitative analysis of DNA chain elongation mechanisms by these DNA polymerases.

(5) Although both DNA polymerases β and γ can synthesize long DNA products on single-stranded templates, their DNA chain elongation mechanisms are different; the mechanism of DNA polymerase β is highly distributive, while that of DNA polymerase γ is highly processive. Therefore, the variation in the amount of primers affects the chain elongation rate and the product number of DNA polymerase β, but not those of DNA polymerase γ.

(6) The purified DNA polymerase α synthesized exclusively short DNA pieces. The size of products (3-5<u>s</u>) is very close to that of the eukaryotic Okazaki pieces. It remains to be seen how DNA polymerase α recognizes the size of products and ceases further elongation.

The results are in agreement with the indirect observations suggesting that DNA polymerase α and γ are involved in replications of nuclear DNA and mitochondrial DNA, respectively. It is likely that the size of the Okazaki piece is regulated by the reaction property of DNA polymerase α, and the continuous elongation of mitochondrial DNA is supported by the high processivity of DNA polymerase γ. Our results also show good agreement with the observations that DNA polymerase γ is involved in the replication of adenovirus DNA, which

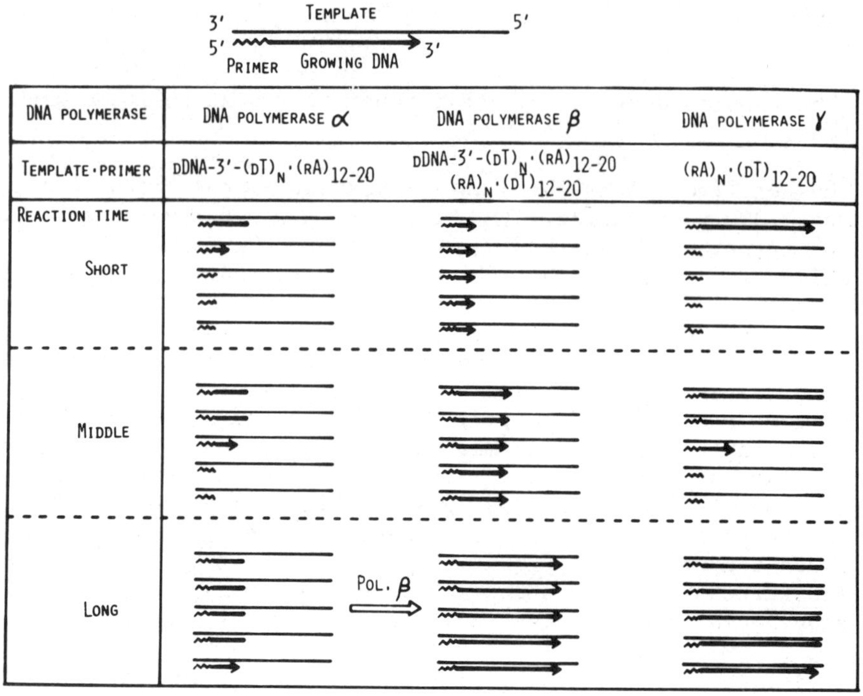

Fig. 17. Models of chain elongation mechanisms of DNA polymerases α, β and γ.

is replicated in a continuous manner. But we have not ruled out the possibility that DNA polymerase β with or without the cooperation of DNA polymerase α is involved in the continuous synthesis of DNA such as adenovirus DNA or the leading strand of the nuclear DNA, in addition to DNA repair.

ACKNOWLEDGEMENTS

We wish to thank Dr. Toshitada Takahashi for his advice in the immunological study, Dr. Katsuhiko Ono for permission to use the enzymatic parameters, and Miss Ikuko Inagaki for technical assistance. These studies were supported in part by a grant-in-aid for Cancer Research from the Ministry of Education, Science and Culture of Japan, a grant from the Research Foundation for Cancer and Cardiovascular Diseases, Osaka, Japan and a grant from The Bio-dynamic Research foundation, Nagoya, Japan.

REFERENCES

1. J. H. Taylor, J. Mol. Biol. 31:579-594 (1968).
2. A. J. Lavine, H. S. Kang and F. E. Bilheimer, J. Mol. Biol. 50: 579-568 (1970).
3. K. Tsukada, T. Moriyama, W. E. Lynch and I. Lieberman, Nature 220:162-164 (1968).
4. G. C. Fareed and N. P. Salzman, Nature New Biol. 238:277-279 (1972).
5. R. M. Fox, J. Mendelsohn, E. Barbosa and M. Goulian, Nature New Biol. 245:234-237 (1973).
6. B. Y. Tseng and M. Goulian, J. Mol. Biol. 99:339-346 (1975).
7. G. Magnusson, V. Pigiet, E. L. Winnacker, R. Abrams and P. Reichard, Proc. Natl. Acad. Sci. U.S.A. 70:412-415 (1973).
8. R. Eliasson and P. Reichard, Nature 272:182-185 (1978).
9. M. A. Wagar and J. A. Huberman, Biochem. Biophys. Res. Commun. 51:174-180 (1973).
10. P. K. Qasba, Biochem. Biophys. Res. Commun. 60:1338-1344 (1974).
11. T. Yamashita, M. Arens and M. Green, J. Biol. Chem. 252:7940-7946 (1977).
12. M. Arens, T. Yamashita, R. Padmanabhan, T. Tsuruo and M. Green, J. Biol. Chem. 252:7949-7954 (1977).
13. H. Kasamatsu, L. I. Grassman, D. L. Robberson, R. Watson and V. Vinograd, Cold Spring Habor Symp. Quant. Biol. 38:281-288 (1973).
14. U. Hübscher, C. C. Kuenzle and S. Spadari, Proc. Natl. Acad. Sci. U.S.A. 76:2316-2320 (1979).
15. M. A. Wagar, M. J. Evans and J. A. Huberman, Nucleic Acids Res. 5:1933-1946 (1978).
16. H. J. Edenberg, S. Anderson and M. L. DePamphilis, J. Biol. Chem. 253:3273-280 (1978).
17. P. C. van den Vliet, and M. M. Kwant, Nature 276:532-534 (1978).
18. H. Krokan, P. Schaffer and M. L. DePamphilis, Biochemistry 18: 4431-4443 (1979).
19. M. Yamaguchi, K. Tanabe, N. Y. Taguchi, M. Nishizawa, T. Takahashi and A. Matsukage, J. Biol. Chem. 255:9942-9948 (1980).
20. L. M. S. Chang, J. Biol. Chem. 248:3789-3795 (1973).
21. T. S.-F. Wang, W. D. Sedwick and D. Korn, J. Biol. Chem. 250: 7040-7044 (1975).
22. K. Tanabe, E. W. Bohn and S. H. Wilson, Biochemistry 18:3401-3406 (1979).
23. D. M. Stalker, D. W. Mosbaugh, and R. R. Meyer, Biochemistry 15: 3114-3121 (1976).
24. K. Ono, A. Ohashi, K. Tanabe, A. Matsukage, M. Nishizawa and T. Takahashi, Nucleic Acids Res. 7:715-726 (1979).
25. J. G. Stavrianopoulos, J. D. Karkas and F. Chargaff, Proc. Natl. Acad. Sci. U.S.A. 69:1781-1785 (1972).
26. G. Brun, F. Rougeon, M. Lauber and G. Chapeville, Eur. J. Biochem. 41:241-251 (1974).

27. M. Yamaguchi, A. Matsukage and T. Takahashi, J. Biol. Chem. 255: 7002-7009 (1980).
28. J. H. Elder, R. A. Pickett, II, J. Hampton and R. A. Lerner, J. Biol. Chem. 252:6510-6515 (1977).
29. Y.-C. Chen, E. W. Bohn, S. R. Planck and S. H. Wilson, J. Biol. Chem. 254:11678-11687 (1979).
30. S. Spadari and A. Weissbach, J. Biol. Chem. 249:5809-5815 (1974).
31. B. J. Lewish, J. W. Abrell, R. G. Smith and R. C. Gallo, Biochim. Biophys. Acta 349:148-160 (1974).
32. A. Matsukage, E. W. Bohn and S. H. Wilson, Biochemistry 14:1006-1020 (1975).
33. K. W. Knopf, M. Yamada and A. Weissbach, Biochemistry 15:4540-4548 (1976).
34. U. Bertazzoni, A. I. Scovassi and G. M. Brun, Eur. J. Biochem. 81:237-248 (1977).
35. S. Yoshida, T. Kondo and T. Ando, Biochim. Biophys. Acta 353:463-474 (1974).
36. A. M. Holmes, I. P. Hesslewood and I. R. Johnston, Eur. J. Biochem. 62:229-235 (1976).
37. A. Matsukage, M. Sivarajan and S. H. Wilson, Biochemistry 15: 5305-5314 (1976).
38. P. A. Fisher and D. Korn, J. Biol. Chem. 252:6523-6535 (1977).
39. N. Nishioka, A. Matsukage and T. Takahashi, Cell Struct. Funct. 2:61-70 (1977).
40. A. Matsukage, N. Nishioka, M. Nishizawa and T. Takahashi, Cell Struct. Func. 4:295-306 (1979).
41. Y. Ono, T. Enomoto and M. Yamada, Gann 69:207-212 (1978).
42. M. Mechali, J. Abadiedebat and A. M. de Rocondo, J. Biol. Chem. 255:2114-2122 (1980).
43. G. R. Banks, J. A. Boezi and I. R. Lehman, J. Biol. Chem. 254: 9886-9892 (1979).
44. G. Villani, B. Sauer and I. R. Lehman, J. Biol. Chem. 255:9479-9484 (1980).
45. E. M. den Tonkelaar and P. van Duijin, Histochemie 4:16-19 (1964).
46. M. Yamaguchi, A. Matsukage and T. Takahashi, Nature 285:45-47 (1980).
47. A. Matsukage, M. Nishizawa and T. Takahashi, J. Biochem. 85:1551-1554 (1979).
48. A. Matsukage, K. Ono, A. Ohashi, T. Takahashi, C. Nakayama and M. Saneyoshi, Cancer Res. 38:3076-3079 (1978).
49. K. Ono, A. Ohashi, A. Yamamoto, A. Matsukage, T. Takahashi, M. Saneyoshi and T. Ueda, Cancer Res. 39:4673-4680 (1979).
50. K. Ono, A. Chashi, K. Tanabe, A. Matsukage, M. Nishizawa and T. Takahashi, Nucleic Acids Res. 7:715-726 (1979).
51. A. Matsukage, M. Nishizawa, T. Takahashi and T. Hozumi, J. Biochem. 88:1867-1877 (1980).
52. S. H. Wilson, A. Matsukage, E. W. Bohn, Y. C. Chen and M. Sivarajan, Nucleic Acids Res. 4:3981-3996 (1977).
53. Y. Kurosawa, T. Ogawa, S. Hirose, T. Okazaki and R. Okazaki, J. Mol. Biol. 96:653-664 (1975).

ADENOVIRUS DNA REPLICATION: MECHANISM AND REPLICATION PROTEINS

Peter C. van der Vliet,* Marijke M. Kwant
Bram G. M. van Bergen and Wim van Driel

Laboratory for Physiological Chemistry
University of Utrecht
Utrecht, The Netherlands

INTRODUCTION

The human adenoviruses, especially type 5 (Ad5) and type 2 (Ad2), can be easily cultivated in HeLa or KB cells. They have been a favorite subject for DNA replication studies in the eukaryotic cell during the last decade. The mature adenovirus genome isolated from virions consists of a linear duplex DNA molecule of about 35,000 base pairs. The DNA contains an inverted terminal repeat of 103 base pairs of which the first 50 nucleotides are very rich in A+T[1,2]. The inverted repeat is rather well conserved in the various adenoviruses and in particular base pairs 9-22 are identical in all human serotypes studied[3].

In addition, each strand contains a virus coded protein with a molecular weight of 55,000 which is covalently bound to the 5'-terminus[4,5,21]. Although the function of this terminal protein is not fully understood, it becomes increasingly clear that it participates in the initiation of viral DNA replication.

Adenovirus DNA replicates very efficiently in the nucleus of permissive cells, with yields of about 10^6 daughter molecules, roughly 10% of which is incorporated into progeny virus. Thus, in infected cells late in infection the amount of viral DNA surpasses the amount of host DNA present. Under commonly used infection conditions,

---

*Dr. P. C. van der Vliet, Laboratory for Physiological Chemistry
Vondellaan 24 s, 3521 GG Utrecht, The Netherlands

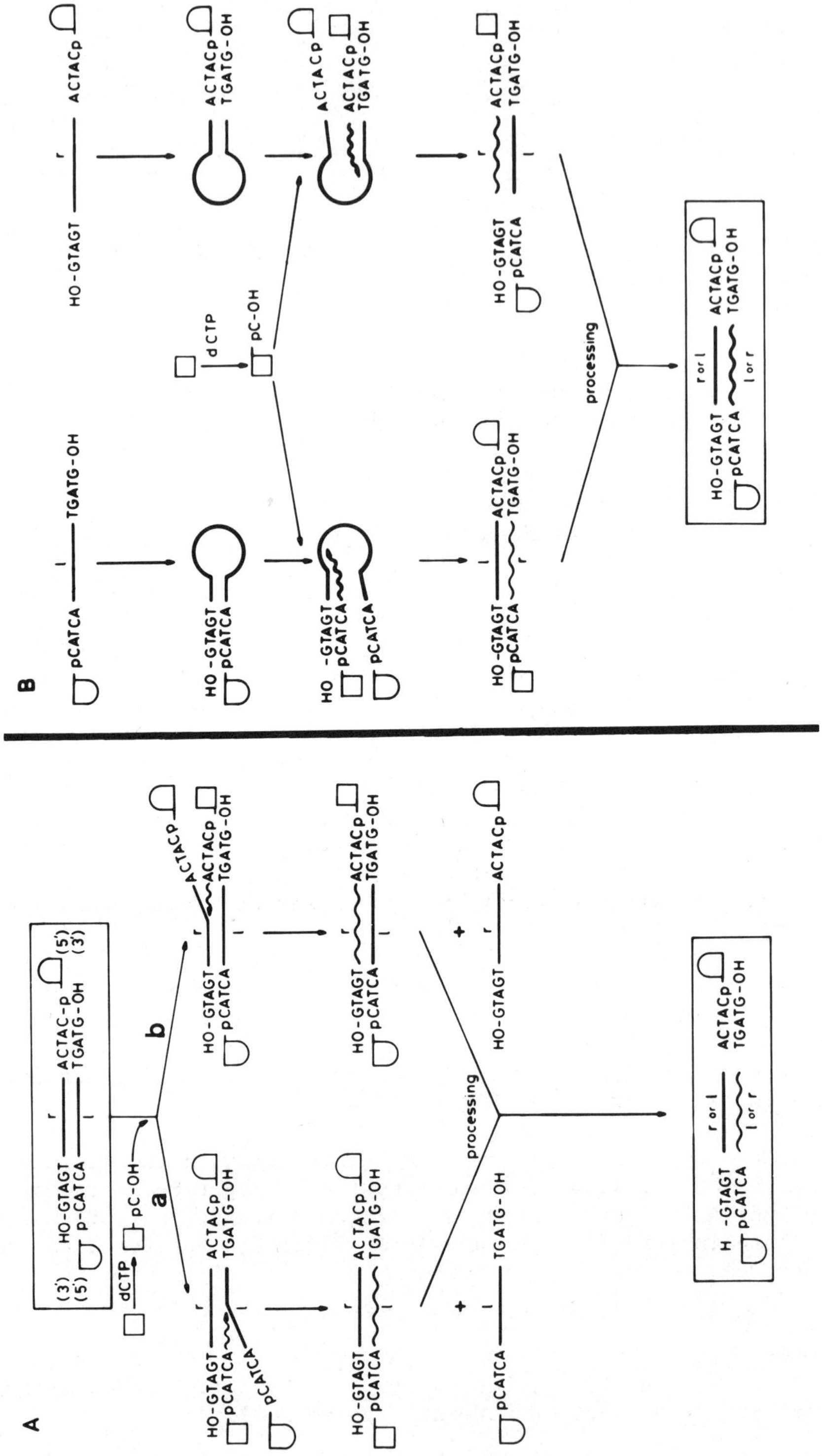
A
B
HO-GTAGT
p-CATCA
ACTAC-p
TGATG-OH
dCTP
pC-OH
a
b
processing
r or l
l or r

adenovirus DNA replication gradually inhibits cellularDNA synthesis[6]. This means that late in the infection cycle DNA synthesis is almost exclusively of a viral nature.

ASYNCHRONOUS REPLICATION MECHANISM

The pattern of DNA replication has been studied in some detail and reviewed[7,8]. Replication initiates at either the right or left end of the viral genome, which are identical. DNA synthesis is unidirectional and proceeds by a strand displacement mechanism. The polymerization reaction is most likely a continuous process. Single-strand fragments[10-12] have been observed in Ad5 after hydroxyurea treatment or in isolated nuclei[10-11], but not under normal infection conditions. These fragments may be derived from an excision repair process rather than being true replication intermediates.

The displaced parental strand may remain single-stranded temporarily before being duplicated. Alternatively, duplication of the displaced strand may start just before, or concomitant with, completion of the displacement synthesis. Replicative intermediates in accordance with both alternatives have been observed.

A model for the replication of adenovirus DNA is outlined in Figure 1. Notable differences with the mechanism of cellular or papovavirus DNA synthesis are:

(1) Initiation does not occur internally but at either end of the molecule, presumably involving protein priming.

Fig. 1. Replication model of adenovirus DNA, based on biochemical and electron microscopic data and adapted from previous models[5,9,12,25]. (A), displacement synthesis. The two routes of single-stranded DNA from l and r strand are produced during replication. Positioning of pC10H may be mediated by protein-protein interaction at the termini. (B), complementary strand synthesis. The displaced single strand may form a panhandle structure which provides ends that are identical to the molecular ends in the parental viral DNA. Although attractive as a hypothesis, these structures have not been observed as intermediates and may be very short-lived. The proposed role of the terminal protein in complementary strand synthesis is hypothetical since this reaction has not been studied *in vitro*. 80-87kD precursor terminal protein, 55kD terminal protein present on mature DNA, parental DNA, newly synthesized DNA. It should be pointed out that the two types of synthesis, drawn as separate events in (A) and (B), might well be coupled *in vivo*.

(2) Elongation is continuous, without evidence for RNA priming or Okazaki fragments.
(3) Synthesis of the two daughter strands is highly asynchronous, producing completely displaced single-stranded molecules as intermediates in the replication process.

INITIATION MECHANISM AND THE ROLE OF THE TERMINAL PROTEIN AS PRIMER

Until recently the mechanism of initiation was poorly understood. Three possible mechanisms have been considered: (a) self priming by hairpin formation, (b) priming by RNA, and (c) protein priming.

(a) A self-priming mechanism as proposed by Cavalier-Smith[13] has been observed in the related adeno-associated viruses and autonomous parvoviruses[14]. Such a mechanism cannot a priori be excluded for adenovirus. However, sequence analysis[1,2] indicates that the hairpin formation, which is required for self-priming, is not possible within the 103 base pairs of the inverted terminal repetition. Formation of longer hairpins as intermediates would give rise to a population of progeny viruses with a scrambled nucleotide sequence at the ends. Such a mixed virus population was never observed. Also, direct evidence against covalent coupling of parental and progeny DNA has been obtained[16,17]. Such a covalent coupling is a prerequisite for a self-priming mechanism.
(b) An initiation mechanism involving RNA priming would require circle or concatemer formation to fill in the single-strand gaps that remain after removal of the RNA primer. In spite of many efforts, such concatemerization has never been observed. If it occurs, it might involve very short-lived intermediates which have thus far escaped detection.
(c) Evidence for priming by the terminal protein: Recently, in vitro systems have been developed that permit initiation of adenovirus DNA synthesis[18-20,23]. In such systems, extracts of infected cell nuclei are prepared and incubated with an exogenous viral DNA template. Legitimate initiation at the ends of the molecule can be observed, but only when adenovirus DNA covalently linked to the terminal protein (DNA-pro) is added to the reaction mixture. Removal of the protein results in aspecific incorporation of nucleotides, randomly distributed over the genome.
An example of such a reaction is shown in Figure 2. In this experiment the viral DNA template was first digested with restriction endonucleases XbaI or XhoI. XbaI produces 5 fragments (A-E) of which C and E are derived from the termini. XhoI produces 7 fragments of which B and C are terminal. After incubation, the reaction product was electrophoresed in agarose gels containing 0.1% SDS and autoradiographed.

Incubation of Ad5 DNA-pro with extracts from Ad5 infected HeLa cell nuclei (lanes 2 and 4) results in preferential labelling of

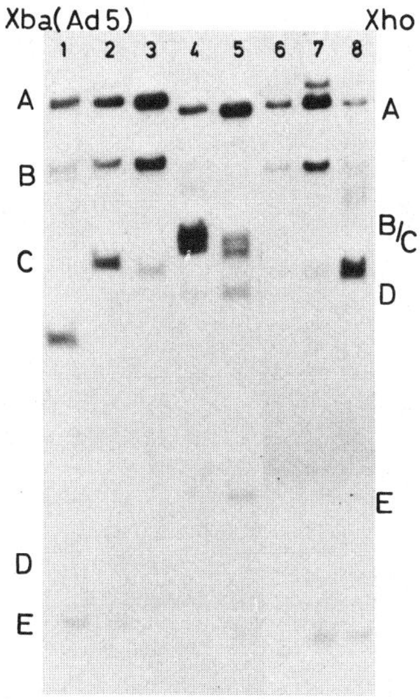

Fig. 2. Initiation of Ad5 DNA replication in vitro. Nuclear extracts from Ad5 infected or uninfected HeLa cells were prepared according to Challberg and Kelly[18]. Restriction enzyme (XbaI or XhoI) digests of DNA-pro or DNA were incubated with the extracts for 30min at 37°C in the presence of a reaction mixture containing $[^{32}P]$-dCTP. The reaction was stopped by addition of 10mM EDTA and 0.1% SDS and the samples were electrophoresed directly in 1% agarose gels containing 40mM Tris-HCl pH 7.8 - 5mM sodium acetate - 1mM EDTA - 0.1% SDS. The gels were dried and autoradiographed. XbaI fragments C and E, and XhoI C and B are derived from the left and right termini, respectively. DNA-pro: DNA containing terminal protein.

Lane:
(1) Ad2 DNA-pro x XbaI + Ad5 extract
(2) Ad5 DNA-pro x XbaI + Ad5 extract
(3) Ad5 x XbaI + Ad5 extract
(4) Ad5 DNA-pro x XhoI + Ad5 extract
(5) Ad5 DNA x XhoI + Ad5 extract
(6) Ad5 DNA-pro x XbaI + uninfected extract
(7) Ad5 DNA-pro x XbaI + Ad5 extract + 20μM ddTTP
(8) Ad5 DNA-pro x XbaI + Ad5 extract + 10μM aphidicolin

fragments containing terminal protein. Although the other fragments are also labelled, termimal fragments contain a 3- to 4-fold higher specific activity (Fig. 3A). With Ad5 DNA as template (lanes 3 and 5) the label is randomly distributed over the fragments (Fig. 3B). This aspecific, random incorporation can be suppressed by aphidicolin (Fig. 2, lane 8). Incorporation into terminal fragments is much more resistant to this drug, indicating that two different polymerizing

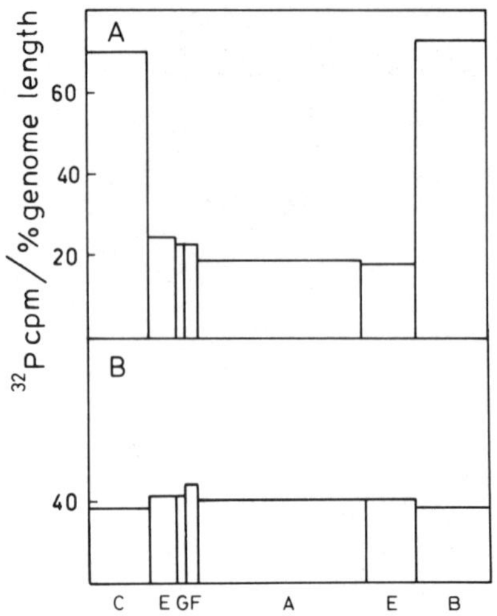

Fig. 3. Specific initiation in vitro of terminal fragments containing covalently bound protein. Radioactive bands from Figure 2, lanes 4 and 5, were cut out, counted and plotted as specific activity (^{32}P-cpm/% of fragment length). No correction for deoxycytidine-content of the restriction fragments was made. (A) = Ad5 DNA-pro; (B) = Ad5 DNA.

reactions may take place during the incubation. The amount of random incorporation in internal fragments varies with different extracts and template preparations. Uninfected cell extracts also produce random incorporation, even with DNA-pro (Fig. 2, lane 6). Only exogenous templates are replicated which can be seen from Figure 2, lane 1, where Ad2 DNA-pro XbaI fragments were used as templates in combination with Ad5 extracts. Only Ad2 fragments become labelled during the incubations. This shows that endogenous DNA, if present, does not participate in the reaction.

These results indicate that an intact terminal protein is of essential importance for initiation at the origin, in agreement with previously published data[18,20]. Newly synthesized adenovirus DNA is covalently bound, both in vitro and in vivo, to a 80-87kD protein which is partially identical in peptide map to the mature 55kD terminal protein and presumably is a precursor of this protein[20-24]. Both forms of the terminal protein are bound to the DNA by a serine-dCMP linkage[25]. This leads to a model for initiation in which the 80-87kD precursor terminal protein binds dCMP and recognizes the mature 55kD protein at the ends of the viral DNA. This may result in a convenient position at the origin which could permit the 3'-OH group of the dCMP residue to function as primer for a DNA polymerase (see model, Fig. 1). In this respect, adenovirus resembles B-subtilis phage ϕ29 for which protein priming has also been proposed[26].

It appears clear that a protein priming mechanism as outlined here will eliminate the need for gap filling of a primer molecule and provides a simple mechanism for termination and completion of the mature progeny DNA. The 80-87kD precursor protein may be processed in a later stage to the mature 55kD form, a reaction performed by a virus coded protease[21,24].

OTHER INITIATION PROTEINS

Although the biochemical evidence for a role of the terminal protein in initiation is rather strong, no genetic support has been obtained due to the absence of mutants affecting the structural gene of the protein. Two or three other viral proteins have been implicated in initiation. They are defined by the different DNA negative complementation groups found in adenovirus (see Table 1), H5ts125 and H5ts36 being the main prototypes.

The H5ts125 mutation affects the 72kD DNA binding protein[27] which is the product of the viral early region E2, located at 0.61-0.66 on the adenovirus map. The protein has been extensively studied. It is a phosphoprotein synthesized in large amounts (about 2 x 10 copies per cell) and it binds both to single-stranded DNA and to the ends of double-stranded DNA[28,29]. The H5ts125 mutation is phenotypically expressed as a defect in the initiation of viral DNA syn-

Table 1. DNA Negative Adenovirus Mutants

Mutant prototype	Region	Map position (% from left)	Mutated protein	Function	Ref.
H5ts125	E2A	60-64	72kD DNA binding protein	initiation elongation protection	6,27
H5ts36	E2B	18.5-22.0	?	initiation	30,33
H2ts111	?	?	?	initiation protection	54,55
H5hrI	E1A	0-4.4	?	indirect□	56
H5hrII△	E1B	6-9.5	?	?	56

□The function of host range I mutants is required for expression of the other early regions.

△Host range II mutants are DNA positive in human embryonic kidney cells, which may provide a helper function not present in cultured cells.

thesis. Upon shift-up, rounds of replication are completed, but new replication starts no longer occur[30]. Viral DNA synthesis in vitro in nuclear extracts is thermosensitive when H5ts125 infected nuclear extracts are used[31] (B.G.M. van Bergen, unpublished results). This defect can be complemented by adenovirus DNA binding protein, but not by DNA binding proteins from bacterial sources, which suggests a specific requirement for the virus coded protein in the initiation reaction. Apart from a function in DNA initiation, the adenovirus DNA binding protein is also required during elongation (see next chapter) and has a function in the control of transcription.

No proteins have been correlated yet to the other DNA negative complementation groups, represented by H5ts36 and H2ts111.The H5ts36 mutation has been mapped in a region which is also responsible for the synthesis of the terminal protein[31,33] (see Table I). It is tempting to speculate that the H5ts36 mutation affects the terminal protein, or its precursor from, but evidence for this is lacking.

PROTEINS INVOLVED IN ELONGATION OF NASCENT DNA STRANDS

Both _in vivo_ and _in vitro_, synthesis of the complete daughter strands takes roughly 20min at 37°C,[34] which is about 30 nucleotides per second. At least two proteins participate in the presumably continuous elongation reaction.

A. Role of the DNA Binding Protein (DBP) in Elongation

The single-stranded DNA originating from the displacement reaction can be recovered from infected cells as a nucleoprotein complex. The major protein present on the single-stranded DNA is the 72kD DNA binding protein. This is demonstrated both by analysis of endogenous replicative complexes[35] and electron microscopic analysis of chromatin spread directly from infected cell nuclei[58].

These findings are in agreement with the abundance of the protein and its stoechiometric requirement in virus multiplication[28,33] and suggest that the protein participates in elongation as well as in initiation. Evidence for an active role of the DNA binding protein was initially suggested from inhibition of the elongation process by anti-DBP-gammaglobulin. These antibodies inhibit viral DNA synthesis up to 50%, both in isolated nuclei[32] and replication complexes[35]. These systems are only capable of elongation of preexisting replication intermediates. A role in elongation was also made plausible from the observed temperature-sensitive elongation reaction in replication complexes from H5ts125 infected cells[31] and by the slow conversion of single-stranded intermediates to double-stranded DNA in H5ts125 infected cells at the non-permissive temperature, compared to the wild type[15].

Although a detailed picture of the function of the protein cannot be given, it may facilitate the unwinding of the parental strands during displacement synthesis, possibly in collaboration with other proteins in the replication fork. Also it may protect single-stranded DNA from nuclease attack and could provide a template configuration which is favourable for the duplication of the displaced single-stranded DNA.

B. Role of DNA polymerase γ

Adenovirus DNA does not code for its own DNA polymerase but relies on a host enzyme for polymerization. To determine which DNA polymerase is required several specific inhibitors have been employed, especially 2'-3'-dideoxynucleoside-5'-triphosphates (ddNTP's) and aphidicolin. ddNTP's are strong inhibitors of DNA polymerase β and in particular DNA polymerase γ, while DNA polymerase α is only affected at high concentrations of the analogues[36,37].

In isolated nuclei, one can easily distinguish between adenovirus DNA synthesis and cellular or SV40 synthesis by adding any one of the four ddNTP's at relatively low concentrations[36], see Figure 4. Under these conditions, cellular DNA proceeds unabated and at least 300-fold higher concentrations of the analogues are required for inhibition of cellular DNA. Similar observations have been made for viral DNA synthesis in replication complexes and in nuclear extracts[31]. During in vitro synthesis in nuclear extracts, ddTTP

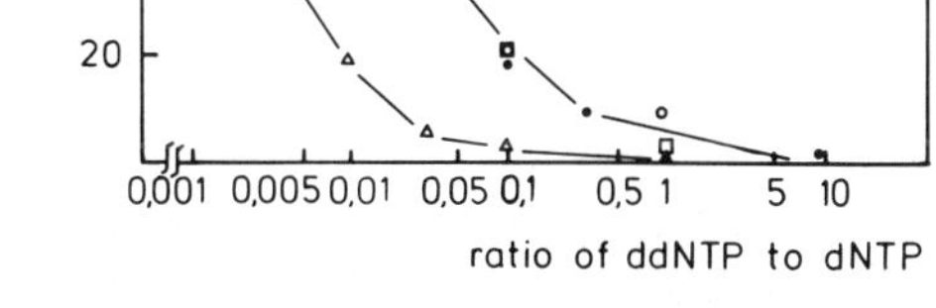

Fig. 4. Specific inhibition of adenovirus DNA synthesis by 2'-3'-dideoxynucleoside-5'-triphosphates (ddNTP's). (——), Ad5 DNA; (---), cellular DNA; (○), ddATP; (●), ddTTP; (□), ddGTP; (Δ), ddCTP. Viral or cellular DNA synthesis was assayed in isolated nuclei in the presence of 50μM unlabelled dNTP, 5μM [^{3}H]-dTTP and ddNTP as indicated. [^{3}H]-dGTP was used and 50μM dTTP when inhibition by ddTTP was tested.

preferentially inhibits the replicative synthesis of fragments containing the terminal protein, while random synthesis, occurring in internal fragments, is not affected (Fig.2, lane 7).

The kinetics of inhibition of adenovirus DNA synthesis in isolated nuclei or in nuclear extracts are most compatible with the inhibition kinetics of isolated DNA polymerase γ, but could also be explained by a function of DNA polymerase β. The latter enzyme is slightly less sensitive than DNA polymerase γ. However, in replication complexes isolated from infected cell nuclei only DNA polymerase α and γ have been found[38-41]. The observed enzymes in the complexes are N-ethylmaleimide sensitive, in contrast to purified DNA polymerase β. An enzyme with the characteristic low sedimentation value of DNA polymerase β was not found. Viral DNA synthesis in these complexes is as sensitive to ddNTP's as in isolated nuclei[40,42].

Therefore these results are best explained assuming a role of DNA polymerase γ in elongation of nascent DNA chains. In agreement with this hypothesis, α-^{32}P-ddTTP at low concentrations is incorporated into Ad5DNA and acts as a chain terminating nucleotide[43]. Under the conditions used for this reaction, only DNA polymerase γ and not DNA polymerase α accepts ddTTP as a substrate. Presumably, DNA polymerase γ is less discriminating with regard to the sugar moiety of its substrate than DNA polymerase α.

Table 2. Adenovirus DNA Replication Proteins

Protein	Origin	Location of structural gene (% from left)	Function
DNA binding protein	viral	61-66	initiation elongation protection
Terminal protein	viral	11-31	initiation
DNA polymerase γ*	cellular	-	elongation

*A possible additional function of DNA polymerase α has been suggested[42,46,47] based upon aphidicolin, phosphonoacetic acid and Ara-CTP inhibition studies (see text). Other proteins like RNA polymerase II and III, DNA ligase, RNase H and a 3' $\rightarrow$ 5' deoxyribonuclease[50] have been found in replication complexes but their function in Ad-DNA replication has not been established.

INHIBITION OF ADENOVIRUS DNA SYNTHESIS BY APHIDICOLIN: AN ADDITIONAL ROLE FOR DNA POLYMERASE α?

Since both DNA polymerase α and γ have been found in replication complexes, it was of interest to study whether DNA polymerase α also has a function in adenovirus DNA synthesis. An additional role of DNA polymerase α, either separately or in a coordinated reaction with DNA polymeras γ, has been suggested from inhibition studies of viral DNA synthesis by phosphonoacetic acid and ara-CTP[40] and by aphidicolin, a specific inhibitor of DNA polymerase α[44,47]. We have investigated the effect of aphidicolin in some detail[48].

When the drug is administered to intact infected or uninfected HeLa cells, cellular DNA synthesis is 300-400 fold more sensitive to aphidicolin than Ad5 DNA replication. A differential sensitivity to aphidicolin is also observed between adenovirus DNA synthesis and SV40 DNA synthesis in monkey cells. These cells can be doubly infected with both viruses, resulting in a combined replication of SV40 DND and adenovirus DNA within the same cell.

Figure 5 shows the result of aphidicolin inhibition in such doubly infected cells. At 3μM aphidicolin more than 99% of the SV40 DNA synthesis is blocked while adenovirus (SA7) DNA synthesis con-

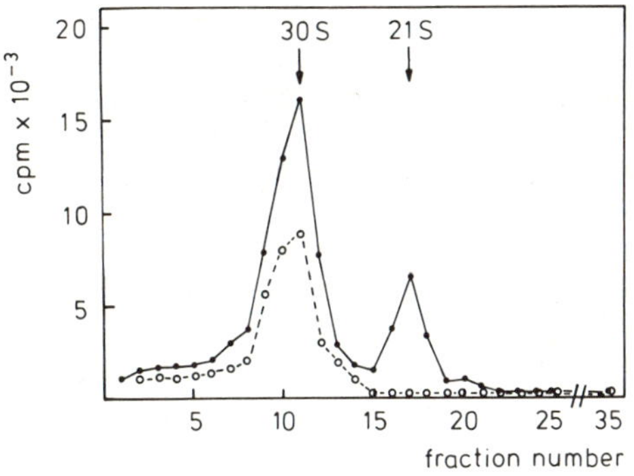

Fig. 5. Differential effect of aphidicolin on adenovirus and SV40 DNA synthesis in doubly infected monkey cells. AGMK cells were infected with SV40 followed by addition of simian adenovirus SA7 after 24hr. Under these conditions both viral DNA's can replicate in the same cell without strong interference. At 16hr after infection with SA7 the cells were labelled with 20 Ci/ml of [^{3}H]-thymidine for 2hr in the absence (●—●) or presence (○—○) of 3μM aphidicolin. Viral DNA was extracted and analyzed by sedimentation. The results of two separate sucrose gradients are shown in one figure. 21 S = SV40 DNA; 30 S = SA7 DNA.

tinues at 55% of the rate observed in the control cells in the absence of the drug. This result makes it less likely that the resistance of adenovirus to aphidicolin is caused by pool size effects or metabolic conversion of the drug. Moreover, the inhibition of Ad5 DNA synthesis in isolated nuclei by aphidicolin does not respond to changes in dCTP concentration, while purified DNA polymerase α is strongly influenced by such changes[48]. The effect of aphidicolin on DNA synthesis in nuclear extracts is also clearly different for randomly synthesized DNA and for replication of fragments containing the adenovirus teminal protein (Fig. 2, lane 8). Label incorporation into the internal fragments A, B and D is much more sensitive than incorporation in the terminal fragments C and E.

Taken together, these results are best explained assuming that DNA polymerase α is not required for adenovirus DNA replication. If this is indeed so, it is difficult to explain why adenovirus DNA synthesis becomes sensitive to aphidicolin at high concentration (< 10μM). Under these conditions, DNA polymerase β and γ, both from infected and uninfected cells, are completely resistant to the drug. It is possible that aphidicolin at high concentrations is reactive towards other components of the adenoviral replication machinery rather than the DNA polymerase. Alternatively, the sensitivity of DNA polymerase γ in a replication fork differs from the sensitivity of the purified DNA polymerase. Another explanation, which we cannot exclude, is that DNA polymerase α is indeed functional in adenovirus DNA synthesis but that its sensitivity towards aphidicolin in the adenovirus replication fork is reduced by the presence of other replication proteins. This may result in a different micro-environment in the replication fork compared to SV40 or host DNA replication. Other approaches to this problem, such as the use of antibodies against DNA polymerases, are clearly required.

CONCLUDING REMARKS

In many ways, adenovirus DNA replication deviates from the host nuclear DNA synthesis. Both initiation and elongation seem to use alternative pathways. With regard to initiation of DNA replication, protein priming has thus far only been observed in the bacteriophage ϕ29 and not in any eukaryotic system[26].

As far as elongation is concerned, there are parallels for displacement synthesis in eukaryotes (Table 3). Both in mitochondiral DNA synthesis[49] and parvovirus[14] replication highly asynchronous duplication of the daughter strands has been observed. Both in the replication of Mt DNA and of the defective adeno-associated virus (AAV) DNA[59], DNA polymerase γ is implicated. In this respect, it is interesting to note that purified DNA polymerase γ can perform processive polymerization in the absence of other replication factors, in contrast to DNA polymerase α and β[53]. Such processiveness would

Table 3. Similarities between Adenovirus and Mitochondiral DNA Synthesis

	Ad	Mt	Cellular/SV40
Mechanism	asynchronous displacement	asynchronous displacement	synchronous (Cairns' type)
Direction	unidirectional	unidirectional	bidirectional
Nucleosome structure	lower number of nucleosomes	no	normal
Cycloheximide	no inhibition	no inhibition	inhibition
Histone synthesis	inhibited	not coupled to DNA synthesis	coupled to DNA synthesis
ddNTP inhibition	strong	strong	no
Aphidicolin inhibition	weak	no	strong

suit a continous mode of DNA chain elongation during displacement synthesis and complementary strand synthesis.

The three systems mentioned above use completely different mechanisms to initiate DNA synthesis, like RNA (Mt-H strand)[60], self-priming (AAV) and protein (adenovirus). Thus, displacement synthesis seems to be independent of the way in which a 3'OH-primer is offered. Recognition and use of DNA polymerase γ may require interaction with initiation proteins that are common to the three systems, or recognition of a specific nucleotide sequence. Alternatively, DNA polymerase γ may recognize an aberrant nucleosome structure. Nucleosomes have not been observed in Mt DNA. During adenovirus infection, histone synthesis is blocked[57]. This leads to a reduced free pool of histones which could explain the low level of nucleosomes found on replicating adenovirus DNA late in infection[34,58]. Such an altered nucleosome structure may be a better template for a nuclear DNA polymerase γ than for DNA polymerase α.

ACKNOWLEDGEMENTS

We thank Drs. H.S. Jansz and J.S. Sussenbach for stimulating discussions, Dr. B.M. Stillman (Cold Spring Harbor) for help in the use of nuclear extracts, and Dr. E. Winnacker (Munchen) for kindly permitting one of us (B.G.M. van Bergen) to work in his laboratory. Aphidicolin was a gift of Drs. S. Spadari (Pavia), A.H. Todd and

B. Hesp (ICI). This work was supported in part by the Netherlands Foundation for Chemical Research and the Netherlands Organization for the Advancement of Pure Research.

REFERENCES

1. P. H. Steenbergh, J. Maat, H. van Ormondt and J. S. Sussenbach, Nucleic Acids Res. 4:4374-4389 (1977).
2. M. Shinagawa and R. Padmanaban, Biochem. Biophys. Res. Comm. 87:671-678 (1979).
3. A. Tolun, P. Alestrom and U. Pettersson, Cell 17:705-713 (1979).
4. A. J. Roninson, H. B. Younghusband and A. J. D. Bellett, Virology 56:54-69 (1973).
5. D. M. K. Rekosh, W. C. Russell, A. J. D. bellett and A. J. Robinson, Cell 11:283-295 (1977).
6. J. Tooze, Molecular Biology of Tumor Viruses 2:443-546 (1980).
7. A. J. Levine, P. C. van der Vliet and J. S. Sussenbach, Curr. Topics Microbiol. Immunol. 73:67-124 (1976).
8. E. L. Winnacker, Cell 14:761-773 (1978).
9. J. S. Sussenbach, P. C. van der Vliet, D. J. Ellens and H. S. Jansz, Nature New Biol. 239:47-49 (1973).
10. J. M. Vlak, Th. H. Rozijn and J. S. Sussenbach, Virology 63:168-175 (1975).
11. E. L. Winnacker, J. Virol. 15:744-758 (1975).
12. R. L. Lechner and T. J. Kelly, Cell 12:1007-1020 (1977).
13. T. Cavalier-Smith, Nature 250:467-470 (1974).
14. P. Tattersall and D. C. Ward, Nature 263:106-109 (1976).
15. S. J. Flint, S. M. Berget and P. A. Sharp, Cell 9:559-571 (1976).
16. B. W. Stillman and A. J. D. Bellett, Nature 269:723-725 (1977).
17. J. S. Sussenbach and M. G. Kuijk, Nucleic Acids Res. 5:1289-1295 (1978).
18. M. D. Challberg and T. J. Kelly, Proc. Natl. Acad. Sci. U.S.A. 76:655-659 (1979).
19. T. Reiter, J. Futterer, B. Weingartner and E. L. Winnacker, J. Virol. 35:662-671 (1980).
20. B. M. Stillman, J. Virol. 37:139-147 (1981).
21. B. W. Stillman, J. B. Lewis, L. T. Chow, M. B. Mathews and J.E. Smart, Cell 23:497-508 (1981).
22. M. D. Challberg and T. J. Kelly, J. Mol. Biol. 135:999-1012 (1979).
23. J. E. Ikeda, T. Enomoto and J. Hurwitz, Proc. Natl. Acad. Sci. U.S.A. 78:884-888 (1981).
24. M. D. Challberg and T. J. Kelly, J. Virol. 38:272-277 (1981).
25. M. D. Challberg, S. V. Desiderio and T. J. Kelly, Proc. Natl. Acad. Sci. U.S.A. 77:5105-5109 (1980).
26. R. P. Mellado, M. A. Penalva, M. R. Inciarte and M. Salas, Virology 104:84-96 (1980).
27. P. C. van der Vliet, A. J. Levine, M. J. Ensinger and H. S. Ginsber, J. Virol. 15:348-354 (1975).

28. P. C. van der Vliet and A. J. Levine, Nature New Biol. 246:170-174 (1973).
29. D. M. Fowlkes, S. T. Lord, T. Linné, U. Pettersson and L. Philipson, J. Mol. Biol. 132:163-180 (1979).
30. P. C. van der Vliet and J. S. Sussenbach, Virology 67:415-426 (1975).
31. L. M. Kaplan, H. Ariga, J. Hurwitz and M. S. Horwitz, Proc. Natl. Acad. Sci. U.S.A. 76:5534-5538 (1979).
32. P. C. van der Vliet, J. Zandberg and H. S. Jansz, Virology 80:98-110 (1977)
33. R. S. Galos, J. Williams, M. K. Binger and S. J. Flint, Cell 17: 945-956 (1979).
34. C. Kedinger, O. Brison, F. Perrin and J. Wilhelm, J. Virol. 26: 364-380 (1978).
35. J. W. Bodnar and G. D. Pearson, Virology 100:208-211 (1980).
36. H. J. Edenberg, S. Anderson and M. L. De Pamphilis, J. Biol. Chem. 253:3273-3280 (1978).
37. P. C. van der Vliet and M. M. Kwant, Nature 276:532-534 (1978).
38. O. Brison, C. Kedinger and J. Wilhelm, J. Virol. 24:423-435 (1977).
39. M. Arens, T. Yamashita, R. Padmanaban, T. Tsurno and M. Green, J. Biol. Chem. 252:7947-7954 (1977).
40. M. M. Abboud and M. S. Horwitz, Nucleic Acids Res. 6:1025-1038 (1977).
41. S. van der Werf, J-P. Bouché, M. Méchali and M. Girard, Virology 104:56-72 (1980).
42. C. H. Shaw, D. M. Rekosh and W. C. Russell, Gen. Virol. 48:213-236 (1980).
43. P. C. van der Vliet and M. M. Kwant, Biochemistry 20:2628-2632 (1981).
44. S. Ikegami, T. Taguchi, M. Ohashi, M. Oguro, M. Nagaro and Y. Mano, Nature 275:458-460 (1978).
45. G. Pedrali-Noy and S. Spadari, Biochem. Biophys. Res. Comm. 88: 1194-1202 (1979).
46. M. Longiaru, J. Ikeda, Z. Jarkovsky, S. B. Horwitz and M. S. Horwitz, Nucleic Acids Res. 6:3369-3386 (1979).
47. H. Krokan, P. Schaffer and M. C. de Pamphilis, Biochemistry 18: 4431-4443 (1979).
48. M. M. Kwant and P. C. van der Vliet, Nucleic Acids Res. 8:3993-4007 (1980).
49. D. C. Robberson, H. Kasamatsu and J. Vinograd, Proc. Natl. Acad. Sci. U.S.A. 69:737-741 (1972).
50. U. Bertazzoni, A. I. Scovassi and M. G. Brun, Eur. J. Biochem. 81:237-248 (1977).
51. U. Hubscher, C. C. Kuenzle and S. Spadari, Eur. J. Biochem. 81: 249-258 (1977).
52. A. Bolden, G. Pedrali-Noy and A. Weissbach, J. Biol. Chem. 252: 3351-3356 (1977).
53. M. Yamaguchi, A. Matsukage and T. Takahashi, Nature 258:45-47 (1980).

54. J. C. D'Halluin, C. Allart, C. Cousin, P. A. Boulanger and G. R. Martin, J. Virol. 32:61-71 (1979).
55. J. S. Sussenbach, personal communication.
56. E. Frost and J. W. Williams, Virology 91:39-50 (1978).
57. G. Tallman, J. E. Akers, B. T. Burlingham and G. R. Reeck, Biochem. Biophys. Res. Comm. 79:815-822 (1977).
58. A. L. Beyer, A. H. Bouton, L. D. Hodge and O. L. Miller, J. Mol. Biol. 147:269-295 (1981).
59. H. Handa and B. J. Carter, J. Biol. Chem. 254:6603-6610 (1979).
60. A. M. Gillum and D. A. Clayton, J. Mol. Biol. 135:353-368 (1979).

TOPOLOGY, TYPE II DNA TOPOISOMERASES AND DNA REPLICATION IN PROKARYOTES AND EUKARYOTES

Patrick Forterre[1], Liliane Assairi[1] and
Michel Duguet[2]

SUMMARY

DNA replication raises several topological questions. i) How did the two parental strands uncoil in spite of the restriction imposed to their rotation either by DNA circularity or by the barriers which segregate the chromosome into topologically independent domains? ii) How is tangling of the two daughter duplexes avoided? Which mechanism generates the interlocked DNA rings found in several organisms? iii) What is the role of negative supercoiling of natural DNAs in their replication if any? iv) How is DNA unwinding compatible with the nucleosomal and the supranucleosomal organization of the chromosome?

The isolation of DNA topoisomerases from prokaryotes and eukaryotes over the last few years should help to answer these questions. For instance, purified preparations of type II DNA topisomerases can force a DNA segment through another DNA segment via a transient double-strand break. Through this mechanism, these enzymes could remove positive supertwists which would have accumulated in front of the replication fork if this supercoiling were the means by which uncoiling of the parental strands were compensated. One such topoisomerase, the bacterial gyrase, can transform these positive supertwists directly into negative ones. Type II DNA topoisomerases

[1]Institut de Recherche en Biologie Moléculaire, Université Paris VII, France.
[2]Institut de Recherches Scientifiques sur le Cancer, B.P. N° 8, 95802 Villejuif Cedex, Université Pierre et Marie Curie, Paris VI, France.

may also decatenate duplex DNA rings thereby separating interlocked products of DNA replication if such catenanes are produced either to compensate unwinding of the parental strands, or from errors during the alignment of the replication complexes, or else from some mechanism occurring at the end of the round of replication.

The involvement of type II DNA topoisomerases in DNA replication has been tested in bacteria in vivo and in vitro using inhibitors of these enzymes and strains carrying thermosensitive mutations in their constitutive subunits.

In the case of the elongation phase of DNA replication, the analysis of all available data leads to the hypothesis that gyrase activity is dispensable for normal progression of the replication fork. Thus, negative supercoiling of the DNA would not help unwinding of the parental strands.

On the contrary, either relaxation of positive supertwists or decatenation is needed for DNA elongation since, when all the bacterial type II DNA topoisomerases are inhibited by drugs or mutations, DNA elongation immediately ceases.

Complications in data interpretation arise from the possibility of DNA topoisomerase molecules to slow or to stop DNA elongation by forcing strong interactions eith the DNA when one or several of their subunits are altered by drugs or mutations. This inhibition mechanism would explain why inhibitors of type II DNA topoisomerases stop DNA synthesis in vitro in spite of the fact that in vitro, the chromosome apparently lacks topological constrains. Indeed, the differences between the mode of action of type II DNA topoisomerase inhibitors in vivo and in vitro suggest that introduction of swivel in the chromosome during the preparation of an in vitro system is sufficient to make DNA elongation independent of any topoisomerase activity. This supports the hypothesis that unwinding of the parental strands is compensated by positive supertwisting, and not by the catenation of daughter molecules. In vivo, this supertwisting would be removed by the swivelase activity of type II DNA topoisomerase.

In contrast, gyrase and/or catenase activities are essential for normal initiation of bacterial DNA replication. Analysis of available data suggests that type II DNA topoisomerases are not required for initiation per se but rather for a process which controls cell divisions, chromosome segregation and later initiation of DNA replication itself. We propose a model in which this process is the formation of the link between the chromosome and the cell membrane.

In eukaryotes, type II DNA topoisomerases are presumably required for initiation of DNA replication but not for elongation. Their role may be to help the decondensation of the chromatin which precedes initiation and/or to promote the initiation process itself. Modifications in the nucleosomal and supranucleosomal organization during the chromatin decondensation would aid DNA topoisomerase and the further progression of the replication fork through the nucleosomal fiber.

The similarities and differences between the chromosomal organization and type II DNA topoisomerases of eukaryotes and prokaryotes are analyzed in terms of evolution. A common origin is postulated for the topological domains which segregate the chromosome in both cell types, and also between the type II DNA topoisomerase-DNA complexes inside these domains which control their superhelicity. A model is proposed for their diverging evolution.

INTRODUCTION

In 1953, Watson and Crick proposed a model for the structure of DNA and for DNA replication[1]. It soon became clear to them that the winding of the two polynucleotide strands around each other postulated by the theory of the double helix, raised an essential topological problem in their scheme of DNA replication[2]. Since semi-conservative replication implied that the two strands come apart to serve as a template for a new chain, their uncoiling was a necessary prerequisite for their separation. At this time, it seemed an impossible task to imagine a mechanism for this uncoiling which would require a very rapid rotation of the DNA molecule. In particular, one could not easily imagine how tangling of the newborn strands could be prevented, a situation that would make their future segregation uncertain.

The difficulties encountered in solving the topological problem appeared so fundamental that Watson and Crick considered them as the main objection to their double-helical model for DNA structure, in their Cold Spring Harbor paper[2] of 1953. Remarkably, in the same paper, these authors appeared much more confident of their scheme for DNA replication, which they never questioned.

They have tried to imagine models for DNA structure in order to avoid the topological problem, as, for example, models in which the two strands are not interlaced, as in a plectonemic coiling, but instead lie side by side, either in a paranemic coiling in which they are not interwound but merely in close opposition to each other, or in a ribbon-like arrangement (see Fig. 1). They also imagined that both right and left handed helices occur sequentially all along the chromosomes. For steric reasons they finally rejected all these models as incompatible with the specific pairing of the bases.

It is interesting to note the pertinence and premonitory nature of Watson and Crick's reasoning for this geometrical standpoint, in the same paper in which they surprisingly suggested the hypothesis of a self-replication mechanism for DNA, independent of any enzymatic

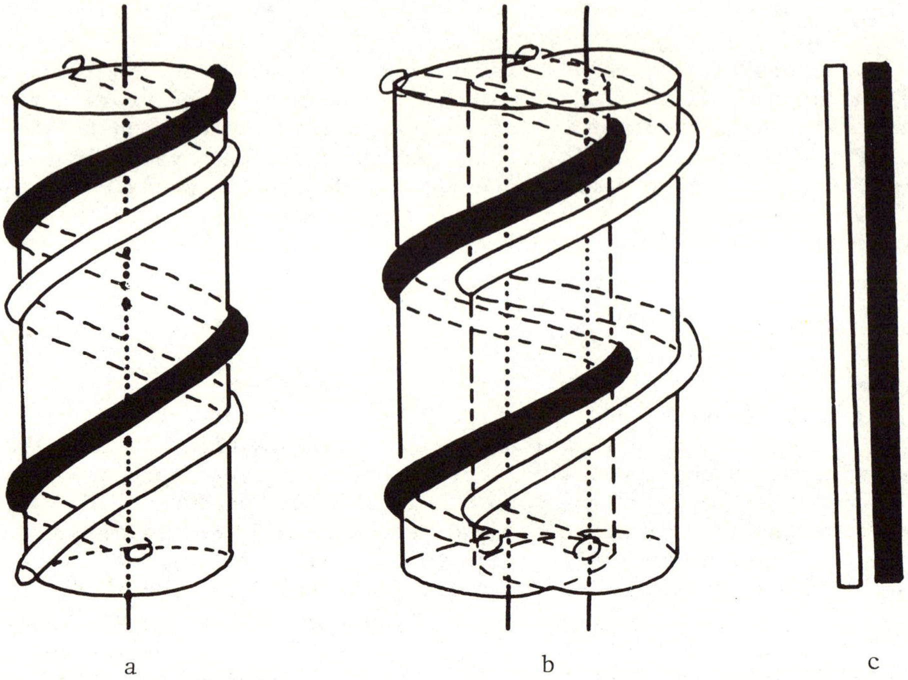

Fig. 1. Several theoretical models for DNA structure discussed by Watson and Crick in 1953. (a) A double helix with plectonemic coiling: the two helices have the same symmetrical axis and are interlaced a number of times which is equal to the number of helix turn. Their separation requires uncoiling. (b) A double helix with a paranemic coiling: the two helices lie side by side but do not have the same symmetrical axis and are not interlaced. They can separate without uncoiling. If it is postulated that each helix turns is formed of alternative right-handed and left-handed sequence, this gives rise to the side by side model as postulated by Rodley et al.[14]. (c) The ribbon-like arrangement: the two strands are not coiled but run in a straight line.

activity. Indeed, they wondered "whether a special enzyme would be required to carry out the polymerization or whether the existing single helical chain could act effectively as an enzyme"[2] and when they thought of a possible swivel mechanism to help uncoiling, they supposed that "one chain of a pair occasionally breaks under the strain of twisting; the polynucleotide chain remaining intact could then release the accumulated twists by rotation about single bonds and following this, the broken ends being still in close proximity, might rejoin".

The disregarding of protein involvement in DNA replication can probably be considered as a late Schroedinger effect[3], with the DNA molecule growing by direct assimilation of free nucleotides being compared to a crystal growing from direct assimilation of ions in a salt solution. It would have been very exciting indeed to imagine the DNA as the ultimate secret of life, possessing one of the main properties of life itself, self-duplication.

In fact, it has become increasingly evident that DNA replication requires many proteins, not only for polymerization but for many other purposes, and in particular, for strand separation.

In recent years, the topological problems arising from the double helical model fell into the domain of enzymology with the discovery of a fascinating new class of enzymes able to modify the topological properties of a DNA molecule, the DNA topoisomerases[4].

It has also become more and more evident that topological problems should be considered not only with respect to naked DNA, but also to the arrangement of the DNA inside the nuclear body of prokaryotes or the chromatin of eukaryotes.

In this paper, we will first try to summarize the various topological problems facing DNA replication, considering the DNA structure at the secondary, tertiary, (circularity and supercoiling) and quaternary (catenation and arrangement inside the chromosome) levels. In the second section, we will briefly review the various type II DNA topoisomerases presently known, which, based on experimental data, appears, at least in prokaryotes, to be the better candidates for solving the topological problems described above. In the third and fourth sections, we will discuss what is presently thought to be the role of these enzymes in DNA replication, first in prokaryotes, and then in eukaryotes. This will be based on results accumulated in vivo and in vitro from studies with inhibitors of these enzymes, or with strains in which these enzymes are thermolabile. Finally, we will compare the eukaryotic and prokaryotic DNA topoisomerases and chromosomal organization in terms of evolution.

I. TOPOLOGY AND DNA REPLICATION

1. Unwinding of the Double Helix: The Topological Problem

The topological problem as first theoretically imagined by Watson and Crick has had to be revised following the discovery of genome circularity in prokaryotes[5]. Circularity implies that the number of interwindings between the two polynucletide strands of a DNA duplex is a topological invariant that remains unchanged whatever the deformation imposed on the DNA molecule, as long as the strands remain intact[6]. Indeed, it is well known that breakage of all the hydrogen bonds linking the two DNA strands of a closed DNA duplex by alkaline or thermal treatment leads to a structure in which the two resulting single strands are interlaced a number of times equal to the previous number of double helical twists in the native duplex. The continuous unwinding of the two DNA strands at the replication fork during the elongation phase of DNA replication would therefore be compensated by a continuous winding of these strands in some other part of the molecule.

Since the incompressibility of interatomic distances precludes shortening of the helix pitch, this winding could only occur in two ways[7]:

(1) supercoiling of the unreplicated portion of the molecule, I.E., coiling of the helical axis of the parental helix with the daughter strands kept aligned by the replication complex (Fig. 2A);
(2) winding of the two daughter molecules around each other, downstream from the replication fork, i.e., winding of the daughter strands about the helical axis of the parental strands (Fig 2B).

In the first situation, each twist which is eliminated when the double helix unwinds (conventionally designated as positive) must be compensated by the creation of one positive supertwist (Fig. 2A). Clearly, some mechanisms should then exist to release positive supertwists as they are introduced, for their accumulation in the unreplicated section of the molecule would hinder separation of the parental strands. Indeed, supertwisting is an energetically unfavorable reaction, and, therefore, the energy required to open the double helix becomes greater as the positive superhelical density increases.

In the second situation, that is, tangling of the two daughter duplexes downstream from the replication fork, some special mechanism should also exist, in this case to decatenate the two daughter strands. Without such a mechanism, one round of replication would result in a quaternary helix incapable of dividing into its two double helical components (Fig. 2B). In fact, this mechanism of decatenation, if it exists, would operate as soon as the uncoiling starts after initiation of DNA replication, since formation of a number of twists in the quaternary helix equal to the number of twists eliminated in the double helix appears impossible for steric and energetic reasons.

In any case, a solution to the topological problem involves the creation of transient breaks in the DNA molecule, since this is the only way to change the number of interwindings between two interlaced closed DNA duplexes (also called the linking number) or to separate two catenated DNA rings. For some time, only the first situation, i.e., positive supertwisting of the unreplicated section of the duplex DNA, was considered. With this in mind, Cairns, in 1962 proposed

a b

Fig. 2. Possible consequences of strand separation during replication of a circular DNA duplex. (a) Opening of the strands at the origin of replication has uncoiled one turn of the duplex and consequently has introduced one positive supertwist in the unreplicated portion of the molecule. (b) Uncoiling of the two daughter molecules. In that figures, several parental duplex turns have been removed but the daughter molecules are only interlaced twice. We should therefore imagine that several interwindings between them have been previously removed as the two replication forks moved in order to avoid the formation of an unstable quaternary helix. Small arrows indicate the direction of the movement of the replication fork; arrows suggest either uncoiling of the parental single-strands or coiling or interlacing of DNA duplex.

the existence of a mechanical swivel allowing free rotation of the two DNA strands around each other in the unreplicated section[5]. For years, the mysterious junction point between DNA and the membrane appeared to be a likely candidate for the location of this mysterious swivel. In the sixties, further discoveries of various DNases and ligases suggested an alternative explanation, i.e., that the swivel is generated by the continuous and concerted action of DNases and ligases ahead of the replication fork, promoting the formation of a transient nick in one of the two parental strands.

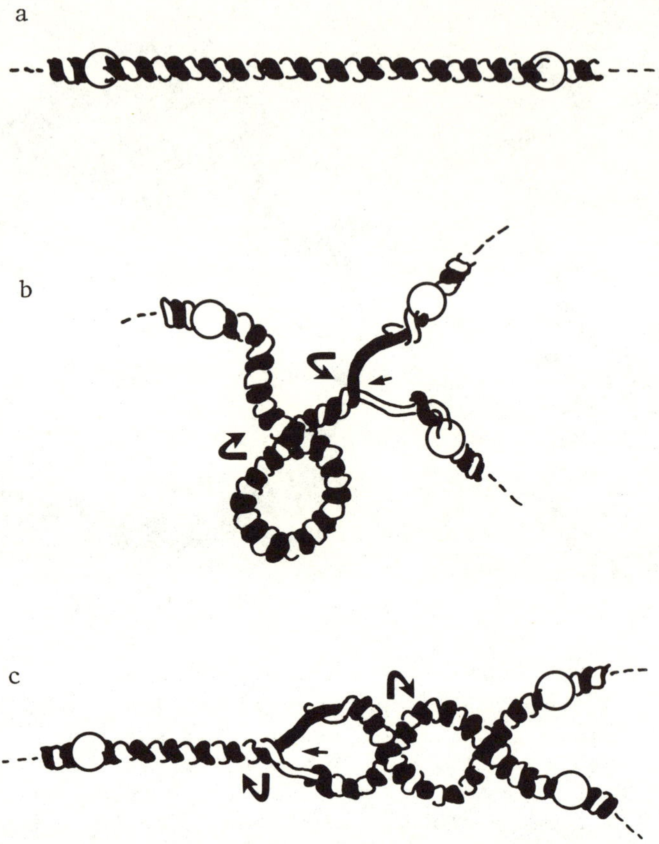

Fig. 3. Consequences of strand separation during replication of a topological domain inside the chromosome. A topological domain is delineated by two topological barriers (a))open circles). When a replication fork moves inside this domain, strand uncoiling in the domain could be compensated by either formation of positive supercoiling of the domain in front of the fork (b) or interlacing of the two daughter molecules behind the fork (c).

The discovery, in the early seventies, of a new enzyme called ω[8], able, in one molecule, to perform both nicking and closing of one strand of a duplex, seemed to put an end to the period of speculation and to firmly establish the swivelase theory. But this theory must now be revised after the isolation of DNA gyrase[9], a protein which may convert relaxed or positively supertwisted DNA into negatively supertwisted DNA at the expense of ATP, and to catenate or decatenate two interlocked DNA rings[4].

The present situation concerning the solution to the main topological problem in bacterial DNA replication is therefore not clear and cannot be deduced from the simple existence of one or another enzyme. A priori, ω (now called DNA Topoisomerase I) is not a good candidate at this level because it can relax only negatively supertwisted DNA and not positively supertwisted DNA. In E. coli, the latter may be either directly converted in negatively supertwisted DNA by the DNA gyrase (also called DNA Topoisomerase II) at the expense of ATP, or relaxed in an ATP-independent reaction by another enzyme related to DNA gyrase, recently isolated, and called DNA Topoisomerase II'. These two enzymes may therefore act as a swivel during DNA replication (Fig. 2a) but DNA gyrase as suggested by Champoux[7] can also decatenate the two daughter double helices if their tangling compensates unwinding of the parental helix (Fig. 2b).

The above topological considerations concerning replication of circular DNA may be extended to eukaryotic chromosomal DNA, although the latter seems to be a linear molecule. Indeed, in both types of organisms, the DNA in the isolated chromosome is segregated into multiple independent topological domains[10,11]. This structural feature has recently been confirmed in vivo in the case of E. coli[2]. The topological independence of these domains means that some unknown structural constraints prevent any modification of the linking number of the two DNA strands inside one domain from changing the linking number of the two DNA strands inside the other domains. Consequently, when the two parental strands pull apart in one domain during replication, the compensatory process, either supertwisting or catenation, must occur inside the same domain and must be performed by a mechanism which is itself located between the two topological barriers which enclose the replication fork (Fig. 3). This occurs in the same way whether the overall DNA structure is linear or circular.

2. The Alignment Problem

The possible interlacing of daughter strands downstream from the replication fork (Fig. 3c) could be envisaged, even if it does not help separation of the parental strands.

Let us imagine the progression of two opposite replication forks in a circular DNA molecule lying within a plane. Unless each replication separation complex proceeds without any rotation with respect to the plane, the result of a round of replication is catenated duplex daughter molecules. This problem was analyzed in 1978 by Pohl and Roberts[13], who estimated that if the complex "makes an error" of 1° per turn with respect to the plane, the number of topological links between the daughter molecules after a round of replication of the E. coli chromosome would be about 1000. Pohl and Roberts argued that tangling of the DNA rings would hinder their separation after completion of DNA replication, and therefore, they formulated what they called the alignment problem as follows: "What is the principle of control of the replication complex, whereby it so places the daughter molecules that are topologically unlinked?" Their answer was that the problem cannot be solved by the double helical theory of DNA, since the fundamental symmetry of the double helix hinders control of the spatial orientation of the replication complex relative to the helical axis, and therefore prevents the replication complex from providing a direction for the separation of the daughter molecules. For that reason, they rejected the double helical model and favored more recent ones, such as the side-by-side model[14] in which the DNA molecule is asymmetric and unlinked (Fig. 1b), with these properties being used by the replication complex to align itself onto the molecule, thereby avoiding tangling.

It is worth pointing out that 25 years after Watson and Crick first encountered this problem[2], topological considerations were again brought up to consider models of the DNA structure different from the model of the plectonemic double helix. Nevertheless, the most recent research on DNA topology supports this latter model rather than the side-by-side model[15]. It is also interesting to note that two years before the discovery of the catenation-decatenation activity of DNA gyrase[16], Pohl and Roberts examined the possible existence of a new enzyme, the "catenase" which would make the alignment problem insignificant[13]. These authors rejected this possibility, however, as they were unable to imagine a way of controlling the mode of action of such an enzyme. Control would be necessary to avoid intramolecular knotting of the chromosome and to drive the overall reaction towards decatenation. On the contrary, they thought that high intracellular DNA concentrations would favor both catenation and knotting.

Does the recent discovery of natural catenase activity solve the alignment problem? We feel that this question should be added to those asked at the end of the preceding section. In other words, do the daughter helices wind around each other downstream from the replication fork? If so, how are DNA topoisomerases involved in the resolution of the catenated products?

If the answer to the first question is affirmative, then the problems raised by Pohl and Roberts concerning the mode of action of catenase remain unsolved: how does the enzyme recognize which daughter molecule should pass through the other, and in which direction? If, conversely, the daughter molecules are indeed aligned, which mechanism organizes this alignment?

Watson and Crick suspected that "the most reasonable way to avoid tangling is to have the DNA fold up into a compact bundle as it is formed"[2]. Indeed, the possibility exists that as the two daughter strands escape from the replication complex, they are immediately packaged to construct the two new chromosomal structures, and therefore are prevented from becoming interwound[17]. Another possibility is that the two replication forks are tied together and uniquely orientated within a binary replication complex[17].

3. Natural Catenated DNA Rings

The natural occurrence of catenated DNA rings was experimentally demonstrated several years ago in the DNA of a wide variety of organisms such as mitochondrial DNA, SV40 DNA, bacterial plasmid DNA and kinetoplast DNA[18]. The role of this catenation process in the life of the cell is unclear, especially in the case of the giant network of interlaced kinetoplast DNA.

Thus far, the main hypothesis which takes into account some of these structures suggests that improper segregation of these molecules occurs at the end of DNA replication. Sundin and Varshavsky suggested that this may arise in the cycle of SA40 DNA replication from the impossibility of the DNA topoisomerase to release positive supertwisting induced by uncoiling of the last duplex twists[17].

This may be the case if, for stereochemical reasons, DNA topoisomerases cannot work on DNA pieces that are too small. Unwinding of the last duplex twists would then directly convert these twists into interlaced duplex rings (Fig. 4).

In our view, the hypothesis proposed by Champoux whereby uncoiling of the parental strands is compensated by winding of the daughter molecules around each other is an extension to the overall elongation mechanism of the proposal of Sundin and Varshavsky which concerns the end of DNA replication. In that case, catenated dimers isolated at the end of replication would be representative of the lag between occurrence of tangling and the action of the catenase. We already pointed out that such a lag cannot occur at the beginning of the elongation process without immediately generating an unstable quaternary helix.

Nevertheless, this difficulty may be circumvented if another kind of mechanism operates at the very beginning of a DNA replication round, such as, for instance, overwinding of the domain containing the origin itself as suggested for bacteriophage T4 DNA replication[19].

4. What About Negative Supertwisting?

We have seen how theoretical considerations lead to the introduction of some topological problems in the theory of DNA replication: the possible formation of positive supertwists or the tangling of the daughter strands. If occurrence of catenated DNA rings during elongation has not yet been observed, the same applies for positive supertwisting which, to date, has not been observed in natural DNA.

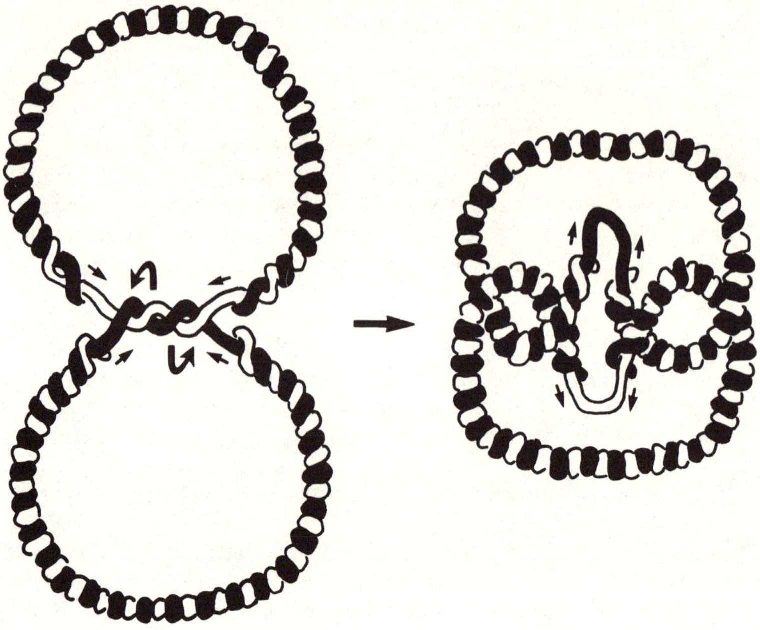

a b

Fig. 4. Catenation of the daughter molecules arising at the end of one replication round. (a) At the end of one replication round one might imagine that two double helical turns remain to be uncoiled preventing the two replication forks from moving onto the last unreplicated portion of the DNA. (b) If the DNA topoisomerases cannot relax positive supercoiling arising in such small portions their uncoiling would directly convert the two replication products into DNA rings interlaced twice. Each ring contains a gap near the terminus of replication which could later be filled by the DNA polymerases. Small arrows indicate the direction of the movement at the replication fork; large arrows indicate uncoiling of the parental strands. In (a) and (b), the linking number between the two parental strands is the same and equal to two.

On the contrary, all the natural circular DNAs isolated so far are negatively supertwisted with about one negative supertwist per 15 double helical twists[20]; this is also true for the DNA inside the chromosomal domains in eukaryotes and in prokaryotes[10,11].

Another question may then be formulated: what, if any, is the role of such negative supertwisting in DNA replication? What is the role, in this process, of an enzyme such as DNA gyrase which is presumably required, at least in prokaryotes, to maintain negative supertwisting?

Partial unwinding of a double helix will introduce positive supertwists in a closed circular DNA duplex, and will therefore remove negative supertwists in a negatively supercoiled duplex. Since the latter reaction is energetically favorable, negative supertwisting of the DNA could help to destabilize the double helix, either for strand separation or for DNA binding of any protein which promoted local denaturation of the molecule[20].

Negative supertwisting could therefore be required for the initation, elongation or segregation step in DNA replication, as well as for other processes such as transcription, repair or recombination.

The recent measurement of the torsional tension in DNA made in vivo[21] indicates that negative supertwisting is accessible to measurement in prokaryotic cells but not in eukaryotic cells. In the latter, torsional tensions are possibly restrained by the winding of the DNA helix both around the nucleosomal core and in the supranucleosomal structure[22].

If negative supertwisting were a prerequisite for DNA chain elongation in prokaryotic cells, this would establish a dramatic difference between the mode of replication in eukaryotes and prokaryotes unless the replicating DNA in eukaryotic cells were locally unrestrained at the level of the replication fork.

Finally, we should notice that negative supertwists should be re-established as they are relaxed by the movement of the replication fork.

5. Unwinding of the Double Helix and Chromatin Structure

The eukaryotic chromatin organization into nucleosomes creates new topological problems for the progression of the replication fork. Nevertheless, these problems remain difficult to define precisely, because too many questions concerning the behavior of the nucleosomes during replication remain unsolved.

It has been shown that elimination of negative supertwisting by ethidium bromide (EtBr), a drug which intercalates between the base pairs of DNA and creates positive supertwists in a covalently closed circular duplex, parallel the removal of the nucleosome core from the DNA in the folded eukaryotic chromosome[11]. If EtBr really dissociates the nucleosome core from the DNA by removing the negative supertwists in the DNA and not by another unknown effect, then one can imagine that positive supercoiling of the DNA related to the progression of the replication fork should also remove the nucleosome ahead of this fork. This mechanism might be used as a driving force to clear the DNA of nucleosomes and thereby prepare it for replication.

Nevertheless, it is not evident that the nucleosomes really leave the DNA when the replication fork reaches them; in a current model, for example, preexisting nucleosomes are conserved and recovered exclusively on the 5'-3' lagging strands[23]. In this case, on the contrary, one should imagine a mechanism which prevents the accumulation of positive supertwists ahead of the fork, avoiding the removal of the nucleosomes located in this region.

The mode of action of the topoisomerases could also be concerned by the presence of nucleosomes if, for instance, eukaryotic type II topoisomerases, like DNA gyrase, wound a long segment of free DNA (about 140 base-pairs) around themselves to perform their reactions. Another complication may arise by the formation by the nucleosomal and/or supranucleosomal organization of strong barriers preventing free rotation of the double helix. Type I DNA topoisomerase, for instance, is thought to provoke the formation of transient break in one of the two DNA strands with the passage of the other strand through the break and the gradual removal of any stress in the DNA[24]. The relaxation wave is therefore initiated at the level of the binding site of type I topoisomerase and further extended to the whole DNA molecule. The problem is whether or not the relaxation wave can travel across the nucleosome succession. This situation is quite similar to that created when nicks are introduced by DNAse in isolated folded chromosomes containing histones. In that case, it has been shown that despite the presence of nucleosomes, the negative supertwists are removed from the DNA loops once the nicks have been introduced[11]. This removal probably results from the rotation of the nucleosomal fiber around the fiber DNA end. However, in these experiments, the nuclear structures do not contain histone H1 and therfore should not be organized in supranucleosomal structures. Therefore, we do not know if the nucleosomal fiber is always free to rotate inside such an organization. If not, the action of type I topoisomerase should be spatially very restricted in the eukaryotic chromosome.

Finally, before leaving the field of nucleosomes, we should point out that the various problems depicted above may not be restricted to eukaryotes, since histone-like proteins have been isolated from prokaryotes[25,26] and nucleosome-like fibers detected in these organisms by electron microscopy[27].

II. TYPE II DNA TOPOISOMERASES

Definition

It is difficult to give a complete definition of type II DNA topoisomerases, since the different functions of these enzymes are only partly elucidated. Basically, type II topoisomerases are enzymes that catalyze changes in the topology of the DNA by using temporary double-strand scissions. This simple property is the key to a number of topological isomerizations in DNA, including supercoiling and relaxation, strand reassociation, knotting and unknotting, catenation and decatenation. The rapid progress which has recently been made in the field of topoisomerases is mainly a consequence of extensive studies of DNA gyrase, a topoisomerase found in prokaryotes.

1. Prokaryotic Type II DNA Topoisomerases

DNA gyrase was initially described[9] as an enzyme which introduces negative superhelical turns into DNA. Gyrase is made up of two subunits, Gyr A (105,000 daltons) and Gyr B (95,000 daltons) which are, respectively, the targets of nalidixic (nal) or oxolinic (oxo) acids and coumermycin A_1 (cou) or novobiocin (novo), four inhibitors of DNA replication in E. coli. By using drug resistance mutants, Gyr A and Gyr B were clearly identified as the protein products of the nal and cou genes, respectively[28,29]. At about the same time, it was shown that DNA gyrase catalyzes several different reactions, wherein both subunits are required in equal molar ratios. These reactions are: (i) The ATP-dependent production of negative superhelical turns (inhibited by the four inhibitors); (ii) Binding to DNA duplex in the absence of ATP (not inhibited); (iii) Relaxation of negative supercoils in the absence of ATP (inhibited by nal and oxo exclusively); (iv) Double-stranded cleavage of DNA in the presence of oxo or nal and sodium dodecyl sulfate (SDS) (not inhibited by novo and cou, but cleavage does not occur with gyrase from an nal^R strain); (v) DNA-dependent ATPase activity on the DNA duplex (inhibited by novo and cou); (vi) ATP dependent catenation or decatenation of duplex DNA rings[16,30] (inhibited by the four drugs). For a review see[4].

To account for these activities, the simplest idea was to consider that the gyrase segregates a circular DNA molecule into two independent topological domains, and in a first step, introduces negative supertwists in one of them and positive supertwists in the other, the linking number of the whole molecule being unchanged. Removal of the positive supertwists by the nicking-closing activity associated with the Gyr A subunits would then introduce net negative superhelicity in the overall molecule.

On this basis, three models have been proposed which differ in the way in which the DNA is segregated and rendered suprecoiled. In the first model, the Gyr B subunits bear an ATP-dependent helicase which would partially melt the DNA bound to the gyrase, introducing positive supercoils into the rest of the DNA molecule[29]. In the second model, the enzyme is firmly bound at the junction of two DNA segments and thus defines two topologically independent loops in the DNA molecule. An ATP dependent rotational movement of the gyrase molecule then accumulates negative supertwists in one loop and positive supertwists in the other[31]. In the third model[32], the DNA is again segregated in two loops by an enzyme-DNA complex located at the junction of two DNA segments; then, positive coiling of the DNA from one segment around the enzyme introduces one negative supertwist in one loop. Following that, an ATP-dependent movement of the gyrase along the DNA displaces the positive twist initially formed around the enzyme into the other DNA loop. These three models however, were not supported by experimental evidence[33]. For instance, Gyr B does not exhibit any helicase activity. Furthermore, these models do not explain why the nicking-closing activity of Gyr A produces transient double-stranded breaks in the DNA. Finally, recent results have shown that the relaxation activity of positive supertwists previously found in gyrase preparations was due to their contamination by another enzyme (see below) and that neither Gyr A nor the whole enzyme are able to relax positive supertwists[34].

In the last two year, new experimental data became available: (i) The DNA duplex is wrapped around the enzyme in such a way that about 140 base pairs are protected from micrococcal nuclease digestion. This complex, of 400,000 daltons which Wang called a "gyrasome"[35], is made up of two Gyr A and Gyr B subunits and has some similarities with a nucleosome core. (ii) The double-stranded cleavage performed in vitro in the presence of oxo occurs as specific sites on the DNA, involving two Gyr A subunits covalently attached to the 5' end of each broken strand[36]. (iii) The use of non hydolysable analogs allows a clarification of the role of ATP in the mechanism of gyrase: ATP hydrolysis does not aid gyrase movement along the DNA as initially proposed by Mizuuchi et al[37] and by Liu and Wang[32], but the binding of ATP and its release (through hydrolysis in ADP) are used to induce conformational changes in gyrase, leading to a cycle of supercoiling[38,39]. New models[38-40] for the gyrase mechanism were generated from the above data and generally agree with the following scheme:

The gyrase binds to the DNA duplex in such a way that a segment of the DNA is wrapped around the protein, forming a complex wherein about 140 base pairs are stabilized in a positive superhelical structure, inducing a negative supertwist in the rest of the DNA molecule. This presumably initiates the cut of both strands by the Gyr A subunits at specific sites near the center of the 140 base pair segment. Two molecules of ATP (an allosteric ligand of gyrase) then bind to the Gyr B subunit, producing a conformational change in the whole molecule and consequently in the DNA (the 5' end of the broken strands remaining covalently linked to each Gyr A subunit). Two models have been proposed based on this scheme; in the first the ATP promoted movement allows the removal of the positive supertwist in the DNA segment bound to the enzyme[33]; in the second model, ATP binding promotes the transport of another DNA segment through the double-stranded break[4,30,40], transforming the positive into a negative supertwist (sign inversion model). Finally, once the break has been closed, ATP hydrolysis would produce a second conformational change releasing the enzyme which returns to its initial configuration.

In the sign-inversion model, the linking number of the circular DNA would change by multiples of two and this prediction has been experimentally demonstrated, making this model very attractive. Furthermore, the strand passing mechanism is the only one which could explain the reversible knotting and catenation activities associated with DNA gyrase. The change in the linking number by multiples of two is now taken as a specific characteristic of type II topoisomerases in contrast to type I topoisomerase which only change the linking number by a step of one[41].

Topoisomerase II'

This E. coli enzyme is made up of two Gyr A subunits associated with two truncated Gyr B (called B' or subunits) lacking the ATP binding sites. Consequently, the enzyme lacks the supercoiling activity and the sensitivity to novo and cou which compete for the ATP binding site in gyrase. Surprisingly, however it conserves the reversible catenase activity of the gyrase, and, uniquely in prokaryotes, is able to relax both positive and negative supercoils[4,34,42].

It is not excluded that this enzyme may be generated by an artificial proteolytic cleavage of the Gyr B subunits; however, topo II' accounts for the 10 fold excess of Gyr A over Gyr B subunits in E. coli, a proportion reproducibly obtained in all gyrase preparations[4]. Thus topoisomerase II' may be in charge of performing a part of the hypothetical functions of the gyrase such as a swivelase activity at the replication fork, with its ability to relax positive superhelical turns. Like gyrase, topo II' stabilizes positive superhelical turns upon stoichiometric binding to the DNA, a property which probably indicates the wrapping of the DNA around the enzyme.

T4 DNA Topoisomerase

About five years ago, the attention of investigators was focused on a group of T4 gene products (genes 39, 52 and 60) which appeared to interact with each other[43,44]. These T4 genes called "DNA delay" genes, were recently found to produce a new ATP-dependent DNA topoisomerase which appeared to be essential for the initiation of T4 replication[19].

Further in vitro studies of this enzyme revealed that it is able to relax both positive and negative supercoils in an ATP-dependent reaction. In the presence of ATP, it is also able to perform knotting/unknotting and catenation/decatenation reactions. As expected for a type II DNA topoisomerase, the T4 enzyme changes the linking number of a circular DNA only by multiples of two[41] but, in contrast with Gyrase, it seems unable to produce negative supercoiling. The T4 topoisomerase is neither inhibited by cou and novo nor by nal and oxo.

2. Eukaryotic Type II DNA Topoisomerases

Type II DNA topoisomerases have also been isolated from higher organisms such as *Drosophila melanogaster*[45], *Xenopus laevis* eggs[46], HeLa cells[47], calf thymus[48] and rat liver[49].

All these eukaryotic enzymes are able to relax the supercoiled DNA in an ATP-dependent reaction which leads to change the linking number by multiples of two, in contrast with eukaryotic type I DNA topoisomerases, which change the linking number by increments of one. However, eukaryotic type II DNA topoisomerases appear unable to perform the supercoiling of relaxed DNA, as does the prokaryotic DNA gyrase. It has been suggested that unlike DNA gyrase, the eukaryotic topo II lack the ability to wrap DNA with a positive twist, and thus are unable to supercoil the DNA, positive wrapping of the DNA around the molecule acting as the orienting factor of the gyration reaction[45].

Highly purified eukaryotic type II DNA topoisomerases can perform, in vitro, the catenation of DNA rings in huge networks in the presence of ATP and if an aggregation substance such as spermidine or histone H1 is present. The catenation activity which is sensitive to proteolysis is also sensitive to ionic strength; a low salt concentration inhibits the catenation activity more strongly than the relaxation activity. Both activities are inhibited by novo, but not by oxo. The dose of novo required to inhibit the eukaryotic Topo II is higher than in the case of the prokaryotic enzyme but inferior to the dose which inhibits purified DNA polymerase α[50]. Finally, a DNA topoisomerase II has also been recently detected in rat liver mitochondria[51].

Table 1 summarizes the main activities of the various type II DNA topoisomerases isolated and characterized so far, and for each enzyme indicates which kind of topological problem, defined in Chapter I, could be theoretically solved.

Table 1. Reactions in vitro by Type II DNA Topoisomerases and their Theoretical Implications in DNA Replication

	Relaxation		Supercoiling		Catenation-Decatenation Knotting-Unknotting	
Enzymes	in vitro	theoretical biological implications in DNA replication	in vitro	theoretical biological implications in DNA replication	in vitro	theoretical biological implications in DNA replication
Gyrase Topo II	only negative supercoils ATP dependent	?	formation of negative supercoils	help to open the double helice maintenance of negative superhelicity, swivel at the fork	yes ATP dependent	solution of the alignment problem
Topo II'	positive and negative supercoils ATP independent	swivel at the fork	no		yes ATP independent	prevent knotting tangling
T4 Topo II	positive and negative supercoils ATP dependent	swivel at the fork	no	help to open the double helix for initiation	yes ATP dependent	decatenation of daughter molecules at the end of replication
Eukaryotic Topo II	positive and negative supercoils ATP dependent	swivel at the fork	no	help to open the double helix for initiation	yes ATP dependent	decondensation of chromosome

III. TYPE II DNA TOPOISOMERASES AND BACTERIAL DNA REPLICATION

A. THE ELONGATION PHASE OF DNA REPLICATION

In the first section we saw that topological problems are related to the movement of the replication fork. Type II DNA topoisomerases, which are able to solve these problems in vitro (Table I) are therefore good candidates as essential components of the elongation mechanism. Indeed, it was found that various activities of the Escherichia coli DNA gyrase (E. coli Topo II) were the intracellular target for several inhibitors of DNA elongation: nalidixic acid (nal), oxolinic acid (oxo), novobiocin (novo) and courmermycin A1 (cou)[28,29,52-55]. Participation of gyrase in the elongation phase of DNA replication was therefore the first aspect of the role of this enzyme in DNA replication to be envisaged.

Two type II DNA topoisomerases isolated from E. coli are sensitive to drugs which stop DNA elongation: DNA gyrase (E.coli Topo II) and the related DNA topoisomerase II' (Topo II'). In the preceding section we saw that these two enzymes have one common subunit, encoded by the gene Gyr A. In Topo II, this subunit is associated with the Gyr B gene product (subunit B) whereas in Topo II', it is associated with a protein related to Gyr B called "B'" or "ν"[34,42]. First, we shall examine the situation in the living cell when the B subunit is altered either by the inhibitors novo and cou, by the related drug chlorobiocin (Chl) or by a thermosensitive mutation, i.e., in the absence of Topo II activities. We will then examine the situation in which the altered protein is the A subunit (in the presence of oxo or nal or by mutation), i.e., in the absence of both Topo II and Topo II' activities. The last part of this section will be devoted to studies performed with in vitro systems.

1. Effect on DNA Elongation in vivo of Alteration in the Gyr B Gene Product

The mode of action of the Gyr B inhibitor cou on DNA synthesis in vivo was first examined by Ryan who showed that the inhibition of DNA synthesis was not followed by DNA degradation[55]. Drlica and Snyder further demonstrated that cou decreases the degree of superhelicity of the chromosome without introducing any swivel, since the drug-relaxed chromosome can always be positively supertwisted by the addition of EtBr[56]. In the presence of cou, the rate of DNA synthesis decreases progressively in parallel with the loss in superhelicity of the chromosome[56]. At first sight, these results suggest that Gyr B inhibitors stop DNA elongation because, by hindering the gyrase activity of Topo II, they lower the negative supertwisting of the chromosome below the level compatible with progression of the replication fork. This hypothesis nevertheless conflicts with the normal continuation of DNA elongation at non permissive temperatures in the Gyr B^{ts} mutant (LE 316) isolated by Orr et al.[57]. Gyrase activity is probably absent in this strain at 40°C for two reasons: (1) this mutant is not leaky, since changes in nuclear morphology can be detected as soon as the temperature is raised, (2) the folded chromosome isolated from LE 316 had a reduced degree of superhelicity[58]. Thus, this mutant apparently replicates its DNA in the absence of gyrase activity and thermoinactivation of the Gyr B subunit does not produce the same effect on DNA synthesis as Gyr B inhibitors. To explain these observations, Orr et al. have suggested that Topo II is not essential for DNA elongation but that Gyr B inhibitors stop DNA elongation by trapping this enzyme in a stable complex on the DNA which would prevent strand separation[57]; in the mutant, at non permissive temperature, Topo II would be released from the DNA with little or no effect on the replication fork movement. In this model, the gradual inhibition of DNA synthesis by cou[56] should correspond to the gradual arrest of the replication forks as they reach the stable Topo II DNA complex formed by the drug. Thus, the parallelism between the inhibition of DNA synthesis and the reduction of the chromosomal superhelicity would be coincidental. Nevertheless, caution should be taken in concluding that negative superhelicity is dispensable for replication fork movement. Indeed, cou relaxes DNA to a greater extent than the ts mutant and Bridges finds a non ts mutant with partially reduced superhelical density that grows well[56,58], the residual superhelicity in LE 316 is therefore perhaps sufficient to sustain DNA elongation, as pointed out by Wright and Bridges: "the data obtained to date do not give any indication as to whether superhelicity is generally decreased throughout the chromosome, or whether certain regions (for instance the domain along which the replication fork moves) present normal superhelicity"[58].

In a second Gyr B^{ts} mutant, N4 177, isolated by Gellert, the rate of DNA elongation is somewhat reduced at non permissive temperature[59]. A priori an explanation for the difference between this mutant and LE 316 might be that the mutation isolated by Gellert affects the protein domain common to B and B' whereas that isolated by Orr is specific of B. In this case, the presence of a normal Topo II' in LE 316, but not in N4 177, could explain why DNA elongation is not altered in LE 316.

Nevertheless, this is apparently not the correct explanation, since it has been reported that the N4 177 mutation does not concern Topo II'[59]. This is not surprising, since both Gyr B^{ts} strains have been isolated by a one mutational step leading to drug resistance and temperature sensitivity. Indeed, the drug binding site is absent in B'; therefore, a mutation which affect drug dependence has a good chance of being located outside the protein domain common to B and B'.

Another hypothesis is that the mutation in N4 177 leads to the formation of a DNA Topo II complex which slows the progression of the replication fork, as does the enzyme-drug complex presumably formed in the presence of cou or novo. Other models could be envisaged, but this behavior stresses the importance of genetic studies in the field of DNA topoisomerases. In particular, it would be very interesting to isolate a mutant of the Gyr B gene product altered in both B and B' subunits. However, such isolation can be successful only if both subunits are encoded by the same gene-a point which is not yet certain. Rapid progress should be made in that field, since a bacteriophage carrying the region of the chromosome including Gyr B is now available[60].

2. Effect on DNA Elongation in vivo of Alteration in the Gyr A Product

Unlike Gyr B^{ts} mutants, the Gyr A^{ts} mutants no longer synthesize DNA after a shift to the non permissive temperature[61]. Therefore, it is likely that when the activities of Topo II are absent, as in the strain LE 316 at high temperature, the reactions of topoisomerization required for DNA elongation are performed by Topo II'.

Dramatic differences exist between the mode of action of nal and oxo on the one hand, and cou and novo on the other hand. Thus, oxo and nal lead to up to 50% degradation of the whole chromosome when the cell remains in contact with the drug for several hours[55]. In addition, and apparently in contradiction with the preceding observation, the folded chromosome isolated after a short period of incubation with oxo is normally supertwisted[62]. This contradiction can be resolved if we assume that the degradation which follows long exposure of the cell to the drug has not been initiated at several sites randomly located along the chromosome, but only at one or a very small number of sites clustered in one or a limited set of the topological domains which segregate the bacterial chromosome. Thus, only these domains are relaxed in the first period of drug action with little effect on the overall chromosomal superhelicity. Indeed nal and oxo rapidly induce local DNA cleavage and some degradation at the level of the replication fork[52,63] as well as the synthesis of rec A and the rapid response of cellular repair mechanisms[59]. Since these drugs which inhibit the gyrase activity of Topo II do not relax the chromosome as does cou, it is likely that the relaxation observed in the presence of cou is performed by an activity dependent on the Gyr subunit of Topo II or Topo II' and not by Topo I[62]. In vivo, the overall superhelicity of the bacterial chromosome would therefore be controlled by a balance between the gyrase and the relaxation activities of Topo II and Topo II', whereas the activity of Topo I would be strongly restrained and closely associated with a particular process like transcription or transposition[64]. The same difference between the effect on chromosomal superhelicity of Gyr A and Gyr B inhibitors has been found in *Bacillus subtilis*[65]. Another difference between the mode of action of these two classes of inhibitors emerges from studies by Snyder, Drlica et al.: oxo arrests DNA synthesis more rapidly than cou[59,62]. Oxo should then stop DNA elongation before the replication forks reach the postulated DNA-Topo II complexes formed in the presence of cou or novo. A likely hypothesis is that DNA elongation is quickly arrested when the cells can no longer solve the topological problem related to uncoiling of the parental strands in the absence of both Topo II and Topo II' activities. Drlica et al showed that oxo inhibition was very rapid even at subsaturing drug concentrations. Although rapid, the inhibition was not complete and stopped after a few minutes; the extent of the residual synthesis was a function of both drug concentration and temperature[59] suggesting that an enzymatic process is required for oxo inhibition. The forks which escaped inhibition continued undisturbed until the end of the replication round. Oxo therefore seems to have two opposite actions occurring sequentially: the first rapidly blocks the replication forks, whereas the second beginning shortly thereafter, saves the forks which have not yet been stopped. The inhibitory action of oxo could be coupled with the introduction of nicks followed by DNA degradation. Crumplin and Smith reported that nal does not stop DNA synthesis when synthesis is measured by pulse labelling but leads to the formation of transient 30 S DNA fragments[63]. They suggested that these fragments are an intermediate step in the joining of the Okazaki pieces to bulk DNA and that nal prevents their sealing. By virtue of the unsealed gaps between them, these 30 S DNA pieces could be substrates for exonuclease action and this would explain the arrest of net DNA synthesis in the presence of nal.

In the light of recent results, these 30 S pieces are more likely to be generated by a nal induced endonuclease activity related to Topo II or II', since nal and oxo trap a complex in which they are covalently bound to a double-stranded DNA break. Treatment of this complex with SDS or alkali cleaves the DNA. The 30 S pieces detected by Crumplin and Smith, then, are perhaps generated during the alkali-sucrose centrifugation of the DNA in the presence of nal. Nicking and cleavage of the DNA by the Topo II-nal or Topo II-oxo complexes may also occur in vivo without SDS or alkali treatment. Nicks or breaks introduced in this way by the drugs would explain the local degradation of the DNA observed in their presence if they form primer sites for exonuclease action; they would also provide a mechanical swivel when Topo II' relaxing activities are blocked in the presence of oxo, thus explaining the insensitivity of the residual synthesis to further incubation with the drug.

The enzymatic process which induces the arrest of DNA replication in the presence of oxo could itself be an endonuclease activity, the accumulation of twisting stress ahead of the fork, or another process which we cannot presently imagine.

Drlica et al found that oxo inhibition was partially released by Gyr B^{ts} strains N4 117 at non permissive temperatures[59]. Therefore Topo II participates in the inhibitory action of oxo and the oxo=Topo

II complex ahould actively inhibit DNA elongation since denaturation of the drug target reduces the degree of inhibition. This result also supports the idea that artificial swivel occurs in the presence of low doses of oxo, since some DNA elongation continues in N4 177 at non permissive temperatures in the absence of normal Topo II gyrase activity (ts mutation) and Topo II' relaxation activity (presence of oxo).

We have previously suggested that the reduction of the DNA elongation rate in N4 177 was caused by an altered Topo II-DNA complex: this altered complex would always be able to introduce swivels inside the chromosome in the presence of oxo but would be less efficient in provoking the fast occurring oxo-mediated reaction responsible for the rapid arrest of DNA synthesis.

Snyder and Drlica have shown that cleavage occurs at 100,000 base pair (100Kb) intervals in vivo when SDS is added to the chromosome of oxo-treated cells[62]. Drlica et al also showed that newly replicated DNA was more susceptible to cleavage induced by SDS than the overall chromosome, i.e., there are more cleavage sites near the replication fork[59]. They concluded that two kinds of oxo targets are present in bacteria, those described at roughly 50 sites all around the chromosome, and those clustered in the vicinity of the replication fork. The equivalence between the number of cleavage sites by the DNA topoisomerases II molecules located throughout the E. coli chromosome and the number of independent topological domains segregating this same chromosome, suggests that each domain contains one binding site for type II DNA topoisomerase. Another possibility is that type II DNA topoisomerase is an essential component of the topological barriers which delineate two neighbouring domains[33].

These sites could also correspond to the location of the barriers which prevent the movement of the replication fork formed when cou or novo are added to the cells. Their inactivation by oxo does not stop progression of the replication fork at low drug concentrations since residual replication occurs even when all these sites are hit by oxo[59]. Nevertheless, at high doses of nal, there is no residual synthesis[66], perhaps because the conformation of the inactivated Topo II at these sites is similar to that formed in the presence of cou. For his part, the rapid arrest of DNA elongation by oxo at subsaturating drug concentration could follow inactivation of the topoisomerases located in the replicating region[59].

3. The Role of Type II Topoisomerases During DNA Elongation in vivo

Certain important questions have been raised from in vivo studies. a) The results with Gyr B^{ts} mutants suggest that negative superhelicity may not play a role in DNA elongation. In other words, the work of the helicases, opening the double helix[67], would be performed without any help from the negative superhelicity of the chromosome in contrast to previous hypotheses (see section I). This has not

been firmly established, as discussed previously, but is in agreement with the data obtained on bacteriophage or plasmid DNA replication. In ϕX 174 RF replication, Col E1 or T4 replication, the DNA elongation step does not seem to require gyrase activity.
b) The topological problem may be solved by Topo II', at least when Topo II activity is blocked by Gyr B inhibitor or by a Gyr B^{ts} mutation. This point agrees with in vitro studies on ϕX 174 DNA replication. The second stage of this replication (RF multiplication) can be performed in vitro with a system of purified proteins in the absence of DNA gyrase[68]. If one takes into account the rolling circle model, this would mean that the free 3'OH end should provide a swivel to compensate for the unwinding of the parental strands. In contrast, when the third stage of ϕX 174 DNA replication (synthesis of single-stranded viral DNA from the RF) is studied in vitro in a system containing several phage-encoded proteins, a topoisomerase activity is required. This activity is inhibited by oxo but not by novo, nor by a thermosensitive mutation in the Gyr B gene product[69]. Therefore, in this case as well, Topo II' can be used as a swivel by the cell. Requirement for a swivel in this reaction may be related to the associatin of the phage prohead with the replication proteins if this association restricts free rotation of the growing strand[69].

It is not yet clear which of the Topo II or Topo II' enzymes is used under normal conditions in vivo. Both enzymes apparently participate in the inhibition by oxo of the replication fork. The gyrase sites located in the vicinity of the replication fork, and revealed by the SDS induced cleavage in the presence of oxo, may be attributed either to Topo II or Topo II'. Location of these sites on the newly made DNA behind the fork could, a priori, be the sign of a decatenase activity and not a swivelase activity, unless this swivelase activity is needed to help the formation of nucleosome-like structure. Nevertheless, this location arises from the preferential binding of the gyrase to relaxed regions of the chromosome that might follow the replication fork and may be the way by which the DNA behind the fork becomes twisted once it is sealed. Furthermore, the experiments described above do not allow the detection of potential Gyr A sites in front of the fork if such sites exist.

We cannot therefore deduce from the experiments assembled in this section whether the topological problem is solved by a swivelase or a catenase activity, nor do we know whether or not the alignment problem must be taken into account by the replication process. Champoux pointed out that the gyrase activity which can directly transform a positive into a negative supertwist with the concomitant hydrolysis of one ATP molecule is less energy-consuming than the formation of the same negative supertwist using decatenation followed by gyration: i.e., two ATP molecules[7]. Nevertheless, the formation in two steps of one negative supertwist using first the ATP independent relaxation activity of Topo II' and, secondly, gyration, is no more energy consuming than the direct formation of this negaitve supertwist by DNA gyrase.

c) There are probably several types of Topo II or Topo II' DNA complexes on the chromosome (some of them perhaps moving with the replication forks). These complexes may be altered in various ways by drugs or by mutation in one of the subunits. Certainly, the four subunits of each topoisomerase molecule (gyrasome) could be exchanged with a pool of free subunits and the pattern of these exchanges could be modified when a mutation changes the affinity of one subunit for another.

One hypothesis is that DNA topoisomerase complexes which normally interact reversibly with the chromosome could, after alteration, dissociate more slowly from the DNA and slow down or even completely arrest DNA replication, depending on the ability of the replication machinery to displace such a block. Indeed, Topo II forms a very stable complex with the DNA in vitro, since it dissociates linear CoI EI DNA with a half-life of a few days at 23°C[4]. Topo II binds less tightly to DNA that is negatively supercoiled than to relaxed DNA[4], and thus relaxation of the chromosomal DNA in the presence of cou may be one explanation for the formation of a barrier to the movement of the replication fork. It is interesting to remember that the hypothesis of drug-DNA complexes altering normal progression of the replication fork along the chromosome had previously been proposed by Ryan and Wells following the first studies on the mode of action of cou on DNA synthesis, and before the discovery of type II DNA topoisomerases[55,70].

4. Mode of Action of Gyr A and Gyr B Inhibitors in vitro: the Stop-Signal Hypothesis

None of the acellular systems established to date for studying DNA replication reflect the true process that occurs in the living cell. In all such systems, one observes the continuation of the elongation process engaged in vivo at the time when the cells were lysed or plasmolysed, but the rate of DNA synthesis is reduced as compared to the in vivo rate, and synthesis occurs only on a short section of the chromosome. The results obtained with such in vitro systems should therefore be extrapolated with caution to understand the in vivo mechanism of DNA elongation. This is particularly true of the topology of DNA replication, since the topological state and the structure of the chromosome may be altered by changes in the ionic environment and by destruction or modification of the cell membrane. DNA elongation in vitro is concerned with the type II DNA topoisomerases, since the inhibitors of these enzymes slow down DNA synthesis in plasmolysed bacteria[52,54,70,71], in lysates concentrated on cellophane disks[72], or in lysates complemented with the isolated folded chromosome[73]. For years, sensitivity to nal was even used as a criterion to distinguish true DNA replication from repair-like DNA synthesis in vitro.

The effects of topoisomerase inhibitors on DNA synthesis in vitro and in vivo are nevertheless somewhat different. For instance, the nal concentration required to inhibit DNA synthesis in acellular systems is greater than the doses which are sufficient to stop bacterial growth[52]. Earlier speculation on intracellular transformation of nal into a more potent form inside the living cell has been discarded after the discovery that in vitro DNA gyrase was inhibited at in vivo doses of the inhibitor[28]. In our opinion, the only logical explanation for this observation is that nal stops DNA replication in vitro by a mechanism other than enzyme inhibition. The same conclusion was reached by Kreuzer and Cozzarelli to explain the fact that nal inhibited bacteriophage T7 DNA replication whereas T7 DNA replication was normal in a Gyr A^{ts} mutant at the non permissive temperature[61]. They imagined that nal does not stop T7 DNA replication through topoisomerases inhibition, because T7 DNA being linear, should not be concerned with topological problems. To explain nal inhibition of T7 DNA replication, they proposed a mechanism similar to that imagined by Orr for the inhibition by novo and cou of E. coli DNA replication, i.e., that, nal induces the formation of a poison when it interacts with the subunit A, forming a stop-signal for DNA elongation[61]. This idea also fits well with the fact that the Gyr A^{ts} mutation renders the phage infected cells insensitive to nal at high temperature. The nal concentration required to block T7 DNA replication was also higher than the concentration which inhibits bacterial DNA replication, once again supporting the idea of different mechanisms for nal inhibition in each case. Interestingly, T7 DNA replication is sensitive to nal concentrations similar to those which inhibit DNA replication in in vitro systems; this suggests a common mechanism of nal inhibition in both cases, i.e., the formation of a stop-signal for DNA elongation. This mechanism could aslo be operational in bacteria in vivo and could explain why only high doses of nal completely suppress DNA replication in vivo[66].

The insensitivity of bacterial in vitro DNA replication to low doses of nal suggests that the bacterial chromosome in vitro, like the T7 chromosome, is free from topological constraints. This could arise from the introduction of nicks or breaks into the bacterial DNA during the preparation of the in vitro systems, providing internal swivels for the progression of the replication fork. DNA breakage probably occurs in the cellophane disk system for DNA synthesis, since no special care is taken to prevent low amounts of DNA damage[72]. Nicking of the DNA inside the folded nucleoid during incubation with a soluble extract is also suggested by the rapid unfolding of this nucleoid in the reaction mixture for DNA replication[73]. Recently, one of us showed that the chromosome inside toluene treated cells (TTC) was probably also nicked when TTC were incubated at 30°C in the reaction mixture for DNA replication[74]. Indeed, when the sedimentary coefficient of the nucleoid isolated from TTC, incubated at physiological temperature in the reaction mixture for DNA elongation was plotted against EtBr concentration, one did not obtain the biphasic curve typical of negatively supercoiled DNA, but rather a linear response typical of nicked DNA. In contrast, a normal response was obtained with chromosome isolated from TTC just after toluenization. This indicates that swivels were introduced in the DNA upon incubation. Therefore, DNA replication *in vitro* has thus far been likely studied on relaxed DNA templates with no topological constraints. This may at least partially explain the discrepancies between *in vitro* and *in vivo* DNA replication.

The explanation of the difference between the nal concentration required to inhibit DNA replication *in vivo* and *in vitro* by the existence of different mechanisms of inhibition in both cases is also supported by the results of Forterre and Kohiyama on the mode of action of Gyr A and Gyr B inhibitors on DNA replication in TTC[71,74]. We have previously mentioned that *in vivo*, oxo inhibition is more rapid than cou inhibition[59]; in contrast, in TTC, the patterns of inhibition of Gyr A and Gyr B inhibitors are very similar: for instance, both nal and novo slowly decrease the rate of DNA synthesis until complete arrest is ensured[71]. The high residual synthesis observed before this arrest has been identified as residual DNA replication and not as DNA repair reaction and its level cannot be lowered by preincubating the TTC with the drugs before DNA replication starts. Therefore, the gradual inhibition of DNA synthesis by nal or novo cannot be explained by some delay in drug action. What is the nature of this residual synthesis which involves initiation, elongation and sealing of Okazaki-like pieces[71]? Kinetic data have shown that DNA replication stops once a definite amount of DNA has been replicated[74]; since the chromosome in TTC is free from topological constraints, this definite amount cannot correspond to the DNA susceptible to be replicated in the absence of swivel or before the complete relaxation of the chromosome. Our hypothesis is that both nal and novo do not stop DNA elongation in vitro through enzyme inhibition, but rather through formation of stop-signals formed by interaction between drugs and type II DNA topoisomerases located at specific sites on the chromosome. In such a model, residual synthesis in the presence of nal or novo stops when all the replication forks have reached a stop-signal. A unique mechanism could then take into account the difference in dose response of DNA replication to nal in vivo and in vitro, the difference between the initiation kinetics in both cases, the phenotype of the Gyr B^{ts} mutants and the inhibition by nal and cou of replication of linear T7 DNA[61,75]. This does not mean that the nature and the characteristics of the stop-signals are always similar; for instance, the inhibition by novo or cou is reversible *in vivo* but irreversible in vitro[55]. Relaxation of the chromosome *in vitro* is likely to modify the interaction between the topoisomerases and the DNA in situ, and stop-signals might perhaps be formed *in vitro* in the absence of any drugs. Indeed, Fonterre and Kohyama have shown that the residual DNA synthesis occurring in TTC in the presence of nal or novo is equivalent to those occurring when ATP in the reaction mixture is replaced by any other rNTP[71]. Perhaps the binding of type II DNA topoisomerases to the relaxed DNA in vitro forms stop-signals for DNA elongation even in the absence of topoisomerase inhibitors and this binding is weakened when ATP is bound to the B subunit. In this case, the main role of ATP in vitro would be to rid the DNA of Topo II complexes in front of the replication fork. This model fits the observation that Topo II binds very strongly to relaxed DNA[4]. A plausible explanation for the reduced rate of DNA replication in vitro compared to the rate observed in vivo may be that even in the presence of ATP, the link between Topo II and the relaxed DNA in vitro is stronger than the link between Topo II and the supercoiled DNA in living cells. Addition of novo hinders the binding of ATP to the B subunit and therefore, in this model, prevents the destabilization of the Topo II-DNA complex in vitro. This would mean that the specific role of ATP in DNA replication in TTC is different from its roles in DNA replication in vivo. It has been shown that GTP as well as ATP sustain the initiation and the elongation of Okazaki pieces in TTC[71], two processes which are strictly ATP dependent in more purified in vitro systems using small bacteriophage DNAs as template[76]. One can imagine that enough ATP is formed in TTC at the expense of GTP to sustain the above processes, but not enough to provide sufficient binding onto the topoisomerase sites to destabilize the enzyme-DNA complexes. This can be explained by the high Km of DNA gyrase for ATP (400μM[38]) which is similar to the Km for ATP (250μM) of DNA replication in TTC[77].

Another interesting question concerning the mode of action of DNA topoisomerase inhibitors in vitro is the difference observed in TTC between the effects of nal and novo on the one hand and of cou and oxo on the other, i.e., between drugs which have the same target. Indeed, cou and oxo are effective at lower doses and inhibit DNA synthesis more rapidly than nal and novo[74]. As a consequence, the residual synthesis is lower in the presence of cou or oxo than in the presence of nal and novo; nevertheless, this residual synthesis always exists even at very high drug concentrations.

We previously reported that in vivo the rate of oxo inhibition was also very high perhaps because of the impossibility for the cells hit by the drug to solve their topological problem. Drlica et al. also showed that this rate was independant of the rate of DNA replication[59]. In contrast, when the rate of DNA synthesis was reduced in TTC (by lowering the dNTP concentration) the rate of oxo inhibition was also reduced[76]. At very low rates, DNA synthesis even became insensitive to the drug. Thus, rapid inhibition by oxo in vivo and in vitro does not reflect a similar mechanism of inhibition. We proposed that, as in the case of novo or nal, cou and oxo do not inhibit DNA synthesis in vitro until the replication forks reach a stop-signal formed by a DNA-Topo II complex in the presence of the drugs. The rapid inhibition by oxo and cou at saturating dNTP concentrations in vitro means that these stop-signals are reached more rapidly in the presence of these drugs than in the presence of nal or novo when the replication forks travel at a normal rate; lowering this rate increases the time needed for a replication fork to reach the nearest stop-signal and finally suppresses the inhibition itself when this time is superior to the time during which the in vitro replication system is active.

One hypothesis to explain that stop-signals are reached more rapidly in the presence of oxo and cou is that their number is greater, at least in the vicinity of the replication fork. This can be interpreted in two ways: either the percent of stop-signals among DNA-DNA topoisomerase complexes is increased or two kinds of stop-signals exist on the chromosome, some of which are inducible only by oxo or cou, and others which are inducible by any of the four drugs.

The second interpretation fits well with the results of Drlica et al., if the two kinds of topoisomerase-DNA complexes detected in vitro correspond to the two levels of oxo action reported in vivo[59]. The size of the DNA replicated in the presence of novo and nal makes it possible that the stop-signals encountered by the replication forks correspond to the 50 DNA topoisomerase sites located throughout the chromosome[62]. In this case, the supplementary stop-signals created by oxo and cou should correspond to the topoisomerase sites detected by Drlica et al. in the vicinity of the replication fork[59].

Unexpectedly, freezing and thawing of the TTC lowers the level of residual synthesis in the presence of nal, novo or an NTP other than ATP to the level of residual synthesis in the presence of cou or oxo[64]. One explanation for this observation may be that freezing and thawing modify the structure of the enzyme-drug complexes which are normally only stop-signals in the presence of oxo and cou, also making them stop-signals in the presence of novo or nal or in the absence of ATP. Another possibility is that freezing and thawing induces a redistribution of the topoisomerase sites on the chromosomes. Such redistribution could be related to the chromosome reorganization induced by chilling which has recently been observed by Edelstein et al. using electron microscopy[78].

The similarities in the mode of action of nal and novo on the one hand, and of oxo and cou on the other, as well as the correlation between ATP specificity and drug sensitivity, indicate that E. coli Topo II and not Topo II' is involved in the mechanism of inhibition of DNA elongation in vitro by the topoisomerase inhibitors, since novo, cou and ATP are only effectors of Topo II. These similarities also suggest that the stop-signals induced by nal and novo are located at the same sites as those induced by oxo and cou.

5. General Discussion on the Mode of Action of Bacterial Type II DNA Topoisomerases in DNA Elongation in vivo and in vitro

The various results and interpretations that we have analyzed in this section are summarized in Table 2. The main difference between the in vitro systems and the living cells concerning DNA elongation may be that in vitro DNA replication is not faced with topological problems since the chromosome is apparently free from topological constraints. This difference would explain why the Gyr A inhibitors do not have the same mode of action in vitro and in vivo. Indeed, the Gyr A inhibitors block both Topo II and Topo II' activities, and, therefore, they would in vivo hinder the resolution of topological problems related to parental strand uncoiling, whereas this problem need not be solved in vitro. If this explanation is correct, it may be an important indication of the mechanism by which the topological problem is resolved in vivo. In the first section, we saw that uncoiling of parental strands may be compensated either by positive supertwisting of the non-replicated section of the DNA, or by interlacing of the daughter molecules. The results obtained in vivo did not allow us to choose between the two hypotheses, since Topo II and Topo II' could remove positive supertwisting as well as decatenate two duplex DNA rings. In contrast, the fact that preparation of the in vitro systems eliminates the topological problem strongly supports the first hypothesis, i.e., transient formation of positive supertwists. Indeed, one can easily understand how damage in the chromosome can provide the swivel for the elimination of these supertwists, whereas one cannot see how they could provide a mechanism for decatenation.

The difference in the topological state of the chromosome in vitro and in vivo perhaps explains the difference in the rate of DNA replication in both cases. We suggest that reduction in this rate in vitro is an indirect effect of the chromosome relaxation in vitro and is due to the formation of stop-signals by Topo II molecules stably bound to the relaxed DNA.

Indeed, if, in vitro the chromosome appears to be free from topological constraints, it is not free of topoisomerases. The reduction in the rate of DNA replication in vitro and the possibility of further reducing it by lowering the substrate concentration may facilitate studies of the interaction between Topo II and the chromosome in situ using in vitro DNA synthesis as a probe to measure the distances between the topoisomerase binding sites. A point which must be established before going any further in this way is whether or not the relaxation of the DNA in vitro creates new DNA binding sites for these enzymes on the chromosome.

Table 3. Relationships Between Alterations in Type II DNA Topoisomerases, Chromosome Superhelicity and the Elongation Phase of Bacterial DNA Replication *in vivo* and *in vitro*

Alteration in type II DNA topoisomerase		Experimental observations: Relaxation of the chromosome	Experimental observations: Inhibition of the DNA elongation
Gyr A inhibitors	low dose	no	yes
	high dose	?	yes
Gyr B inhibitors		yes	yes
Gyr A^{ts} mutation		?	yes
Gyr B^{ts} mutations	LE 316	yes	no
	N 4177	?	partially
Gyr A	low dose	the chromosome is even relaxed *in vitro*	no
	high dose		yes
Gyr B inhibitors			yes
Substitution of ATP by another rNTP			yes
Gyr A^{ts}, Gyr B^{ts} mutations			never checked

Hypotheses	
Inhibition of DNA elongation through hindering resolution of the topological problem	Inhibition of DNA elongation by the formation of stop-signals on the chromosome
yes	
yes	yes
no (the topological problem is solved by Topo II')	yes
yes	yes
no (the topological problem is solved by Topo II')	yes
no (the chromosome is free from topological constrains: therefore no topological problem to be solved)	yes
	yes
	yes

Various effects dependent on the nature of the topoisomerases-DNA complexes formed when one subunit is thermolabile; possibility of stimulation if the enzyme dissociates from the DNA.

B. THE INITIATION PHASE OF DNA REPLICATION

Theoretically, the initiation of bacterial DNA replication does not involve topological problems distinct from those related to DNA elongation. A specific involvement of DNA topoisomerases at this stage of DNA replication has therefore been deduced only from experimental observation. The first originated from studies of bacteriophage DNA replication: an in vitro system for the second stage of ϕX 174 DNA replication (RF multiplication) synthesized DNA in the absence of Topo II only when the exogenous template was a supercoiled RF (RFI)[79]. When the system was supplied with a relaxed RF, replication only occurred if an active DNA gyrase was present to supercoil the template. In that case, negative superhelicity of the RF was required for the bacteriophage gene A product to cleave the viral strand near the replication origin. The product of this specific cleavage reaction was a relaxed RF with the Cis A protein covalently bound to the 5' end of the viral strand[80]. One can suppose that the energy freed by relaxation of the supercoiled RF helps the formation of the covalent bond, since the energy freed during the cleavage reaction cannot be fully recovered for ligation. As for ϕX174 DNA, Itoh and Tomizawa reported that in vitro replication of a supercoiled Col E1 plasmid DNA occurs in the absence of Topo II[81]. Nevertheless, addition of Topo II to the system increased the length of the DNA synthesized.

Orr and Standenbauer, using another in vitro system, later found that Col E1 DNA replication was completely inhibited even at the permissive temperature when this system was prepared from extracts of the Gyr B^{ts} mutant LE 316[82]. They noticed that absence of plasmid replication was correlated with the rapid relaxation by these extracts of the exogenous supercoiled Col E1 DNA. We can therefore deduce that for Col E1 DNA replication, as for ϕX174 DNA replication, the initiation phase specifically requires a negatively supertwisted DNA substrate, and that DNA gyrase is required either to counterbalance the relaxation of this DNA by one or several topoisomerases present in the extract, or to supertwist the substrate when the latter is added in a relaxed form.

Initiation of plasmid Col E1 DNA replication does not require the cleavage of one of the two DNA strands, but rather the formation of an RNA primer synthesized by the bacterial RNA polymerase[83]. The requirement for this process of a superhelical template suggests that partial melting at the origin needed for transcription of the primer is facilitated by relaxation of the supercoiled template. Experimental observations have indeed shown that both annealing of a single stranded RNA stretch to a duplex DNA and extension of such primer by a DNA polymerase requires a supertwisted duplex[4]. In vivo, this mechanism is thought to help the binding of RNA polymerase to some bacterial promotors and should explain the requirement of DNA gyrase for their activation[84] as well as stimulation of RNA synthesis in a mutant defective in type I DNA topoisomerase[64].

A third prokaryotic system in which initiation of DNA replication but not elongation requires active DNA topoisomerases is the replication of bacteriophage T4. Indeed, mutations in one of the three genes which code for T4 DNA topoisomerase (the so called "DNA delay" gene) result in a reduction of the initial phase of T4 DNA replication, without changing the rate of fork movement[19,85]. In this case, the phage coded type II DNA topoisomerase, once purified, does not exhibit gyrase activity but an ATP dependent relaxing activity (see section II). The hypothesis of Alberts et al., to explain its role during initiation is that, in vivo, this enzyme introduces negative supertwists at the expense of ATP, but only in a limited portion of the DNA containing the replication origin[19]. In this model the binding of T4 gyrase on the linear T4 DNA creates a topologically circular structure able to be supertwisted to help unwinding of the double-helix at the origin.

The involvement of Topo II in the initiation of bacterial DNA replication was demonstrated two years ago in E.coli studies on Gyr B^{ts} mutants. In the preceding section we have reported that, unexpectedly the thermo sensitive strain LE 316 isolated by Orr et al., continued to synthesize DNA normally at the non-permissive temperature, at least until the end of one replication round[57]. Nevertheless, Orr et al., observed a gradual reduction in the accumulation of DNA after longer times. Since this reduction was more rapid than the decrease in the rate of RNA synthesis under the same conditions, they concluded that the alteration of the Gyr B subunit resulted in a specific reduction of the frequency of DNA initiation which was not related to the decrease in transcriptional activity in the absence of gyrase activity.

In fact, the pattern of inhibition of DNA synthesis in LE 316 does not correspond exactly to that of an initiation mutant; for instance, the decline in DNA synthesis was less rapid at the non-permissive temperature than in the presence of DNA initiation inhibitors such as rifampicin or chloramphenicol[57].

A similar result was obtained by Drlica et al. during their studies on the mode of action of oxo in vivo: when the forks which survive the inhibitory phase of oxo action (see section III) have terminated one round of replication, new initiation events may take place in the continuous presence of the drug, since addition of chloramphenicol or rifampicin reduces the level of the residual DNA synthesis[62]. These results suggest that initiation of DNA replication per se is not directly affected by a defect in Topo II. Rather, initiation may be gradually hindered in the absence of topoisomerase activity because some cellular process upstream of DNA initiation itself is dependent on topoisomerase activity. We have previously noticed that this process could not be transcription, since RNA synthesis decreases less rapidly in the mutant than does DNA synthesis. Plausible hypotheses are that, Topo II is required for the formation of an initiation complex preceding DNA initiation itself or that decatenation of the daughter chromosomes by Topo II is required for further initiation. Orr et al. have shown that cell division in

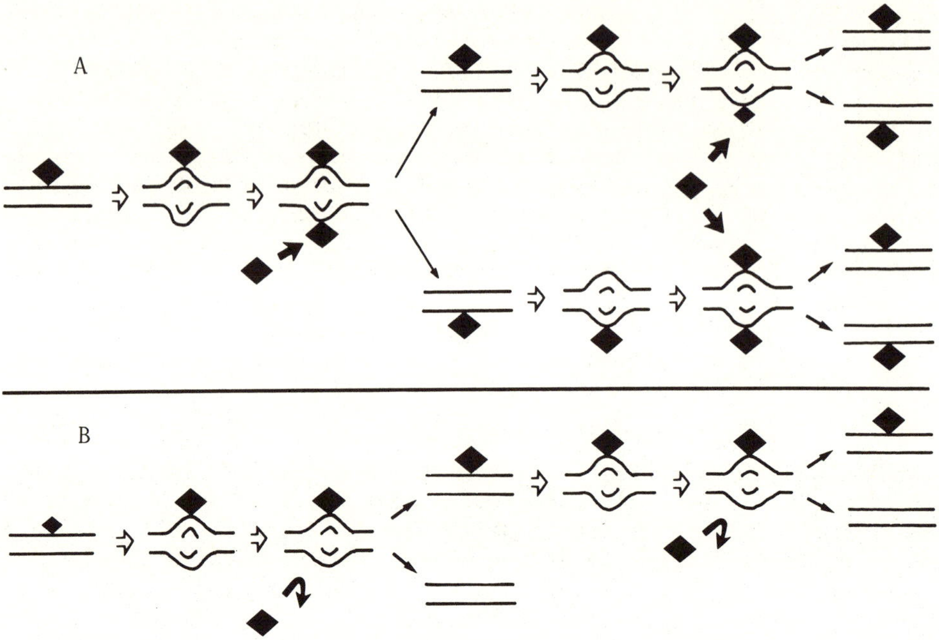

Fig. 5. Model for the relationship between the chromosome and the membrane during two generations. In normal conditions when the DNA is negatively supercoiled (A), one DNA strand of the chromosome is bound to the membrane via one or several DNA binding proteins (black losenges). After opening of the duplex and the start of DNA replication, the unlinked strand becomes attached to the membrane. After completion of one replication round the two daughter molecules are bound to the membrane by one of their two strands and can both serve of template for a new replication round. In the absence of gyrase activity, when the chromosome is relaxed B, the unlinked strand cannot be attached to the membrane; therefore, after completion of one replication round, only one of the daughter molecules can be used as a template for DNA replication. If cell division requires the presence of at least one DNA-membrane bond one can see that in normal conditions, the two replication rounds lead to three divisions and four viable cells whereas in the absence of gyrase activity they lead to only two divisions and one viable cell.

the LE 316 mutant continued after a shift to the non-permissive temperature, but at a linear rate, and that no increase in the viable population could be dectected[57]. This reduction in the cell division rate is not related to the effect of the drug on DNA synthesis since, in contrast, to other DNA replication inhibitors, chlorobiocin (a drug related to novo and cou) also affects cell division when added to cells in the D period, i.e., when they are not replicating their DNA[66]. At the non permissive temperature, and one hour after the shift, Orr et al. also noticed the appearance in LE 316 of a new population of anucleated cells with a smaller average cell volume[66]. These features were also observed in chlorobiocin-resistant strains with an altered Gyr B subunit and a chromosome with reduced superhelicity[58,66]. Therefore, a defect in E. coli Topo II alters both cell division and chromosome segregation. According to the replicon hypothesis, these two processes are coordinated by a physical link between the DNA and the cell membrane[86]. Possibly this linkage is altered in some way in the Gyr B^{ts} mutant. In Figure 5 we propose a model which can explain the observations made in E. coli in terms of the relationship between DNA topoisomerase and DNA initiation (Fig. 5), and which is based on the model for DNA strand segregation previously proposed by Perucci and Zuchovski[87].

In this model, at the beginning of a replication round, only one of the two parental strands is bound to the cellular membrane via one or several DNA binding proteins near the origin of replication[88]. The second strand becomes bound after unwinding of the DNA at the end on one replication round; both chromosomes are then linked to the cellular membrane by one of their two DNA strands and therefore are ready to be segregated (Fig. 5A). Furthermore, we suppose that negative supercoiling of the DNA is required for the formation of the DNA-membrane link but that, once this link has been made, it is stable even if the chromosome is later relaxed. Finally, we hypothesize that DNA initiation can only take place when the chromosome is associated with the cell membrane via the origin.

In this model, when the chromosome is relaxed in the absence of gyrase activity, the first DNA initiation taking place after relaxation occurs normally, but is not followed by the association between the second parental strand and the membrane. Therefore, at the end of the replication round, only one of the two newly replicated chromosomes is attached to the cell membrane and can be used as a matrix for further initiation. This could explain why the frequency of DNA initiation decreases in the presence of Chl or in the LE 316 mutant at non-permissive temperatures (Fig. 5B). If we make the further assumption that cell division only takes place in cells which contain at least one membrane-bound DNA, we can see that the end of the first replication round after chromosome relaxation is followed by a division with random nuclear segregation for one of the two chromosomes and production of only one viable cell capable of further division. This could explain why the cell division rate in the Gyr B^{ts} mutant is linear and why the number of viable cells does not increase after the temperature shift. Random segregation during cell division would produce either one longer cell with high nuclear content and one anucleate cell with a smaller volume, or two cells with normal nuclear content, but only one of them being able to divide further. Clearly, this ideal pattern might be complicated by the existence of more than one genome per cell and by the pleiotropic effect of the absence of gyrase activity, especially on transcription and thereafter on protein syntheis. A specific requirement for the decatentation activity of E. coli Topo II at the terminus of DNA replication can also be considered as an alternative or as a complement to the above model.

Ogasawara et al. have found that novo and nal inhibited the formation of an inhibition potential in Bacillus subtilis when these drugs were added at the onset of germination of a thymine-requiring mutant grown in a medium lacking thymine[89]. The presence of an initiation potential was measured in their experiments by the ability of the cells to initiate and complete one round of replication in the presence of thymine but in the absence of RNA and protein synthesis. They have also shown that initiation of chromosomal replication occurred in the presence of topoisomerase inhibitors when the drug was added after the initiation potential had been fully formed. In their experiments, the initiation potential was destroyed by prolonged incubation with novo in the absence of thymine and the rate of destruction paralleled the reduction in the degree of supercoiling of an endogenous plasmid DNA under the same conditions. Destruction of the initiation potential is therefore probably related to relaxation of the chromosome. This was supported by the fact that nal which, as in the case of E. coli, does not cause relaxation of the chromosome also does not destroy the initiation potential after it has been formed[89].

These findings, together with the observation that the addition of EtBr affects both the initiation potential, reversibly, and the RNA-DNA complex containing the origin of replication of Bacillus subtilis (S-complex)[90], suggest that formation of the initiation potential corresponds to the binding of the supertwisted chromosome to the cellular membrane at the onset of germination.

It is also interesting to note that when the drugs are added after the initiation potential has been fully formed, initiation per se is not prevented, but chromosomal replication initiated in their presence ceases after a fragment of approximately 15MD (15 x 10^6 daltons) has been replicated[91]. This observation might represent a confirmation that neither bacterial initiation of DNA synthesis itself, nor DNA elongation requires a supercoiled DNA, but that DNA replication in the presence of novo ceases once the fork encounters a topoisomerase binding site locked by the drug. The fragment of 15MD detected in the laboratory of Yoshikawa may, therefore, correspond to the distance which separates the replication origin from the first type II DNA topoisomerase binding site on the B. subtilis chomosome.

DNA TOPOISOMERASES AND EUKARYOTIC DNA REPLICATION

A. THE INITIATION PHASE OF DNA REPLICATION

Several lines of evidence have led to the conclusion that the initiation of eukaryotic DNA replication, like the initiation of prokaryotic DNA replication, is dependent on DNA supercoiling.

The requirement for superhelicity in initiation has been shown in permeable Chinese hamster ovarian cells, where the presence of single stranded breaks in the DNA produced by X-ray irradiation is sufficient to block initiation of all replicons within a cluster[92]. Other experiments have also demonstrated that small concentrations of EtBr stimulate DNA synthesis, but that higher concentrations of EtBr, which lead to a positive supercoiled DNA, inhibit DNA synthesis[93]. Thus, it has been postulated that a replicon cluster must be in a specific supercoiled conformation before replication can be initiated.

It has been shown that novo, which is an inhibitor for eukaryotic type II DNA topoisomerase, inhibits DNA synthesis in Chinese hamster ovary cells[94,95]. The continued presence of novo in the medium is required to maintain this inhibition, novo seems therefore to be an inhibitor of the initiation step, this has also been shown for DNA replication of adenovirus infected cell[96] and SV40(50). This drug was shown to affect the sedimentation properties of nucleoids consequent to a reduction in negative supertwists found in the DNA of Chinese hamster cells[94]. However, if the experiments using novo suggest that inhibition occurs at the initiation step rather than at the elongation step, it is difficult to discriminate between chromatin decondensation and the initiation level itself.

a) Chromatin Decondensation

The complex and compact organization which is seen on the eukaryotic chromosome must first be released to promote the replication mechanism, and it has been suggested that type II DNA topoisomerases could play a role in this reversible packaging of the DNA. The partially reversible inhibitory effect of novo on chromosome decondensation was indeed reported during the study of DNA repair after UV irradiation of mitotic HeLa cells[94].

The different levels of chromatin organization have been revealed by electron microscopy[98-100] and nuclease digestion studies, and additional information has been obtained on the organization of chromatin using histone depleted metaphase chromosomes[101]. These structures consist of a central "scaffold core" surrounded by a halo of many DNA loops. Each loop shows a maximum length of 90 kilobases anchored in the "scaffold core" at its base and may correspond to the topological domains which have been detected in isolated nuclear structures (see the first section). The loops, which are composed of nucleohistone fibers, can be further twisted to form structures of 500-600 Å in diameter[102].

Different chemical modifications of histones are reportedly involved in conformational changes in the nucleosome, which becomes more open and accessible to the transcription process. These modifications could take place during DNA replication at the level of chromatin decondensation; models involving modifications of histones during the elongation step here appear too complicated.

Postranslational modifications of five histones have been correlated with changes in structural properties of the genome. Acetylation and phosphorylation reversibly modify histones; conversely, methylation is an irreversible process. Since these modifications occur on amino acids conserved during evolution, they might be of major biological significance. The enzymes (acetylase, deacetylase, kinase, phosphorylase, methylase) involved in histone modification have now been isolated from different organisms. Acetylated histones are preferentially released during early stages of DNase I digestion, showing a destabilization of DNA-histone interaction. Other histone modifications, such as phosphorylation and dephosphorylation of histones H1 and H3, are associated with the condensation and decondensation of chromosomes during mitosis[103].

Other non-histone chromosomal proteins are thought to be involved in the destabilization process of the nucleosome. They include the HMG proteins, which have been isolated from several organs. Some HMG proteins such as HMG14, HMG17 and H6 are thought to be involved in the transcription process, whereas HMG1 and HMG2 would be involved in replication.

The involvement of HMG2 in DNA replication has been suggested by observations wherein the level of HMG2 follows the proliferative activity in tissues[104]. Nucleosomes can be reconstituted with HMG1 and HMG2[105]. However, HMG2 can be detected only in nucleosomes which have lost histone H2A, suggesting that HMG2 replaces histone H2A in the nucleosome core. This replacement of histone H2A is facilitated by the relatively weak interactions[106] existing between histone H2A and other core histones. HMG1 binds strongly to histone H1, whereas HMG2 binds only weakly to histone H1[107-109].

These modifications in nucleosomal and supranucleosomal organization which could occur at the stage of decondensation probably modify and facilitate the work of the DNA topoisomerases during decondensation, and thereafter during elongation. In section I, for example, we suggest that activity of type I DNA topoisomerase if facilitated in the absence of histone H1, and that of type II DNA topoisomerases is facilitated when the size of DNA free from histone is increased.

b) Initiation at the Origin of the Replicon

In the preceding chapter, we saw that supercoiling of the DNA which locally generates single-stranded regions might play a role in recognition of DNA by the enzymes and proteins involved in the initiation or bacterial plasmid and viral replication. Liu[47] suggested that the eukaryotic ATP-dependent type II DNA topoisomerases could function as origin specific DNA gyrases. This suggestion was based on the similar characteristics found for both type II DNA topoisomerases isolated from phage T4 and eukaryotic cells.

However, differences exist between prokaryotic and eukaryotic origins of DNA replication. Eukaryotic replicating replicons have been visualized by electron microscopy as loops or eyes along the DNA fiber. Several such structures have been observed on the same fiber, demonstrating that bidirectional DNA replication is initiated simultaneously at many sites throughout the eukaryotic chromosomes[110,111].

Eukaryotic replication units are of variable sizes, most of them falling into the range of 50-300 kilobase pairs. The cell type and growth conditions influence the size of the replication units: small replication units predominate in cells undergoing fast rates of DNA synthesis. This variability in the size of the replication unit has led Callan to suggest two models for the initiation of the replication. One model supposes the existence of different classes of initiation sites recognized by different initiators. The other model supposes one class of initiation sites, but proposes that their use could be determined by the availability of initiator molecules or, alternatively, by the precise molecular configuration at those sites[112].

Little is known about the structure of the site of origin of replication of the eukaryotic chromosome. Nevertheless, a specific origin for the initiation of replication has been found in eukaryotic viruses, and yeast. Individual replication origins have been cloned from yeast and from certain isolated DNA segments which confer the ability to replicate on non-replicating plasmids and can serve as sites for the initiation of DNA replication[113].

To understand the mechanism of eukaryotic replication, several groups have chosen to study the replication of SV40 which replicates as a minichromosome in the host nucleus. SV40 DNA replication starts at a unique site[114] composed of about 75 base pairs[115-117]. The minichromosome is made up of 24 nucleosomes; however, one nuclease sensitive region located between the origin of replication and the region coding for the late leader RNA[118-120] appears to be free of nucleosomes[121]. It has been suggested that the nucleosome free region, with a maximum length of 385 base pairs, could play a role in replication, transcription, or virion assembly.

The presence of such a region has not been observed in eukaryotic chromosomes; however, it could correspond to a region where the negative supercoiling may accumulate unrestrained. A nucleosome free region could be recognized by an enzyme such as a DNA gyrase able to introduce supplementary superhelicity, leading to the destabilization of the double helix necessary for the interaction with the protein involved in the initiation of replication.

B. THE ELONGATION PHASE OF DNA REPLICATION

Chromosome decondensation leads to the release of the fibril structure composed of aligned nucleosomes. In the first chapter we saw that the nucleosomal structure raises some problems related to progression of the replication fork. We shall successively examine the kinds of modifications which could occur in the nucleosome leading to its destabilization or to effects upon its reconstitution.

The nucleosome consists of the exterior winding of 146 base pairs of DNA in 1 3/4 left-handed (or negative) superhelical turns around the histone core. Spectral studies have demonstrated that the nucleosomal DNA adopts the B conformation and that its major groove is largely accessible and could be the binding site for the non-histone proteins[122]. The periodicity of the double helix is thought to be reduced from 10.4 base pairs to about 10.0 base pairs. A zigzag ribbon (250 A^o diameter) of stacked nucleosomes has been proposed[122] by recent observations suggesting that a repeat unit is made up of two nucleosomes connected by a relaxed spacer DNA.

The presence of such nucleosomal structures could appear as a barrier for the progress of the replication fork. Different models can be considered for the progression of the DNA polymerase along the chromosomal DNA. 1) Conservation of the nucleosomal structure, 2) separation of the two DNA strands, one of which remains associated with nucleosomes, 3) removal of the nucleosome as a unit, or its complete disintegration.

The structure of the replicating fork has been reviewed by de Pamphilis and Wassarman[123]. Some experimental arguments are in favor of the removal of the nucleosmal structure before the synthesis of the Okazaki fragments. Indeed, 90% of the Okazaki fragments present in replicating SV40 chromatin are not associated with histones[124,125]. It has also been reported that the DNA ligase cannot seal interruptions in nucleosomal duplex DNA[126,127]. This leads to the hypothesis that the synthesis of Okazaki fragments, as well as ligation, precede nucleosome association. Preferred sites are involved in initiation, and these "initiation zones" could be determined by the nucleosome spacing[123]. Nucleosome reassembly rapidly follows the complete synthesis of Okazaki fragments.

We have seen in the first part of this chapter, that chemical modifications in histones, as well as the effect of HMG proteins, lead to the destabilization of the nucleosome, but not to the removal of the nucleosomal structure from the double helix. This removal is only one possibility; the other is that nucleosomes are conserved through the passage of the replication fork. If the histone cores are indeed removed in front of the fork, this would lead to the presence of negative supertwists which could help the progression of the DNA polymerase.

Benyajati and Worcel[11] have shown that, at high salt concentrations (0.9M NaCl), EtBr treatment displaces the four histones in equimolar concentrations from the isolated folded genome of Drosophila melanogaster. Dissociation of the four histones occurs at the equivalence point of 2μg of EtBr per ml. a concentration which corresponds to those required to completely relax the negative superhelicity of natural DNAs. Thus, EtBr which modifies the topology of the DNA introducing positive supertwists can remove the nucleosomal structure. It is possible that formation of positive supertwists related to the replication fork movement produces the same effect in vivo and that removal of the nucleosomes is either prevented by elimination of the positive supertwists in front of the fork, or is used to prepare the DNA for replication according to the behavior of the nucleosomes at the replication fork (see chapter I). Benyajati and Worcel have also shown that DNase treatment which, like EtBr relaxes the folded chromosome of Drosophila, is not accompanied by a displacement in the histones[11]. In fact, DNase treatment only relaxes the superhelicity induced by the high ionic strength, in addition to supercoiling of the nucleosome. Relaxation by introduction of nicks in the DNA, therefore, cannot remove the nucleosomal structure. This is in contrast to topological changes induced by EtBr. At first glance, the specific inhibitor effect of novo at the initiation step in eukaryotic cells[94,95] suggests that a type II DNA topoisomerase is not involved in the elongation phase. If this is the case, it may be an important difference with respect to the prokaryotic cells, whose elongation phase requires the activity of a type II DNA topoisomerase (either Topo II or Topo II' in E. coli), probably for removing positive supertwisting ahead of the replication fork. In contrast to their bacterial counterparts, eukaryotic type I DNA topoisomerases may remove positive supertwisting and therefore can be used as swivel during the elongation phase of DNA replication. Nevertheless, up until now no experimental evidence has been obtained for the participation of eukaryotic type I DNA topoisomerase in DNA replication.

Nucleosome Reconstitution

The synthesis of the five principal classes of histones (H1, H2A, H2B, H3 and H4) involved in the nucleosomal structure is restricted to the S phase.

A functional relationship between the expression of histone genes and DNA replication was suggested by the rapid shut off of histone synthesis in the presence of cytosine arabinoside[128], hydroxyurea[129] or amethopterin, three inhibitors of eukaryotic DNA replication.

A preferential association of newly synthesized histones with nascent DNA has been demonstrated for chick myoblasts[130], Drosophila melanogaster[131] and SV40[132]. During DNA replication, the old histones from the parental double helix, as well as newly synthesized histones, must assemble into nucleosomes wrapped by the daughter double helices. The assembly and segregation of the nucleosomes have been investigated in different organisms; both random and non random mechanisms for the segregation of the chromosomal proteins have been reported[133-137]. Some experiments have demonstrated that assembly and segregation of the histone core are not random; thus, the new nucleosomes which are composed only of new histones segregate conservatively and are preferentially localized next to each other[138]. Nucleosomes prepared from cultures of chick myeloblasts during the course of replication in the presence of dense and radioactive amino acids have beeen cross-linked and analyzed by density gradients. The conservative mode of nucleosome assembly was demonstrated when cells grown in heavy medium were chased in light medium. The sedimentation of the histone octamer suggests that old and new histones are not mixed in one nucleosome. In addition, the results have shown that most of the new histone octamers are preferentially located near other new histone octamers, but this is not an absolute rule.

The in vitro reconstitution of the nucleosomal structure from histones H2A, H2B, H3, H4 and circular covalently closed SV40 DNA requires high ionic strength followed by lengthy dialysis, i.e., non-physiological conditions. However, the assembly of chromatin (SV40) occurs under physiological conditions if a nicking-closing activity is present in addition to histones[139-143]. Type I DNA topoisomerases could remove the positive supertwists occurring during nucleosome reconstitution. In addition, it has been found that acidic polypeptides replace the type I DNA topoisomerases and increase the rate of nucleosome reconstitution[144]; this last observation poses the problem of the real participation of the type I DNA topoisomerases in nucleosome assembly.

Once the nucleosome is reconstituted, the new nucleosomal fibril becomes compact through histone H1 binding. It has been suggested that DNA compaction occurs through the phosphorylation and dephosphorylation of histone H1. The fiber (solenoid) is then packaged into a higher ordered structure which is the native chromatin conformation.

The protein requirement for chromatin condensation is as yet unknown. However, it is possible that proteins as well as enzymes could participate with histone H1 in condensation of a higher level of organization. As was suggested in the first section, type II DNA topoisomerase could be involved in chromatin decondensation and condensation.

IV. EUKARYOTES, PROKARYOTES: TOPOLOGY AND EVOLUTION

Table 3 summarizes some of the similarities and differences observed between the prokaryotic and eukaryotic chromosomes and DNA topoisomerases. What, then, can we learn about evolution? Similarities are certainly the sign of a common origin. In our opinion, it is false to think that, because of its relative simplicity, the present prokaryotic organization is the ancestor of its present eukaryotic counterpart. It is better to imagine the existence of a common ancestral organization from which both types derived. We might hypothesize, for instance, that the organization of the chromosome into nucleosomal and supranucleosomal structures, in both kinds of cell types, played a common role in packaging the DNA, and this from the beginning of their diverging evolution. In eukaryotes, this organization became the basis for the present sophisticated mechanisms controlling gene expression and cell differentiation. In such an evolution, the association between DNA and histones became very strong, and negative superhelicity was restrained[21], unable to control promotor activity. On the other hand, we can suppose that negative superhelicity became an essential tool for the precise control of overall gene expression in prokaryotes with the concomitant destabilization observed in the links between the DNA and nucleosome-like structures. In eukaryotes, the selection pressure led to the formation of multicellular and complex organisms with large and differentiated cells growing at various but usually slow rates in a well defined intercellular environment. In contrast, in prokaryotes, the selection pressure led to small unicellular organisms with rapid growth rates and the ability to readily adjust their metabolism to environmental changes.

From that point of view, it is perhaps futile to search for gyrase activities in eukaryotes similar to those of prokaryotes if the appearance of the latter paralleled the development in bacteria of topological control of gene expression. Indeed, negative supertwisting of the eukaryotic chromosome does not prove the existence of a gyrase activity in eukaryotes since, in vivo, the removal by any topoisomerase of the positive supertwists generated by the negative winding of the DNA around the nucleosome core leads to negative supertwisting of this DNA, once it is isolated and deproteinized.

The reduction in the degree of negative superhelicity of the eukaryotic chromosome in the presence of novo could therefore simply mean that removal of the positive supertwists generated during the formation of chromatin is performed by the eukaryotic type II DNA topoisomerase, an enzyme which has been purified and which is known to be a target for these drugs.

Table 3. Comparison between Structure, Topology and Replication of Eukaryotic and Prokaryotic Chromosomes

	Eukaryotic chromosome	Prokaryotic nucleoid
Histone or histone like proteins	five histones Hi, H2A, H2B, H3, H4	HU and protein H, the latter cross-react with H2A
Nucleosome or nucleosome like structure	stable, have been isolated	unstable, have only been vizualized, never isolated
Reconstitution of nucleosome or nucleosome like structure	realized with the four histones, stable	realized with only HU unstable over 0.1M salt
Negative supertwists	one for 15 twists restrained in vivo	one for 15 twists unrestrained in vivo maintained by gyrase activity
Domains of superhelicity	several per chromosome perhaps one per chromosomal loop and per replication unit	several per chromosome 50-100kb
Origin of replication	several per chromosome variable number perhaps one per loop	one origin
Effect of type II DNA topoisomerase inhibitors on DNA replication	stop initiation of DNA replication not elongation	stop both initiation and elongation of DNA replication
DNA topoisomerase activities discovered up to now	-relaxation of positive supertwists (Topo I and II) -relaxation of negative supertwists (Topo I and II) -catenation, decatenation (Topo II)	relaxation of positive supertwists (Topo II') relaxation of negative supertwists (Topo I, Topo II and II') catenation decatenation (Topo II and II') formation of negative supertwists (Topo II)

The fact that the same inhibitors affect bacterial and eukaryotic type II DNA topoisomerase indicates that these enzymes are closely related. One wonders why eukaryotic type II DNA topoisomerases, which like their prokaryotic counterparts contain an ATP binding site, are not able to perform gyrase activity or, in other terms, why bacterial gyrase is able to do so?

One hypothesis, previously made by Liu et al., may be that eukaryotic type II DNA topoisomerases introduce supertwists in the DNA, but only when they are bound to specific DNA sequences located, for instance, near the various eukaryote replication origins[47]. This would explain the role of type II topoisomerases in initiation of eukaryotic DNA replication in the same way that Alberts hypothesis for the role of the bacteriophage T4 type II topoisomerase in T4 DNA replication. In this context, directional binding of a specific DNA sequence around the "gyrasome" may be the main process permitting orientation of the overall reaction by the type II topoisomerase towards the gyrase reaction. This hypothesis could be tested by checking eukaryotic type II DNA topoisomerase for gyrase activity using plasmid DNA containing the autonomously replicating sequences (ars elements) isolated in eukaryotic DNA and which allows colinear DNA to replicate in yeast[145]. In the preceding chapter, we saw that the chromosome of eukaryotic cells is organized in many tandem replicons, each one containing one origin for DNA replication. It has been proposed that these replication units may be equivalent to the topological domains which segregate the eukaryotic chromosome[95]. We suggest that each of these replication units also contains one binding site for type II topoisomerase. We have also seen that the size of the replication unit varies, yet the average spacing between adjacent initiation sites is not too different from the size of the topological domains which divide the prokaryotic chromosome, and, in both cases, this size varies with growth conditions. Therefore, if each topological domain in the prokaryotic chromosome also contains one topoisomerase binding site as previously suggested (see chapter III), there is perhaps a lineage between eukaryotic and prokaryotic topoisomerase binding sites, and also a common function in chromosomal organization. The ars elements have not been found in E. coli, but the type II topoisomerase binding site in eukaryotes controlling the initiation of replication may be located, not in the ars element, but in a domain containing it. Type II DNA topoisomerase binding sites may have diverged somewhat in their function between eukaryotes and prokaryotes, depending on the existence or absence of a replication origin in the topological domains under their control. During the diverging evolution of prokaryotes, the ancestors of the ars elements might have conceded their functions as origins of replication to a unique origin located in the vicinity of the membrane and DNA junction. The ars elements would have been eliminated by natural selection, but the adjacent topoisomerase binding sites may have been preserved to allow control by this enzyme of the chromosome superhelicity with an improvement which is the appearance of the gyrase activity. In this case, the present bacterial topoisomerases II and II' could have been generated from an ancestor eukaryotic type II DNA topoisomerase to provide the cell with two proteins, each specialized in a different function. This kind of model would explain why, in contrast to the situation occuring in bacteria, the replication fork in eukaryotes does not, during the elongation phase, encounter any type II DNA topoisomerase sites locked to the DNA in the presence of topoisomerase inhibitors.

It is possible to determine with at least some degree of confidence whether, as suggested in the above model, the main features of chromosome organization in the common ancestor are more similar to those of the present eukaryotic chromosome (for example, the presence of the ars elements), or to those of the present bacterial chromosome? Or, are they very different from both?

A tentative answer can perhaps be given taking into account the particular status of mitochondria. There are two main hypotheses to explain their origin. In the first, mitochondia were generated from intracellular invagination of cytoplasmic membranes[146]. In the second, mitochondira originated from symbiotic bacteria of protoeukaryotic cells[147].

Arguments exist in favor of both hypotheses. In our opinion, strong evidence against the pure symbiotic hypothesis is the fact that the genetic organization of the mitochondrial genome displays some features in common with its eukaryotic nuclear counterpart (such as the intron-exon organization of genes) which cannot have been generated inside a prokaryotic-like structure by mimetism. We suggest that mitochondria were generated inside protoeukaryotic cells by an endogenous mechanism (such as invagination of the cell membrane) during a period of evolution when the divergencies between prokaryotes and eukaryotes first occurred. This would explain why the diverging evolution of the mitochondria inside the cell has conserved some features of their common ancestor which may resemble certain characteristics of present day bacteria. Such a mechanism seems more probable than the idea that mitochondria are symbiotic bacteria or bacteria autonomous mitochondria. If this suggestion is correct, it would mean that the genetic organization of present eukaryotes is more similar to the organization of the common ancestor than that of the present prokaryotes. This might also be the case for the structure and replication of the chromosome. With this in mind, it will be interesting to learn about the mode of supercoil formation in mitochondria. To date, neither histone, nor DNA gyrase have been detected in these organelles, but the existence of a negative supercoiling has been demonstrated[148] and mitochondrial DNA replication is inhibited by cou[149]. Recently, the isolation of a mitochondrial type II DNA topoisomerase has been reported[51].

The fundamental restructuration which occurred in the prokaryotic genome after its divergence from the common ancestral genome could be explained by the rapid growth and therefore mutation rate of prokaryotes. These fast propagating and simple organisms have now invaded all the ecological niches by skillfully using the negative superhelicity of their DNA to adjust their growth rate and the expression of their genes to changing environmental conditions. Therefore, they might be considered as the most recent and best achievements in biological evolution, whereas the trend of the eukaryotic organisms towards increasing complexity may reflect a fundamental weakness in these old-fashioned forms of life.

ACKNOWLEDGEMENTS

We would like to thank Drs Anne-Lise Haenni and Suzan Elsevier for their suggestions during the preparation of the manuscript. We would like to thank Anne-Marie Solomiac and Richard Schwartzmann for the presentation and the illustration of the text.

REFERENCES

1. J. D. Watson and F. H. C. Crick, Nature 171:737-738 (1953).
2. J. D. Watson and F. H. C. Crick, Cold Spring Harbor Symp. Quant. Biol. 18:123-131 (1953).
3. E. Schroedinger, in:"What is life, the physical aspect of the lung cell, Cambridge (1944).
4. N. R. Cozzarelli, Science 207:953-960 (1980).
5. J. Cairns, Cold Spring Harbor Symp. Quant. Biol. 28:43-46 (1963).
6. F. H. C. Crick, Proc. Natl. Acad. Sci. U.S.A. 73:2639-2643 (1976).
7. J. J. Champoux and M. D. Been, in:"Mechanistic Studies of DNA Replication and Recombination"m B. M. Alberts and C. F. Fox, eds., pp. 809-815, Academic Press, New York (1980).
8. J. C. Wang, J. Mol. Biol. 55:523-532 (1971).
9. M. Gellert, K. Mizuuchi, M. H. O'Dea and H. A. Nash, Proc. Natl. Acad. Sci. U.S.A. 73:3872-3876 (1976).
10. A. Worcel and E. Burgi, J. Mol. Biol. 71:127-147 (1972).
11. C. Benyajati and A. Worcel, Cell 9:393-407 (1976).
12. R. R. Sinden and D. E. Petitjohn, Proc. Natl. Acad. Sci. U.S.A. 78:224-228 (1981).
13. W. F. Pohl and G. W. Roberts, J. Math. Biology 6:383-402 (1978).
14. G. A. Rodley, R. S. Scobie, R.H.T. Bates and R. M. Lewitt, Proc. Natl. Acad. Sci. U.S.A. 73:2959-2963 (1976).
15. J. C. Wang, Trends Biochem. Sci. 5:219-221 (1980).
16. K. N. Kreuzer and N. R. Cozzarelli, Cell 20:245-254 (1980).
17. O. Sundin and A. Varshavsky, Cell 21:103-114 (1980).
18. H. Kasamatsu and J. Vinograd, Ann. Rev. Biochem. 43:695-720 (1974).
19. L. F. Liu, C. C. Liu and B. M. Alberts, Nature 281:456-461 (1979).

20. W. R. Bauer, Ann. Rev. Biophys. Bioenerg. 7:287-313 (1978).
21. R. R. Sinden, J. O. Carlson and D. E. Petitjohn, Cell 21:773-763 (1980).
22. A. Worcel, S. Strogatz and D. Riley, Proc. Natl. Acad. Sci. U.S.A. 78:1391-1465 (1981).
23. I. M. Leffak, R. Grainger and H. Weintraub, Cell 12:837-845 (1977).
24. O. P. Brown and N. R. Cozzarelli, Proc. Natl. Acad. Sci. U.S.A. 78:843-847 (1981).
25. J. Rouvière-Yaniv, M. Yaniv and J. E. Germond, Cell 17:265-274 (1979).
26. V. Hubscher, H. Lutz and A. Kornberg, Proc. Natl. Acad. Sci. U.S.A. 77:5097-5101 (1980).
27. J. D. Griffith, Proc. Natl. Acad. Sci. U.S.A. 73:563-567 (1976).
28. M. Gellert, K. Mizuuchi, M. H. O'Dea, T. Itoh and T. Tomizawa, Proc. Natl. Acad. Sci. U.S.A. 74:4772-4776 (1977).
29. A. Sugino, C. L. Peebles, K. M. Kreuzer and N. R. Cozzarelli, Proc. Natl. Acad. Sci. U.S.A. 74:4767-4771 (1977).
30. K. Mizuuchi, L. M. Fisher, M. H. O'Dea and M. Gellert, Proc. Natl. Acad. Sci. U.S.A. 77:1847-1851 (1980).
31. K. Mizuuchi, M. H. O'Dea and M. Gellert, Proc. Natl. Acad. Sci. U.S.A. 75:
32. L. F. Liu and J. C. Wang, Proc. Natl. Acad. Sci. U.S.A. 75:2098-2102 (1978).
33. P. Forterre, J. Theor. Biol. 82:255-269 (1980).
34. P. O. Brown, C. L. Peebles and N. R. Cozzarelli, Proc. Natl. Acad. Sci. U.S.A. 76:6110-6114 (1979).
35. L. F. Liu and J. C. Wang, Cell 15:979-984 (1978).
36. A. Morrison and N. R. Cozzarelli, Cell 17:175-184 (1979).
37. K. Mizuuchi, M. H. O'Dea and M. Gellert, Proc. Natl. Acad. Sci. U.S.A. 75:5960-5963 (1978).
38. A. Sugino, M. P. Higgins, P. O. Brown, C. L. Peebles and N. R. Cozzarelli, Proc. Natl. Acad. Sci. U.S.A. 75:4838-4852 (1978).
39. A. Morrison, M. P. Higgins and N. R. Cozzarelli, J. Biol. Chem. 255:2211-2219 (1980).
40. J. C. Wang, R. I. Gumport, K. Javaherian, K. Kirkegaard, L. Klevan, M. L. Kotewicz and Y. C. Tse, in:"Mechanistic Studies of DNA Replication and Recombination", B. M. Alberts and C. F. Fox, eds., pp.769-784, Academic Press, New York (1980).
41. L. F. Liu, C. C. Liu and B. M. Alberts, Cell 19:697-707 (1980).
42. M. Gellert, L. M. Fisher and M. H. O'Dea, Proc. Natl. Acad. Sci. U.S.A. 76:6289-6293 (1979).
43. W. B. Wood and H. R. Revel, Bact. Rev. 40:847-868 (1976).
44. G. Stetler, J. K. Gretchen and Wai Hun Huang, Proc. Natl. Acad. Sci. U.S.A. 76:3737-3741 (1979).
45. T. S. Hsieh and D. Brutlag, Cell 21:115-125 (1980).
46. M. I. Baldi, P. Benedetti, E. Mattocia and G. P. Tocchini-Valentini, Cell 20:461-467 (1980).

47. L. F. Liu, in:"Mechanistic Studies of DNA Replication and Recombination", B. M. Alberts and C. F. Fox, eds., pp. 817-831 Academic Press, New York (1980).
48. L. M. Assairi, unpublished results.
49. M. Duguet, unpublished results.
50. H. J. Edenberg, Nature 286:529-531 (1980).
51. F. J. Castora, G. C. Brown and M. V. Simpson, in:"The organization and expression of the mitochondrial genome", A. M. Kroon and C. Saccone, eds., Amersterdam, Elsevier, in press (1981).
52. N. R. Cozzarelli, Ann. Rev. of Biochem. 46:641-668 (1977).
53. D. H. Smith and B. O. Davis, J. of Bacteriol. 93:71-79 (1967).
54. W. L. Staudenbauer, J. Mol. Biol. 96:201-205 (1975).
55. M. J. Ryan, Biochemistry 15:3769-3777 (1976).
56. K. Drilica and M. Snyder, J. Mol. Biol. 120:145-154 (1978).
57. E. Orr, N. F. Fairweather, I. B. Holland and A. H. Pritchard, Mol. Gen. Genet. 177:103-112 (1979).
58. B. Van Wright and B. A. Bridges, J. of Bacteriol. 146:18-23 (1981).
59. K. Drlica, E. C. Engle and S. H. Manes, Proc. Natl. Acad. Sci. 77:6879-6883 (1980).
60. F. Hansen and K. von Meyenburg, Molec. Gen. Genet. 175:135-144 (1979).
61. H. N. Kreuzer and N. R. Cozzarelli, J. Bacteriol. 140:424-435 (1979).
62. M. Snyder and K. Drlica, J. Mol. Biol. 131:287-302 (1979).
63. G. S. Crumplin and J. T. Smith, Nature 260:643-644 (1976).
64. R. Sternglanz, S. Dinardo, K. A. Woelkel, Y. Nishimura, Y. Hirota, K. Becherer, L. Zumstein and J. C. Wang, Proc. Natl. Acad. Sci. 78:2747-2751 (1981).
65. N. Ogasawara, M. Seiki and H. Yoshikawa, Mol. Gen. Genet. 181:332-337 (1981).
66. M. F. Fairweather, E. Orr and I. B. Holland, J. of Bact. 142:-53-161 (1980).
67. M. Abdel-Monem and H. Hoffmann-Berling, Trends in Bioch. Sci. 5: 128-130 (1980).
68. K. Arai, N. Arai, J. S. Schlomai and A. Kornberg, Proc. Natl. Acad. Sci. U.S.A. 77:3322-3326 (1980).
69. R. H. Hamatak, R. Mukai and M. Mahashi, Proc. Natl. Acad. Sci. U.S.A.
70. M. J. Ryan and R. D. Wells, Biochemistry 15:3778-3782 (1976).
71. P. Forterre and M. Kohiyama, Europ. J. Biochem. 90:537-546 (1978).
72. H. Schaller, B. Otto, V. Nusslein, J. Huf, R. Herrmann and J. Bonhoeffer, J. Mol. Biol. 63:183-200 (1972).
73. T. Kornberg, A. Lockwood and A. Worcel, Proc. Natl. Acad. Sci. U.S.A. 71:3189-3193 (1974).
74. P. Forterre and M. Kihiyama, unpublished observations.
75. T. Itoh and J. Tomizawa, Nature 270:78-79 (1977).
76. S. Wickner, Proc. Natl. Acad. Sci. U.S.A. 74:2815-2819 (1977).
77. P. Forterre and M. Kohiyama, in:"Mechanism and Regulation of DNA replication", pp. 22-35, Plenum Press, New York (1973).

78. E. Edelstein, L. Parks, H. E. Tsan, L. Daneo Moore and M. L. Higgins, J. of Bacteriol. 146:798-803 (1981).
79. R. J. Marians, J. E. Ikeda, S. Schlagman and J. Hurwitz, Proc. Natl. Acad. Sci. U.S.A. 74:1965-1968 (1977).
80. S. Eisenberg, J. Griffith and A. Kornberg, Proc. Natl. Acad. Sci. U.S.A. 74:3198-3202 (1977).
81. J. Tomizawa, in:"DNA synthesis present and future," pp.797- Plenum Publishing Corporation, New York (1978).
82. E. Orr and N. Staudenbauer, Mol. Gen. Genet. 181:51-56 (1981).
83. T. Itoh and J. Tomizawa, Cold Spring Harbor Symp. Quant. Biol. 43: 409-413 (1978).
84. B. Sanzey, J. Bact. 138:40-47 (1979).
85. D. McCarthy, C. Minner, H. Bernstein and C. Bernstien, J. Mol. Biol. 106:963-981 (1976).
86. F. Jacob, S. Brenner and F. Cuzin, Cold Spring Harbor Symp. Quant. Biol. 28:329-348 (1963).
87. O. Pierucci and C. Zuchowski, J. Mol. Biol. 80:477-503 (1973).
88. A. Jacq, H. Lother, W. Messer and M. Kohiyama, in:"Mechanistic Studies of DNA Replication and Genetic Recombination," B. M. Alberts and C. F. Fox, eds., pp. 189-197, Academis Press, New York (1980).
89. N. Ogasawara, M. Seiki and H. Yoshikawa, Mol. Gen. Genet. 181: 332-337 (1981).
90. H. Yoshikawa, N. Ogasawara and M. Seiki, Mol. Gen. Genet. 179: 265-276 (1980).
91. M. Seiki, N. Ogasawara and H. Yoshikawa, Nature 281:699-701 (1979).
92. L. F. Pouirk and R. B. Painter, Biochim. Biophys. Acta. 432:267-272 (1976).
93. M. R. Mattern and R. B. Painter, Biochim. Biophys. Acta 563:293-305 (1979).
94. M. R. Mattern and R. B. Painter, Biochim. Biophys. Acta 563:306-312 (1979).
95. M. R. Mattern and D. A. Sandiero, Biochim. Biophys. Acta 563:248-258 (1981).
96. J. C. D'Halluin, M. Milleville and P. Boulanger, Nucleic Acid. Res. 8:1625-1641 (1980).
97. A. Collins and R. Johnson, Nucleic Acids Res. 7:1311-1320 (1979).
98. A. L. Olins and D. E. Olins, Science 183:330-332 (1974).
99. C. L. F. Woodcock, J. P. Safer and J. E. Stanchfield, Exp. Cell. Res. 97:101-110 (1976).
100. C. L. F. Woodcock, H. E. Swwean and L. L. Erado, Exp. Cell. Res. 97:111-119 (1976).
101. J. R. Paulson and U. K. Laemmli, Cell 12:817-828 (1977).
102. Y. Daskal, M. L. Mace, J. R. W. Wray and M. Busch, Exp. Cell. Res. 100:214-212 (1976).
103. L. R. Gurley, J. A. D'Anna, S. S. Barham, L. L. Deaven and R. A. Tobey, Eur. J. Biochem. 84:1-15 (1978).
104. S. M. Seyedin and W. S. Kisler, J. Biol. Chem. 254:11264-11271 (1979).

105.K. Mita, M. Zama, S. Ichimura and K. Hamana, Biochem. Biophys. Res. Commun. 98:330-336 (1981).
106.J. A. D'Anna and I. Isenberg, Biochemistry 13:499214997 (1974).
107.P. Cary, K. Shooter, G. Goodwin, E. Johns, J. Olayemi, P. Hartman and E. Bradbury, Biochemistry 183:657-662 (1979).
108.M. J. Smerdon and I. Isenberg, Biochemistry 15:4242-4247 (1976).
109.S. H. Yu and T. G. Spring, Biochem. Biophys. Acta 492:20-28 (1977).
110.J. A. Huberman and A. D. Riggs, J. Mol. Biol. 32:327-341 (1968).
111.J. A. Huberman and A. Tsai, J. Mol. Biol.75:5-12 (1973).
112.H. G. Callan, Phil. Trans. Roy. Soc. B 181:19-41 (1972).
113.C. Chan and B. K. Tye, in:"Mechanistic Studies of DNA Replication and Genetic Recombination," B. M. Alberts and C. F. Fox, eds., pp. 347-358 (1980).
114.G. Fareed and D. Davoli, Annual Rev. Biochem. 46:471-522 (1977).
115.H. Van Heuversyn and W. Fiers, Eur. J. Biochem. 100:51-60 (1979).
116.V. B. Rudy, B. Thimmappaya, R. Dhar, K. V. Subramanian, B. Zain, J. Pan, P. Gosh, M. Celma and S. Weissman, Science 200:494-502 (1978).
117.W. Fiers, R. Contreras, G. Haegman, R. Rogiers, A. Van de Voorde, H. Van Heuversyn, J. Van Herreqeghe, G. Volckaert and M. Ysebaert, Nature 273:113-120 (1978).
118.A. J. Varshavsky, O. H. Surdin and M. J. Bohn, Nucl. Acid. Res. 5: 3469-3478 (1978).
119.W. Scott and D. Wigmore, Cell 15:1511-1518 (1978).
120.W. Waldeck, B. Fohring, K. Chowdhury, K. Gruss and G. Sauer, Proc. Natl. Acad. Sci. U.S.A. 75:5964-5968 (1978).
121.S. Saragosti, G. Moyne and M. Yaniv, Cell 20:65-73 (1980).
122.O. Goodwin, J. Vergne, J. Brahms, N. Defer and J. Kruh, Biochemistry 18:2057-2064 (1979).
123.M. DePhamphilis and P. Wassarman, Annual Rev. Biochem. 49:627-666 (1980).
124.M. DePhamphilis, S. Anderson, R. Bar-Shavit, E. Collins, H. Edenberg, T. Herman, B. Katas, G. Kaufmann, B. Krokan, E. Shelton, R. Su, D. Tapper and P. Wassarman, Cold Spring Harbor Symp. Quant. Biol. 43:679-692 (1978).
125.T. Herman, M. DePamphilis and P. Wassarman, Biochemistry 18:4563-4571 (1979).
126.S. B. Zimmerman and C. J. Levin, Biochemistry 14:1761-1677 (1975).
127.S.B. Zimmerman and C. J. Levin, Biochem. Biophys. Res. Cancer 62: 357-361 (1975).
128.E. Robbins and T. Borun, Proc. Natl. Acad. Sci. U.S.A. 57:409-416 (1967).
129.S. Takai, T. Borun, T. Muchmore and I. Lieberman, Nature 219:860-861 (1968).
130.M. Leffak, R. Grainger and H. Weintroub, Cell 12:837-845 (1977).
131.A. Worcel, S. Hans and M. Wong, Cell 15:969-977 (1978).
132.C. Cremisi, A. Chestier and M. Yaniv, Cell 12:947-951 (1977).
133.H. Weintroub, Cold Spring Harbor Symp. Quant. Biol. 38:247-256 (1973).
134.R. Tsanev and G. Russev, Eur. J. Biochem. 43:257-263 (1974).
135.R. Seale, Proc. Natl. Acad, Sci. U.S.A. 73:2270-2279 (1976).

135. R. Seale, Proc. Natl. Acad. Sci. U.S.A. 73:2270-2279 (1976).
136. V. Jackson, D. Granner and R. Chalkley, Proc. Natl. Acad. Sci. U.S.A. 73:2266-2269 (1976).
137. E. Freedlender, L. Taichman and O. Smithies, Biochemistry 16:1802-1808 (1977).
138. M. Leffak, R. Grainger and H. Weintroub, Cell 12:837-845 (1977).
139. R. Laskey, A. Mills and R. Morris, Cell 10:237-243 (1977).
140. R. Laskey, B. Honda, A. Mills and J. Finch, Nature 275:416-420 (1978).
141. J. Germond, B. Hirt, P. Oudet, M. Gross-Bellard and P. Chambon, Proc. Natl. Acad. Sci. U.S.A. 72:1843-1847 (1975).
142. J. E. Germond, J. Rouvière-Yaniv, M. Yaniv and D. Brutlag, Proc. Natl. Acad. Sci. U.S.A. 76:3779-3783 (1979).
143. A. Ruiz-Carillo, J. Jarcano, G. Eder and R. Luy, Proc. Natl. Acad. Sci. U.S.A. 76:3284-3288 (1979).
144. A. Stein, J. Whitelock and M. Bina, Proc. Natl. Acad. Sci. U.S.A. 76:5000-5004 (1979).
145. D. T. Stinchcomb, M. Thomas, J. Kelly, E. Selker and R. W. Davis,
146. R. A. Ruff and H. R. Mahler, Science 177:575-582 (1972).
147. L. Margulis, in: "Origin of Eukaryotic Cells," Yale Univerity Press (1970).
148. D. Bogenhagen and D. A. Clayton, J. Mol. Biol. 119:69-81 (1978).
149. F. J. Castora and M. V. Simpson, J. Biol. Chem. 254:11193-11195 (1979).

PART II

REGULATION OF DNA REPLICATION

PROTEIN PHOSPHORYLATION AND CHROMATIN REPLICATION:

STUDIES ON THE INTERACTION OF TWO NUCLEAR PROTEINS WITH DNA

Hans Stahl, Herbert König and Rolf Knippers

Fakultät für Biologie
Universität Konstanz
D-7750 Konstanz, FRG

SUMMARY

1. Eukaryotic single strand specific DNA binding protein was phosphorylated *in vitro* by a specific nuclear protein kinase. The untreated and phosphorylated protein have similar DNA binding properties. The phosphorylated binding protein, however, fails to stimulate DNA polymerase α.

2. A growth dependent histone H1 specific protein kinase *in vitro* adds an average of three phosphate groups to one mole of isolated histone H1. Under appropriate ionic conditions, unmodified H1 binds cooperatively to DNA, while *in vitro* phosphorylated H1 binds distributively to DNA.

INTRODUCTION

Typical preparations of chromatin proteins, histones plus nonhistone chromatin proteins, contain approximately one percent phosphorus by weight, corresponding to 4-5 phosphate groups per hundred amino acids[1]. Recent determinations in our laboratory showed that, in resting as well as in mitogen stimulated lymphocytes, about 95% of the phosphate groups are covalently bound to serine and about 5% to threonine side chains. Phosphotyrosine in lymphocyte chromatin was less than 0.2% of total recovered phosphoamino acids. In mouse Ehrlich ascite cell chromatin, about 93% of all recovered phosphate groups were found on serine, about 7% on threonin and about 0.05% on tyrosine residues[2].

The phosphate groups on chromatin proteins turn over quite rapidly. The average half life of protein bound (^{32}P) phosphate has been estimated to be about 6 hours while the protein half life is at least 4-5 times longer[3]. The phosphate turnover and the rate of protein phosphorylation vary with the physiological condition of the cell. For example, when resting lymphocytes are stimulated by Concanavalin A to proceed from a dormant to a metabolically more active phase, nuclear protein phosphorylation is one of the earliest reactions detectable in the nucleus leading to a higher phosphate content of non-histone proteins. The protein phosphorylation rate remains approximately constant during the G1 phase and increases rapidly again at the beginning of the DNA replication phase, remaining high until the end of the S-phase. During the S-phase, histones as well as non-histones are phosphorylated[4,5].

Since temporal correlations exist between an increase in the rate of protein phosphorylation and genetic events like transcription and DNA replication, many authors have speculated on a regulatory role of nuclear protein phosphorylation[1,6,7]. It has been suggested, for example, that the introduction of negative charges such as phosphate groups into proteins might weaken the interaction of proteins with DNA[8]. This straightforward concept, however, is probably an oversimplification. We describe below some experiments which were performed with two different nuclear proteins, both in an un- or underphosphorylated form and as phosphoproteins. We shall demonstrate that *in vitro* phosphorylated DNA binding proteins interact quite efficiently with DNA, but have some properties which are different from those of untreated proteins.

First, we would like to summarize our current knowledge of protein phosphorylating enzymes in the cell nucleus.

Nuclear Protein Kinases

Protein kinases are enzymes which transfer the γ-phosphate group from ATP to amino acid side chains of proteins. The two best characterized protein kinases[9] are cAMP dependent and occur in high activities in the cytoplasm of the cell. These are also the most abundant soluble protein kinase activities in the nucleus[10].

There is also a group of cAMP independent protein kinases in the nucleus. This group is usually found in association with an insoluble structure, most probably chromatin. After solubilization in 0.4 M NaCl and column chromatography, we could distinguish two subgroups[11].

Histone specific protein kinases, including (a) at least two H1 specific protein kinases, one of which is detectable in resting cells while a second H1 kinase is most active in proliferating cells, for example, after Concanavalin A activation of lympho-

cytes ("growth-dependent H1 kinase"[12]); (b) a H2b specific kinase, and (c) a kinase activity which phosphorylates H2b, H3 and H4; it is possible that this activity consists of two or even three distinct enzymes[13,14].
Non-histone specific protein kinases, including (a) an enzyme, sedimenting with about 4S through sucrose gradients ("4S kinase"), and (b) a second enzyme, sedimenting with 8S ("8S kinase"). We find[11] 3-4 times more active 8S kinase in proliferating than in resting lymphocytes. The 4S kinase appears to be equally active under both conditions. Both protein kinases phosphorylate specific sets of nuclear non-histone proteins. (Their activity is tested in the *in vitro* assay with caseine or phosvitin).

Among the phosphate accepting substrates for the 8S kinase, but not for the 4S kinase, is a single strand specific DNA binding protein. We shall describe some properties of the phosphorylated form of this protein.

Effects of Phosphorylation on a Single Strand Specific DNA Binding Protein

Single strand specific DNA binding proteins (DBP) appear to occur in many, if not all, eukaryotic cells. Some of the DBP preparations stimulate the activity of DNA polymerase α but not of DNA polymerase β or of prokaryotic DNA polymerases[15]. A DBP, prepared from meiotic plant cells[16], and a similar protein from mammalian liver cells[17], seem to possess two phosphate groups. When completely dephosphorylated *in vitro* by alkaline phosphatase, these proteins lose their specificity for single stranded DNA and bind equally well to single stranded and to double stranded DNA[16]. Removal of one phosphate group did not change the binding specificity of the DBP[16]. Attempts to restore the single strand specificity by *in vitro* phosphorylation of the totally dephosphorylated DBP with a cAMP dependent protein kinase were unsuccessful. However, treatment with a nuclear cAMP independent protein kinase, probably related to the 8S kinase (see above), restored the binding specificity of the protein by introduction of two phosphate groups/molecule DBP[16].

We have previously shown that a DBP of molecular weight of about 30000, prepared from proliferating mouse ascite cells, is phosphorylated *in vitro* by the nuclear 8S kinase which transfers maximally one mole phosphate/mole DBP (unpublished). (We have not yet determined the original phosphate content of the ascite DBP preparation. Since DBP preparations from meiotic plant cells[16] and from deer liver cells[17] maximally contain two phosphate groups/molecule it seems reasonable to conclude that the ascite cell DBP can also aquire up to two phosphate groups/molecule, suggesting that our original DBP preparation had one mole phosphate/mole protein).

In vitro phosphorylation of the ascite cell DBP causes an increase in the affinity for single stranded DNA. Filter binding experiments show that the phosphorylated DBP prefers single stranded over double stranded DNA, even at low salt concentrations when untreated DBP shows little specificity (Fig. 1).

An interesting difference which we find between untreated and in vitro phosphorylated DBP is that the latter does not stimulate DNA polymerase α (Fig. 2). Since both untreated and phosphorylated DBP bind under the conditions of the polymerase assay to single stranded DNA with similar efficiencies, we think that the observed difference must be due to a change in DBP-polymerase interaction. Our previous attempts to demonstrate a direct DBP-polymerase interaction were unsuccessful[18]. We have therefore tried an indirect approach and found that untreated, but not in vitro phosphorylated, DBP partially protects DNA polymerase α from heat inactivation (Table 1), suggesting a less intimate interaction of the phosphorylated DBP with polymerase.

Table 1. Heat inactivation of DNA polymerase α.

	Percent of control activity			
	0	2	5	10
	min at 50°C			
DNA polymerase α, control	100	13	2	3
DNA polymerase α + DBP	100	70	33	22
DNA polymerase α + P-DBP	100	14	2	2

About 0.1 unit each of the partially purified DNA polymerase α preparation, used in the experiment in Fig.2, were kept at 50°C in 0.05ml volume reaction buffer, containing 1 mg/ml bovine serum albumin. After several minutes at 50°C, the samples were transferred to an ice bath. Activated calf thymus DNA and deoxyribonucleoside triphosphates, including (^{3}H) dTTP, were then added before the DNA polymerase activity was assayed by incubation at 30°C for 60 min. To test the effect on heat inactivation 2.5μg of untreated and of in vitro phosphorylated DBP (P-DBP) were added to the DNA polymerase samples.

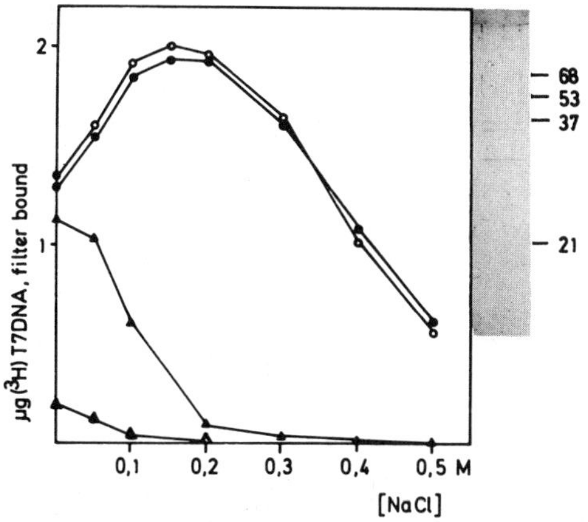

Fig. 1. The interaction of a DBP preparation from mouse ascite cells with single and double stranded DNA.

About 2µg of a DBP preparation were added to 5µg (^{3}H) T7 DNA (spec. radioactivity: ca. 5 x 10^3 cpm/µg) in 0.2 ml binding buffer (50mM Tris-HCl, pH 7.8; 1mM EDTA; 0.1 mM dithioerythritol and 200µg/ml bovine serum albumin), containing NaCl in the concentrations indicated in the graph. After 10 min at room temperature, the mixture was passed through nitrocellulose filters[15].

Native phage T7 DNA (triangles) and heat denatured single stranded T7 DNA (circles) were used in different experiments.

Closed (●,▲) and open (o,Δ) symbols represent data obtained with untreated and with in vitro phosphorylated DBP, respectively.

For in vitro phosphorylation, 50µg of DBP were incubated for 60 min at 30°C in the presence of 1mM γ (^{32}P) ATP with 50 units of a partially purified 8S kinase[11]. The phosphorylated DBP was then precipitated by ammonium sulfate, dialysed and used for the filter binding assay. Control experiments had shown that the kinase preparation was free of DNA binding activity.

The insert shows an analysis of the DBP preparation by polyacrylamide gel electrophoresis in the presence of SDS. We find three bands of apparent molecular weights of 36 000, 33 000 and 31 000 as estimated from the molecular weight of the following standards: bovine serum albumin (Mr. 68 000), glycerinaldehyde dehydrogenase (Mr. 53 000) glutamate dehydrogenase (Mr. 37 000) and trypsin inhibitor (Mr. 21 000). The position of the stained bands of these standard proteins (as determined in a different slot on the same gel) are indicated by the vertical bars.

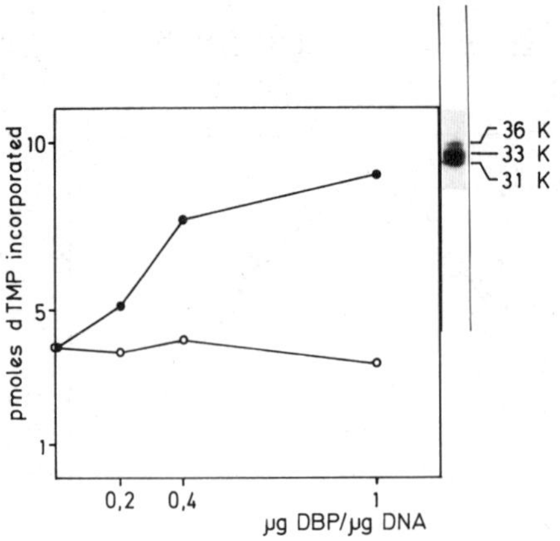

Fig. 2. Stimulation of DNA polymerase α by untreated DBP.

DNA polymerase α was partially purified from mouse ascite cells[15]. The polymerizing activity was assayed as described in[18] using 1µg heat denatured calf thymus DNA as primer template and (^{3}H) dTTP, dCTP, dGTP and dATP as monomeric substrates. Untreated (●) and in vitro phosphorylated (o) DBP were added to reaction mixtures in amounts indicated in the graph. Incubation was at 37°C for 60 min.

The insert shows an autoradiogram of in vitro phosphorylated DBP, after SDS gel electrophoresis, demonstrating that the protein is not degraded during the 60 min incubation with the protein kinase. Note that the gel of Fig. 1 contained 7.5% polyacrylamide and the gel used for autoradiography 15% polyacrylamide.

From the observations summarized above, we conclude that:

- unphosphorylated DBP binds to DNA, but does not show the preference for single stranded DNA as expected for a helix destabilizing protein[16,17];
- an underphosphorylated (monophosphorylated) form of DBP specifically binds to a single stranded DNA and stimulates DNA polymerase α;
- an in vitro phosphorylated (diphosphorylated) DBP form also binds specifically to single stranded DNA, even at low NaCl concentration when the single strand specificity of underphosphorylated DBP is less pronounced; this presumably diphosphorylated DBP does not stimulate the DNA polymerase.

We do not know whether and how the various phosphorylation reactions regulate the chain elongation events at the replication fork. It would be interesting to find out, for example, whether changes in the degree of DBP phosphorylation occur at the beginning of the DNA replication phase of the cell cycle.

Binding of in vitro phosphorylated histone H1 to DNA

In the preceding section, we described the effects of a growth dependent non-histone protein kinase ("8S kinase") on the properties of a specific protein. Here we would like to discuss the DNA binding properties of histone H1, phosphorylated in vitro by the growth dependent H1 kinase[11,12].

Remarks on the in vivo phosphorylation of H1

Histone H1 is phosphorylated during the DNA replication phase of the cell cycle[19], acquiring about three phosphate groups on serine residues in the C-terminal part of the molecule. In the late G2 phase and during mitosis, additional phosphate groups are introduced into the C-terminal as well as into the N-terminal part of the H1 molecule on both serine and threonine side chains. It has been suggested that the latter modification reaction is required for the initiation of mitosis[20].

To learn more about the relationship between H1 phosphorylation and DNA replication, we investigated the appearance of phosphorylated histone H1 on Simian Virus 40 (SV40) chromatin in vivo. Viral chromatin in lytically infected cells is a popular model for studying several aspects of eukaryotic DNA replication[21].

It is known that, except for the initiation of replication which is under the control of the viral gene A product, SV40 chromatin replication is very similar to cellular DNA replication as far as the molecular mechanisms of chain elongation are concerned[21]. Mature and replicating SV40 chromatin can be separated by sucrose gradient centrifugation. Under the conditions used in our laboratory, we estimate a sedimentation rate of 75 S for mature and 95 S for the major fraction of late replicating SV40 chromatin.

In the experiment shown in Fig. 3, we labeled cultures of SV40 infected monkey cells for 15 min and for 60 min with (^{32}P) phosphate. The viral chromatin was then extracted and sedimented through sucrose gradients. Fractions of these gradients were then investigated by SDS-polyacrylamide gel electrophoresis followed by autoradiography. After a 15 min (^{32}P) phosphate pulse, we found most radioactivity on a core histone, probably H2b, and very little on histone H1. After a 60 min pulse, the radioactivity on H2b had not significantly increased, while most (^{32}P)-phosphate was now recovered on H1. Although it is difficult to quantitatively evaluate an experiment of the type shown in Fig. 3, it appears safe to conclude that the binding of phosphorylated H1 to SV40 chromatin is a rather late event in chromatin replication, probably occurring at the end of a replication cycle (about 15 min for SV40 chromatin). This conclusion is supported by a more detailed analysis of (^{32}P) pulse labeled SV40 chromatin after sucrose gradient sedimentation (Fig. 4): phosphorylated H1 was detected in nucleoprotein complexes which sediment like mature 75 S SV40 chromatin; some (^{32}P) radioactivity was detected on core histones in the leading shoulder of 95 S replicating SV40 chromatin but very little, if any, on H1. (The sucrose gradient, shown in Fig. 4, also contains partially matured virus particles sedimenting with 240 S. These particles contain phosphorylated viral proteins, VP 1 and VP 3, in addition to some residual phosphorylated H1. It is likely that, during the maturation process, phosphorylated H1 is gradually replaced by viral proteins).

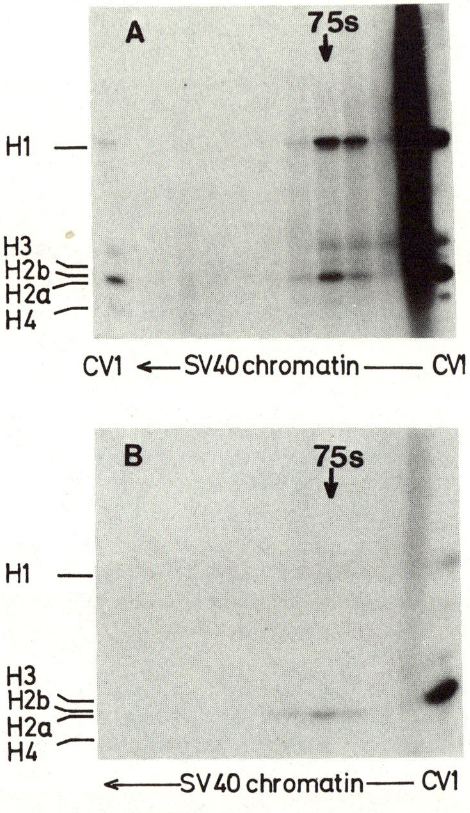

Fig. 3. Autoradiograms of (^{32}P) phosphate labeled chromatin proteins in SV40 nucleoprotein complexes.

CV-1 cells, about 40 hours after infection with SV40, were labeled for several hours with 50μCi/ml (^{3}H) thymidine and then starved for 1 h in phosphate-free medium before 0.2 mCi (^{32}P) phosphate was added in 1 ml phosphate-free medium. The cells were harvested 60 min (A) and 15 min (B) after addition of (^{32}P) phosphate. Viral nucleoprotein was prepared and isolated by sucrose gradient centrifugation as described[22,28]. Fractions of the sucrose gradient were precipitated by trichloroacetic acid. The precipitates were washed with acetone and resuspended in a buffer suitable for SDS polyacrylamide gel electrophoresis. After electrophoresis through 12.5% polyacrylamide, the gel was dried and autoradiographed. The vertical arrow indicates the position of the 75S viral nucleoprotein as identified by the (^3H) thymidine label (not shown). The horizontal arrow shows the direction of sedimentation.

The autoradiograms also show the (^{32}P) phosphate label in cellular histones ("CV1"), prepared by acid extraction from nuclei of the infected cells.

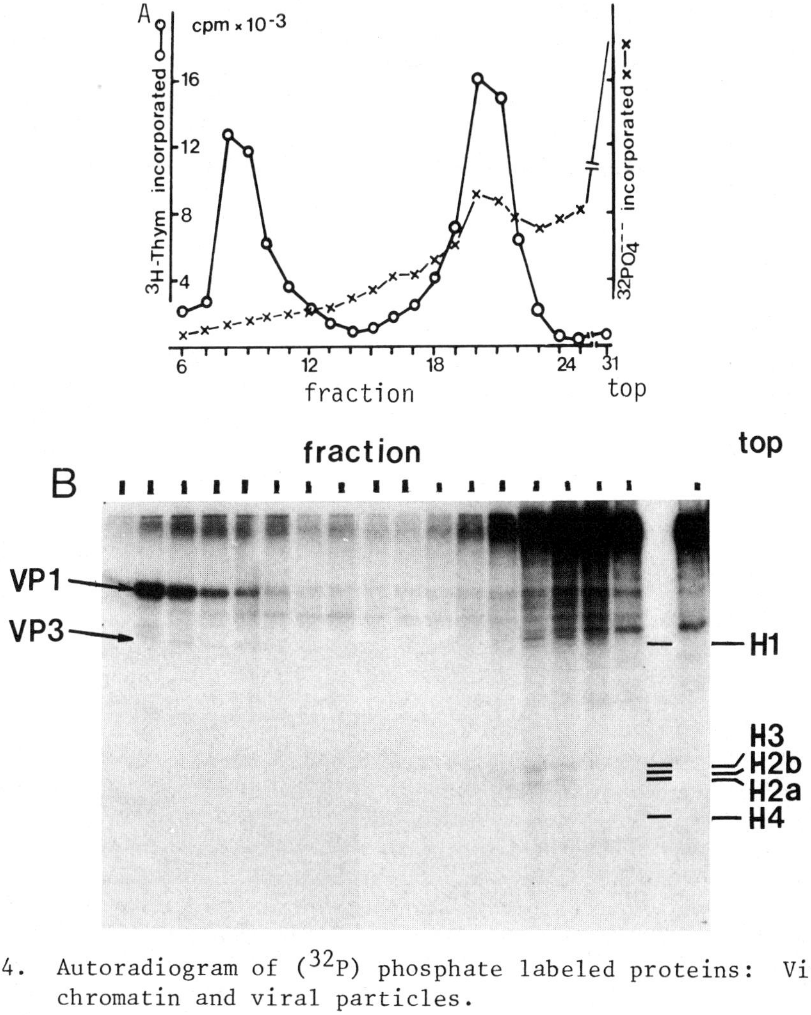

Fig. 4. Autoradiogram of (^{32}P) phosphate labeled proteins: Viral chromatin and viral particles.

SV40 infected CV-1 cells were labeled for 1 h with (^{3}H) thymidine and (^{32}P) phosphate as in Fig. 3. Viral nucleoprotein was prepared and sedimented through sucrose gradients. Two peaks of (^{3}H) labeled material were identified: a) fast sedimenting structures (about 240S), containing partially matured virus particles (fractions 7-10); and b) a 75S structure of non-replicating viral chromatin (fractions 17-22).

Fractions of this sucrose gradient were precipitated by trichloroacetic acid, washed with acetone and investigated by SDS-polyacrylamide electrophoresis as in Fig. 3. The dried gel was then used for autoradiography. (A) Aliquots from fractions 6-24 of the sucrose gradient (total fraction number: 31) were precipitated with trichloroacetic acid to determine the distribution of total radiography.

Note: the leading shoulder of the peak around fractions 20 and 21 contains the 95S replicating viral chromatin. (B) The remainder of fractions 7-23 and fraction 25 was prepared for gel electrophoresis and autoradiography as described in Fig. 3. The horizontal lines indicate the position of the five standard histones as identified on the stained gel.

Enzymatic phosphorylation of H1

For our *in vitro* studies, histone H1 was prepared from calf thymus chromatin. This H1 preparation contained no detectable endogenous phosphate groups. We incubated this H1 preparation with the growth dependent H1 kinase in the presence of γ (^{32}P) ATP and found a transfer of an average of three moles of phosphate to one mole of H1. About 80% of the (^{32}P) phosphate groups were recovered in tryptic peptides containing the serine residues 160, 176 and 199 (numbering according to the H1 sequence published by Jones et al[23]. [The remaining (^{32}P)-phosphate was found on serine residues 36 and 114] (G. Keil and R.K. unpublished). Thus, *in vitro* phosphorylation of histone H1 leads to a modification reminiscent of the *in vivo* phosphorylation during the DNA replication phase of the cell cycle.

Binding of in vitro phosphorylated histone H1 to DNA

To compare their DNA binding properties, we added unmodified and *in vitro* phosphorylated H1 to (^{3}H) labeled supercoiled plasmid DNA. After 10 min at room temperature, the mixtures were passed through nitro-cellulose filters which, under the experimental conditions, retain DNA-protein-complexes but not free DNA. We found that about 10 times more unmodified H1 than phosphorylated H1 was required for filter binding of identical amounts of DNA (Fig. 5). This result does not necessarily mean that phosphorylated H1 has a higher affinity to DNA than unmodified H1. In fact, the experiment, shown in Fig. 6,

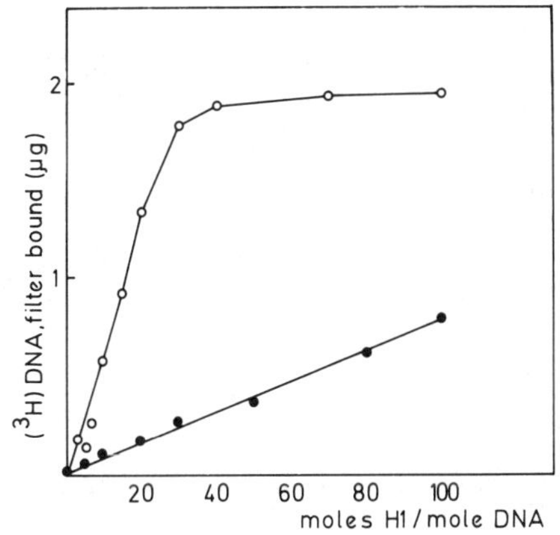

Fig. 5. Filter binding of H1-DNA-complexes.

Samples of 2µg (^{3}H) pBR 322-DNA (spec. radioactivity: 12 000 cpm/µg) in 0.5 ml binding buffer (20mM NaCl, 10mM Tris HCl, 1mM EDTA, 100µg/ml bovine serum albumin, pH 7.6) were mixed with untreated (•) and in vitro phosphorylated (o) histone H1. After 10 min at room temperature, the mixture was passed at a rate of about 1 ml/min through nitrocellulose filters which had been presoaked in binding buffer. The filters were then washed twice with 5 ml each of binding buffer, dried and counted.

clearly excludes this possibility: after sedimentation through sucrose gradients, similar amounts of untreated and of phosphorylated H1 were found to be associated with DNA (Fig. 6). When these complexes were investigated in the filter binding assay, we again found that on nitrocellulose filters, phosphorylated H1 retains more than 90% of the DNA while unmodified H1 retains only 10-20% (data not shown, see ref[24]). These results suggest a different distribution on DNA of phosphorylated and unmodified H1. Under the conditions of the assay, most unmodified H1 binds to a small fraction of the DNA molecules ("cooperative binding", see ref[25], while phosphorylated H1 is distributed more or less randomly on all DNA molecules present in the reaction mixture ("distributive binding"; for a full presentation and discussion of the data, see ref[24]).

Speculations concerning the physiological function of H1 phosphorylation during chromatin replication.

Free H1 is phosphorylated by nuclear protein kinases but DNA bound H1 is not[26]. However, during chromatin replication, both new and old histone H1 are modified by phosphorylation[27].

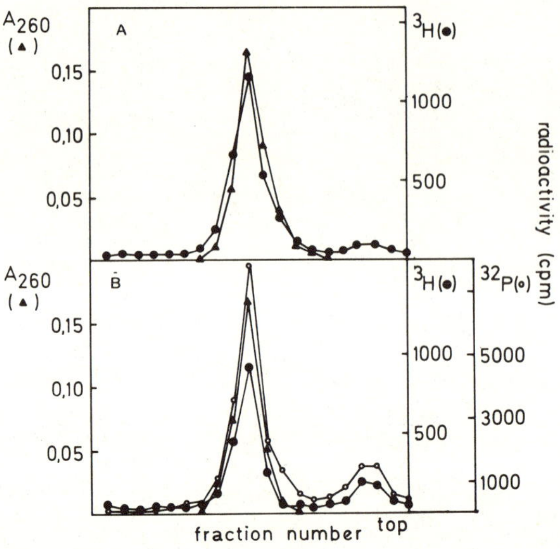

Fig. 6. Sedimentation of H1-DNA-complexes through sucrose gradients.

Non-radioactive pBR 322-DNA and (^{3}H) labeled histone H1 were used in these experiments. (^{3}H) labeled H1 was extracted from lymphocytes, proliferating after mitogen stimulation in the presence of (^{3}H) lysine. (A) About 20μg pBR 322 and 4μg (^{3}H)-H1 were mixed in binding buffer. (B) (^{3}H)-H1 was phosphorylated in vitro with γ (^{32}P) ATP and H1 kinase before addition to pBR 322-DNA. After 20 min at room temperature, the samples were centrifuged through 25% to 5% linear sucrose gradients, buffered by binding buffer.

The DNA was identified in individual fractions by absorbance of ultraviolet light (▲). ^{3}H (●) and ^{32}P (o) label were determined in trichloroacetic acid precipitates.

Does "parental" H1 dissociate from the unreplicated DNA stem, and does it reassociate again with the replicated DNA some distance behind the replication fork?

Several observations are compatible with such a possibility.

1. Chromatin containing pulse labeled DNA (i.e. freshly replicated), is rapidly degraded to monomeric nucleosomes by low concentrations of micrococcal nuclease when a nuclease attack on bulk chromatin is barely detectable[28], showing that the linker region between adjacent nucleosomes is much more exposed in nascent than in nonreplicating DNA as expected if nascent DNA did not contain H1:

2. Monomeric nucleosomes, containing (^3H) thymidine pulse labeled DNA, travel in agarose gels, exactly like core nucleosomes composed of a 145 base pair long DNA piece plus all four inner histones, but without H1 (unpublished data, E.-J. Schlaeger, pers. comm.):
3. Studies on chromatin assembly *in vivo* show that H1 is added to replicated chromatin at some time after the formation of nucleosome cores[29], indicating that the deposition of H1 is a late event in chromatin replication. (See: Fig. 3 and 4).

If, as we suppose, "old" H1 dissociates ahead of the replication fork and reassociates together with "new" H1 as one of the last steps in chromatin maturation, it seems reasonable to conclude that one of the functions of H1 phosphorylation may be connected with the deposition of H1 on newly replicated DNA. For example, phosphorylation could prevent a clustering of H1 molecules on DNA, thereby facilitating the correct interaction of H1 with the nucleosome core and with linker DNA.

CONCLUSION

We have discussed two nuclear DNA binding proteins, a single strand specific protein and histone H1. Both are phosphorylated by specific kinases which are most active during the DNA replication phase. We have shown that the DNA binding properties and other functions of both proteins are affected by phosphorylation. There is no evidence, however, that the increase in negative charges on the protein leads to a decrease in protein-DNA interaction.

ACKNOWLEDGEMENT

This work was supported by Deutsche Forschungsgemeinschaft (SFB 138). We thank Dr. B. Otto for discussions and suggestions.

REFERENCES

1. L. J. Kleinsmith, *in* "The Cell Nucleus", Vol. VI ("Chromatin", Part C) (H. Busch, ed.) Acad. Press, New York, 1978.
2. M. Heil, Diploma-Thesis, Univ. Konstanz, 1981.
3. C. S. Teng, C. T. Teng and V. G. Allfrey, *J. Biol. Chem.* 246, 35-97, 1971.
4. W. Sons, H. J. Unsöld and R. Knippers, *Eur. J. Biochem.* 65, 263-269, 1976.
5. R. Levy, S. Levi, S. Rosenberg and R. Simpson, *Biochemistry*, 12, 224, 1973.
6. G. S. Stein, T. C. Spelsberg and L. J. Kleinsmith, *Science*, 183, 817-824, 1974.

7. V. G. Allfrey, in "Chromatin and Chromasome Structure" (H. J. Li and R. Eckhardt, eds.), Acad. Press, N. Y. 1977.
8. L. J. Kleinsmith and V. G. Allfrey, Biochim. Biophys. Acta 175, 123-135, 1969.
9. E. G. Krebs and J. A. Bearo, Ann. Rev. Biochem. 48, 923-959, 1979.
10. J. Schlepper and R. Knippers, Eur. J. Biochem. 60, 209-220, 1975.
11. H. Stahl and R. Knippers, Biochim. Biophys. Acta. 614, 71-80, 1980.
12. R. S. Lake and N. P. Salzman, Biochemistry 11, 4817-4823, 1972.
13. R. A. Masarachia, B. E. Kemp and D. A. Walsh, J. Biol. Chem.252, 7109-7117, 1977.
14. J. P. Whitlock, R. Augustine and H. Schulman, Nature 287, 74-76, 1980.
15. B. Otto, M. Baynes and R. Knippers, Eur. J. Biochem. 73, 17-24, 1977.
16. J. Szopa and H. Janska, Hoppe-Seyler's Z. Physiol. Chemie 361, 1235-1241, 1980.
17. Y. Hotta and H. Stern, Eur. J. Biochem. 95, 31-38, 1979.
18. A. Richter, R. Knippers and B. Otto, FEBS Lett. 91, 293-296, 1978.
19. P. Hohmann, R. A. Tobey and L. R. Gurley, J. Biol. Chem. 251, 3685-3692, 1976.
20. E. M. Bradbury, R. J. Inglis, H. R. Matthews and T. A. Laugan, Nature 249, 553-556, 1974.
21. M. L. DePamphilis and P. M. Wassarman, Ann. Rev. Biochem. 49, 627-666, 1980.
22. E. Fanning and I. Baumgartner, Virology 102, 1-12, 1980.
23. G. M. T. Jones, S. C. Rall and R. D. Cole, J. Biol. Chem. 249, 2548-2553, 1974.
24. R. Knippers, B. Otto and R. Böhme, Nucl. Ac. Res. 5, 2113-2131, 1978.
25. M. Renz and L. Day, Biochemistry 15, 3220-3228, 1976.
26. J. Böhme, G. Keil and R. Knippers, Eur. J. Biochem. 78, 251-266, 1977.
27. V. Jackson, A. Shires, N. Tanphaichitr and R. Chalkley, J. Mol. Biol. 104, 471-483, 1976.
28. K. H. Klempnauer, E. Fanning, B. Otto and R. Knippers, J. Mol. Biol. 136, 359-374, 1980.
29. A. Worcel, S. Hans and M. L. Wong, Cell 15, 969-977, 1978.

STRUCTURE AND REPLICATION OF SV40 CHROMATIN

M. Yaniv*, S. Saragosti*, and G. Moyne** †

*Department of Molecular Biology
Institut Pasteur
25, rue du Dr. Roux
75015 Paris

**Institut de Recherches Scientifiques sur le Cancer
B.P. n° 8, 94802 Villejuif Cedex

1. HIGH RESOLUTION ELECTRON MICROSCOPY STUDIES OF CHROMATIN

The DNA of SV40 is associated with cellular histones in a chromatin structure in the virions and in the nuclei of infected cells[1]. Biochemical and electron microscopy studies have shown that viral chromatin, and not pure DNA is the template for the replication machinery in vivo[2-7]. The process of SV40 replication should thus involve replication of the DNA double strand found in a nucleosomic structure and the doubling of the histone content in the daughter minichromosomes. With the aim of obtaining more information on the structure of the SV40 chromatin, we undertook high resolution electron microscopy studies of viral chromatin[8]. SV40 minichromosomes were prepared from nuclei of infected cells by the method of Varshavsky et al.[9] in the presence of Triton X-100. After purification on sucrose gradients, the chromatin was adsorbed to activated carbon films by the technique of Dubochet et al.[10]. The samples were briefly stained with aqueous uranyl acetate and observed directly, either by conventional dark field microscopy or with scanning transmission electron microscopy at a magnification of up to x500,000. Figure 1 demonstrates the quality of images obtained by these techniques. Under the conditions employed, the uranyl acetate stains mainly the DNA and not the proteins. It is obvious that the DNA is found on the outside of the nucleosome, in agreement with neutron and X-ray diffraction data[11-13]. The stain in the periphery of the nucleosomes is thicker than in the internucleosomal segments, suggesting that the DNA is wrapped twice around the histome octamer core. The high quality images that we

† Gilles Moyne died accidentally in September 1981.

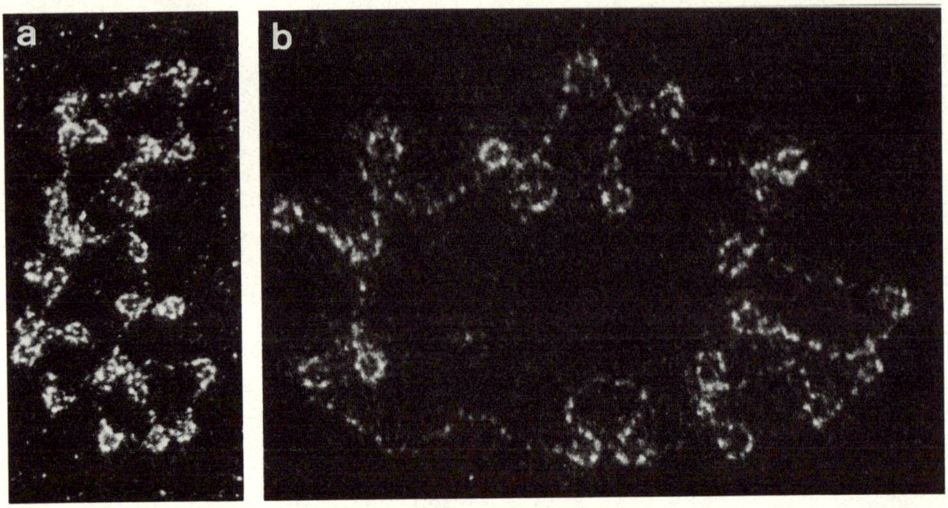

Fig. 1. SV40 minichromosomes displaying or not displaying a stretch of DNA Devoid of Nucleosomes (Gap).
(A) Example of minichromosome with no visible gap. (B) Minichromosome displaying an obvious gap. The gap is always single, with a variable length. In this case, the gap is 225 bp long. Magnification: A. x600.000 B. x426.000

obtained and the knowledge of the exact number of nucleotides in SV40 DNA[14], permitted a more accurate determination of the DNA content of the nucleosomes than previously obtained. Two approaches were used to calculate this value. In the first, we measured the contour length of the internucleosomal linker DNA, deduced it from the total length of SV40 DNA spread under the same conditions, and divided the value obtained by the number of nucleosomes. A value of 162±8 b.p. per nucleosome was obtained for minichromosomes observed by DF, and a value of 163±13 b.p. for minichromosomes observed by STEM. In the second approach, we measured the length of the internucleosomal DNA in a randomly selected population of minichromosomes. In a parallel manner, the mean number of nucleosomes per minichromosome was determined in a population that permitted the precise count of the number of nucleosomes. The knowledge of these two values and that of the number of nucleotides in SV40 DNA permitted us to calculate a mean value of 163 b.p. per nucleosome. This value is in good agreement with that calculated by the first method[8]. In all these studies, we measured the DNA content in the particles defined by electron microscopy from the entry point to the exit point of the DNA. Most frequently, these two points coincided. This structure may, in fact correspond to the core particle plus the 20 b.p. which are involved in the interaction with histone H1 and which are partially resistant to micrococcal nuclease even in its absence[15-19]. This structure is different from the nucleosome which represents the repeat unit of the chromatin, including the linker DNA as deduced by nuclease digestion. The particles we observed are ovoid with a long axis of 8.9nm and a small axis of 7.9nm[8].

2. NON RANDOM DISTRIBUTION OF HISTONES ON SV40 DNA

Careful examination of electron micrographs of SV40 chromatin revealed that about 20% of these complexes contained a nucleosome free DNA segment comprising 250-400 b.p.[20,21]. The thickness of the DNA of this segment did not exceed that of free plasmid DNA observed under the same conditions. The mean number of nucleosomes (24±0.2) on complexes with or without free segment (gap) was identical[20]. Thus, the free DNA segment is not generated by a loss of histones during the purification, nor by the "melting" of a nucleosome. To verify whether this structure may have a unique position on SV40 DNA, we cleaved the minichromosomes with different restriction endonucleases that cut SV40 DNA once (see figure 2). The linear molecules generated were observed by electron microscopy and the location of the gap determined. After cleavage with Bam H1, which cleaves SV40 DNA at 0.14 map units, the gap was observed in the middle of the linear chromatin structure. Cleavage with BglI, which cleaves SV40 DNA at the origin (0.67), generated linear molecules with free DNA at one end. The nucleosome free region is hence located in close proximity to the origin of replication, but is non-symmetrically relative to that point[20]. To more precisely localize this region, and to confirm its existence in situ, we examined the DNAase I sensitivity of the SV40 chromatin in nuclei isolated from infected cells. Following mild digestion of the nuclei for increasing time, the total infected cell DNA was purified and fractionated on agarose gels before and after cleavage with different restriction enzymes. After transfer to DBM paper, SV40 DNA was detected by hybridization with viral DNA labeled with $|^{32}P|$ by nick translation. During DNAase I digestion, the superhelical closed circular viral DNA is conve-ted into nicked circles, and then into linear molecules before being degraded to shorter fragments by the DNAase I. Cleavage of the linear material with a restriction enzyme like Eco R1, which cleaves SV40 once at 0.0 coordinates, will generate discrete fragments if the initial double strand cleavage was in a unique site. Indeed, we found fragments of 67% and 33% genome length, confirming that the region in proximity to the origin of replication is sensitive to DNAase I in the nuclei[20]. Cleavage with several restriction enzymes (Hpa II, Taq 1, Bgl I) which cut closer to the gap permitted a more accurate mapping of the DNAase I sensitive sites. This region starts at about nucleotide 5210 (30 b.p. to the early side of the origin and extends to nucleotide 370 on the late side of the origin, altogether 400 b.p.[22] This value and the location of the DNAase I sensitive region are in good agreement with those observed by electron microscopy. The DNAase I sensitive region is not homogenous; it contains several hypersensitive sites where the frequency of cleavage is higher (Figure 3). These sites are interrupted by relatively resistant segments. The

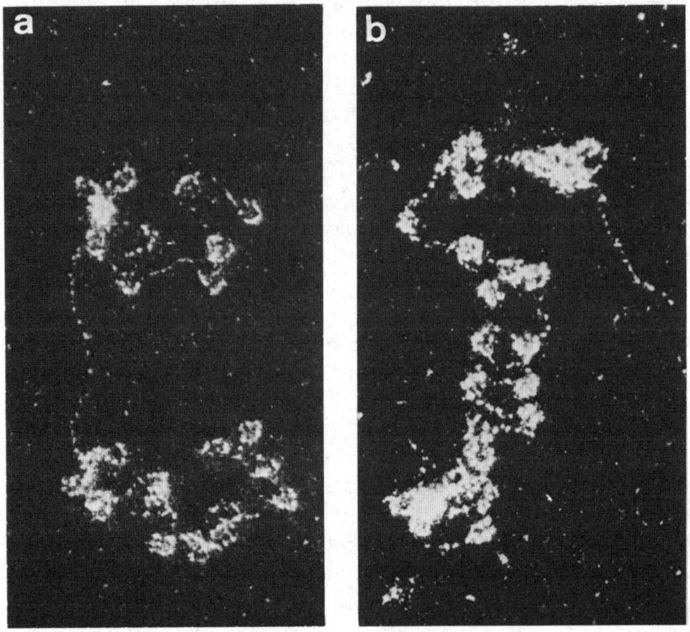

Fig. 2. Minichromosomes Digested with Single-Cut Restriction Endonucleases.
(A) Chromatin digested with Bam HI. The gap occupies a central position. (B) Chromatin digested with Bgl I. The gap is in a distal position.

product of the A gene of SV40, the large T antigen, was shown to bind to three sites on SV40 DNA: at the early side of the origin, on the origin, and on the late side of the origin[23]. The three binding sites are included in the DNAase I sensitive segment.

Our present findings do not permit us to specifically link the nucleosome free region to the process of replication. In fact, in the case of polyoma virus infected cells, we also found a DNAase I sensitive region that was localized between the origin of replication and the beginning of the sequence coding for the late viral protein VP2[24]. The origin is just barely included in this sensitive segment. Further studies on the mechanism of the initiation of replication of SV40 or polyoma DNA are necessary in order to understand the possible role of the nucleosome free DNA segment in the control of viral DNA replication. However, our present studies show that in certain cases, histone may be assembled on specific sequences and not at random. We do not know what the constraints are which keep nucleosomes out of this region of the SV40 DNA. It is possible that nonhistone cellular proteins are associated with at least a part of this DNA segment. These proteins may be responsible for the existence of DNAase I

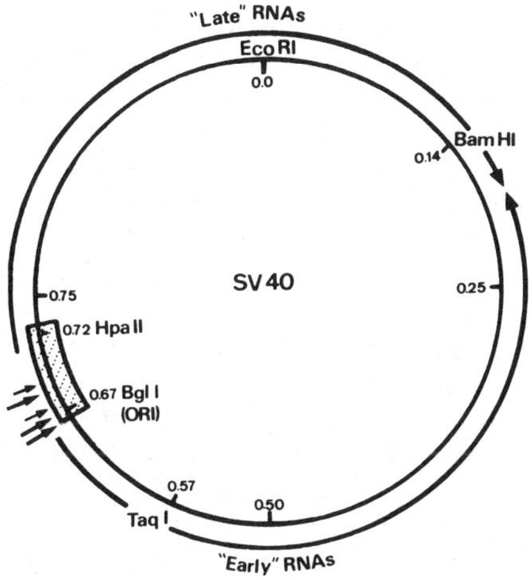

Fig. 3. Physical Map of SV40 DNA and Localization of the Hypersensitive Sites.
The origin of replication (ORI) is coincident with the Bgl I site. The dotted box indicates the extent of the sensitive region, and the arrows the localization of the hypersensitive sites.

hypersensitive sites in this region. They would participate in the processes of transcription and (or) replication, eventually via attachment to the nuclear matrix[25].

3. NUCLEOSOME STRUCTURE AND REPLICATION FORK PROGRESSION

The process of DNA replication in the eukaryotic organisms requires at least the partial dissociation of the nucleosomal structure to permit the progression of the replication fork. Extensive evidence has been accumulated showing that histones remain associated with the DNA during this process: eg. we have shown that the density of replicative intermediates is not very different from that of mature viral chromosome[2]. EM studies have shown that replicative intermediates contain nucleosomes on both the parental and the newly replicated segments[4,26]. SV40 or polyoma nucleoprotein complexes extracted from infected cells contain replicative intermediates that can be elongated <u>in vitro</u>[5-7]. Replication in the presence of protein synthesis-inhibitors such as puromycin, generates complexes with a mean number of 12 nucleosomes - suggesting that parental nucleosomes are distributed between the two daughter molecules[4].

Furthermore, the work of Seidmann et al.[27] has shown that the parental histone octamers are segregated with the DNA strand that is replicated continuously. The folding of two turns of DNA around the histone core creates physical constraint on the progression of the replication machinery. A process that will unfold the nucleosome without dissociation of the histones will facilitate the replication process. Covalent modification of the histones like acetylation or phosphorylation changes the net positive charge of the N-terminal segment of these proteins and should affect the stability of the nucleosome structure[28]. After short pulses of H^3 acetate[29] or of P^{32}-phosphate[30], we isolated the viral minichromosomes and compared the level of modification of the replicative and the mature chromatin. We could not show preferential incorporation of the radioactivity in the replicative intermediates. Under the same conditions, roughly 50% of [^{3}H]-thymidine was incorporated in the replicative intermediates[3]. Although these experiments do not exclude the possible role of covalent modification of histones during replication, they render them unlikely. One can speculate that either the replicating complex can unfold the nucleosomes during fork progression or alternatively, that other nuclear proteins are responsible for this unfolding.

4. GENERATION OF NONCOMPACTED STRUCTURES FROM SV40 CHROMATIN

Several methods were used in the past to dissociate SV40 or polyoma virions, and to show that the viral DNA is associated with histones in a chromatin-like structure. Most of these techniques were performed at slightly basic pH (9.5-10.5); the viral chromatin thus obtained can be separated from the capsid proteins by sedimentation on a sucrose gradient[31-34]. Recently, milder conditions were developed by Brady et al.[35,36]. They require the treatment of virions with EGTA and DTT, to remove calcium, an essential constituent of the virions, and to reduce the S-S bands forming the network of the major capsid protein VP1[37]. When we performed the dissociation under these conditions at pH 9.5, we obtained a minichromosome structure devoid of H1 sedimenting at 70 S and a peak of free viral capsid proteins at the top of the gradient. On the contrary, when the dissociation was performed at pH 7.5, a unique species with an S value of 110 S, containing all the components of the virions, was detected[38,39]. This complex differs from intact nontreated virions that sediment at 240 S. Both EGTA and DTT were required for the formation of this complex. Previously, similar complexes were described for polyoma and SV40 by Brady et al.[35,36]. We next observed these structures by EM (Figure 4). The 70 S complex obtained at pH 9.5 was similar to minichromosomes containing histones and DNA extracted from infected cells. Surprisingly the complexes obtained at pH 7.5 did not show any nucleosomal structure, they appeared as relaxed circular structures containing 70-80 particles. These particles were smaller than the nucleosomes ($\sim$70nm) in diameter, they did not show the peripheral uranyl acetate staining typical of the nucleosomes. The contour length of these

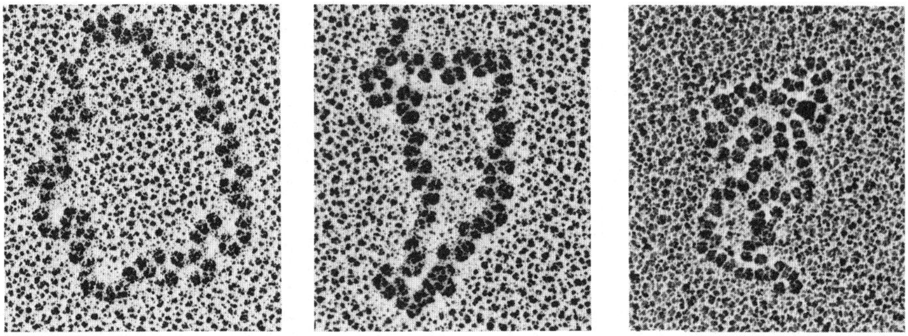

Fig. 4. Nucleoprotein Complexes Obtained by Mild Dissociation of SV40 Virus Particles.
Electron microscopic observations show a bead-on-a-string structure comprising about 70 particles 7nm in diameter along the viral DNA. No particle of nucleosome size was visible. The compaction of DNA in this complex is probably close to unity.

structures was roughly equal to that of free DNA, thus showing no compaction of the DNA. In this respect, these particles are markedly different from half nucleosomes observed previously after treatment of SV40 chromatin with very low ionic strength[40]. Furthermore, the treatment of SV40 chromatin isolated from infected cells under identical conditions (DTT, EGTA, pH) did not generate half nucleosomes. Careful observations of the circular structures generated from the virions show that the particles are arranged along the DNA. These particles seem homogenous and resemble the capsomers of the virion. The number of particles is roughly equal to that of capsomers in the virion[42]. It is clear from our studies of the protein content of these complexes that the full complement of the viral proteins is present in these complexes (results not shown). We deduce that the nucleosomes were unfolded by interaction with the reduced viral proteins in the absence of calcium ions. The unfolded structure obtained may be a better template for transcription and replication. Recent studies by Brady et al. have shown that a similar complex obtained from SV40 was a much better template for the E. coli RNA polymerase than the viral minichromosomes[36]. Upon infection, the environment in the cell may reduce the disulfide bands between the viral capsomers, remove the calcium, and generate a structure that will be efficiently transcribed and replicated in the nuclei.

The properties of the viral capsid proteins observed here may be shared by other non-histone proteins found in the nuclei. This family of proteins can interact with segments of the chromatin during transcription or replication, and permit the transient generation of unfolded structures. In fact, several reports on EM studies of ribosomal genes have suggested a non-nucleosomal expanded structure for these genes slightly before and during transcription[41,42].

CONCLUSIONS

Analysis of the lytic cycle of the circular double stranded papovaviruses like SV40 or polyoma is an extremely useful tool for the study of the chromatin and the processes of transcription and replication. In the present report, we have shown that high resolution electron microscopy can give more details on the structure of the viral chromatin. A segment of DNA of 250 to 400 bp starting from the origin and extending towards the late region is not assembled into nucleosomes. This segment exhibits a unique DNAase I sensitivity pattern that suggests possible interaction with other nuclear components (matrix eg.). This interaction may be required for its transcription and replication.

Our studies on the protein-DNA complexes obtained by mild treatment of the virions in the presence of EGTA and DTT at neutral pH showed that chromatin unfolding can be induced by the viral capsid proteins. We suggest that similar interactions between the cellular chromatin and other nuclear proteins may be involved in the processes of transcription and replication in the eukaryotic cell.

REFERENCES

1. J. Tooze (ed.), "The molecular biology of tumor viruses", part 2 DNA tumor viruses, Cold Spring Harbor Laboratory, Cold Spring Harbor, New York, 1980.
2. C. Cremisi, P. F. Pignatti, O. Croissant and M. Yaniv. J. Virol. 17, 204, 1976.
3. C. Cremisi, A. Chestier and M. Yaniv. Cell 12, 947, 1977.
4. C. Cremisi, A. Chestier and M. Yaniv. Cold Spring Harbor Symp. Quant. Biol. 42, 409, 1977.
5. R. T. Su and M. L. DePamphilis. Proc. Natl. Acad. Sci. U.S.A. 73, 3456, 1976.
6. H. J. Edenberg, M. H. Wagar and J. A. Huberman. Nucl. Acids Res. 4, 3083, 1977.
7. W. Waldeck, U. Spaeren, G. Mastromei, R. Eliasson and P. Reichard. J. Mol. Biol. 135, 675, 1979.
8. G. Moyne, S. Saragosti and M. Yaniv. J. Mol. Biol. in press.
9. A. J. Varshavsky, S. A. Nedospasor, V. V. Schmatchenko, V. V. Bakayer, P. M. Chumackov and G. P. Georgiev. Nucl. Acids Res. 4, 3303, 1977.
10. J. Dubochet, M. Ducommun, M. Zollinger and E. Kellenberger. J. Ultrastruct. Res. 35, 147, 1971.

11. R. P. Hjelm, J. J. Kneale, P. Suau, J. P. Baldwin, E. M. Bradbury, and K. Ibel. Cell 10, 139, 1977.
12. J. T. Finch, L. C. Lutter, D. Rhodes, R. S. Brown, B. Rushton, M. Levitt and A. Klug. Nature 269, 29, 1977.
13. J. F. Pardon, D. L. Worcester, J. C. Wooley, R. I. Cotter, D. M. J. Lilley and B. M. Richards. Nucl. Acids Res. 4, 3199, 1977.
14. H. Van Heuverswyn and W. Fiers. Europ. J. Biochem. 100, 51, 1979.
15. G. Felsenfeld. Nature 271, 115, 1978.
16. R. T. Simpson. Biochemistry 17, 5524, 1978.
17. A. Ruiz-Carrillo, J. L. Jorcano, G. Elder and R. Lurz. Proc. Natl. Acad. Sci. U.S.A. 76, 3284, 1979.
18. W. O. Weischet, J. R. Allen, G. Riedel and K. E. Van Holde. Nucl. Acids Res. 6, 1843, 1979.
19. J. Allen, P. G. Hartman, C. Crane-Robinson and F. X. Aviles. Nature 288, 675, 1980.
20. S. Saragosti, G. Moyne and M. Yaniv. Cell 20, 65, 1980.
21. E. B. Jakobovits, S. Bratosin and Y. Aloni. Nature 285, 263, 1980.
22. S. Saragosti and M. Yaniv, manuscript in preparation.
23. R. Tjian. Cell 13, 165, 1978.
24. P. Herbomel, S. Saragosti, D. Blangy and M. Yaniv. Cell, in press.
25. A. J. Buckler-White, G. W. Humphrey and V. Pigiet. Cell 22, 37, 1980.
26. M. M. Seidman, C. L. Garon and N. P. Salzman. Nucl. Acids Res. 5, 2877, 1978.
27. M. M. Seidman, A. J. Levine and H. Weintraub. Cell 18, 439, 1979.
28. A. J. Luoie, E. P. M. Candido and G. H. Dixon. Cold Spring Harbor Symp. Quant. Biol. 38, 803, 1973.
29. A. Chestier and M. Yaniv. Procl. Natl. Acad. Sci. U.S.A. 76, 46, 1979.
30. I. Iceckson and M. Yaniv. unpublished observations.
31. A. J. Louie. Cold Spring Harbor Symp. Quant. Biol. 39, 259, 1975.
32. G. Christiansen, T. Landers, J. Griffith and P. Berg. J. Virol. 21, 1079, 1977.
33. B. A. J. Ponder and L. V. Crawford. Cell 11, 35, 1977.
34. E. S. Huang, M. K. Esles and J. B. Pagano. J. Virol. 9, 923, 1972.
35. J. N. Brady, V. D. Winston and R. A. Consigli. J. Virol. 27, 193, 1978.
36. J. N. Brady, C. Lavialle and N. P. Salzmann. J. Virol. 35, 371, 1980.
37. G. Walter. Cold Spring Harbor Symp. Quant. Biol. 39, 255, 1974.
38. G. Moyne, F. Harper, S. Saragosti and M. Yaniv. C. R. Acad. Sci. Paris 292, 749, 1981.
39. G. Moyne, F. Harper, S. Saragosti and M. Yaniv, in preparation.
40. P. Oudet, C. Spadafora and P. Chambon. Cold Spring Harbor Symp. Quant. Biol. 42, 301, 1977.
41. V. E. Foe. Cold Spring Harbor Symp. Quant. Biol. 42, 723, 1977.
42. W. W. Franke, U. Scheer, M. Tredelenburg, H. Zentgrat and H. Spring. Cold Spring Harbor Symp. Quant. Biol. 42, 755, 1977.

CHROMATIN STRUCTURE, DNA SEQUENCES AND REPLICATION PROTEINS: SEARCHING FOR THE PRINCIPLES OF EUKARYOTIC CHROMOSOME REPLICATION

Melvin L. DePamphilis, Michael E. Cusick, Ronald T. Hay
Cynthia Pritchard, Lois C. Tack, Paul M. Wassarman and
David T. Weaver

Department of Biological Chemistry
Harvard Medical School
25 Shattuck Street
Boston, Massachusetts 02115 USA

INTRODUCTION

Our objective has been to understand, at the molecular level, the replication of eukaryotic chromosomes as a prerequisite to investigating the regulation of chromosome replication and its relationship to cell proliferation. Chromosomes from eukaryotic cells, simian virus 40 and polyoma virus consist of a series of nucleosomes, each containing about 200bp of DNA coiled around an octamer consisting of two each of the four core histones that gives chromatin its stable repeating structure. Additional structure is provided by histone H1 and nonhistone chromosomal proteins. However, chromatin is not homogenous. The chromatin structure of quiescent genes differs from that of active genes, and chromosomal regions differ in their higher-ordered structure. For example, chromatin can be euchromatic or heterochromatic, 10nm thick fibers or 30nm thick coils, folded into loops or not folded. Therefore, replication of eukaryotic chromosomes, in contrast to prokaryotic chromosomes, requires accurate duplication of the structure of both chromatin and DNA. Furthermore, both chromatin structure and DNA sequence are likely involved in controlling replication by modifying the interaction of proteins with DNA. For example, nonuniformly spaced clusters of replication origins are activated at different times during S phase, but the same origin is never activated twice, and some origins that are active in embryonic cells are not used in somatic cells[9]. Once a replicon is activated, replication proteins must be directed to replication forks and termination of replication must result in complete separation of sibling chromatids to allow normal mitosis.

We have focused our attention on native replication forks with the expectation that a detailed analysis of their life cycle will reveal the principles of replication. To this end, we have asked three basic questions: (i) What is the sequence of structural and catalytic events that occur at native replication forks? (ii) How is chromatin structure and assembly related to DNA replication? (iii)How are events in replication modulated by DNA sequences? To answer these complex questions, the simian virus 40 (SV40) chromosome was utilized as a relatively simple, but appropriate, model for mammalian replicons.

SV40 replicates in the nuclei of African Green monkey kidney cells as a small circular chromosome containing covalently-closed duplex DNA organized into nucleosomes whose structure and histone composition are indistinguishable from those of its host[9,38,39] (and M. Yaniv, this volume). The SV40 genome contains 5243 bp of completely sequenced DNA[50,53]. With the exception of initiation, all subsequent steps in viral DNA replication and chromatin assembly appear to be carried out by the host cell[50]. Furthermore, the final stages in replicon maturation also appear to be the same for both virus and cell since the topological problems in separating two sibling viral chromosomes are analogous to the merger of two adjacent replicons; the ability to rotate one DNA strand about the other is as restricted in an "infinitely" long linear DNA molecule as it is in a circular, covalently-closed molecule (Figure 1).

This report summarizes our current view of replicating SV40 chromosomes, emphasizing the relationships between chromatin structure, DNA sequence, replication proteins, and the sequence of events that takes place at native replication forks.

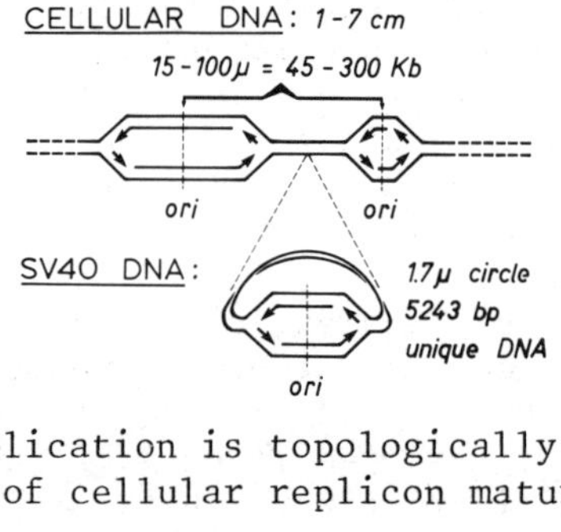

Fig. 1. SV40 DNA replication is topologically equivalent to the final 2-10% of cellular replicon maturation.

RESULTS

SV40 DNA Replication Cycle (Figure 2)

During the period of maximum viral DNA synthesis, replication is initiated at a genetically defined site (ori) on the covalently-closed,

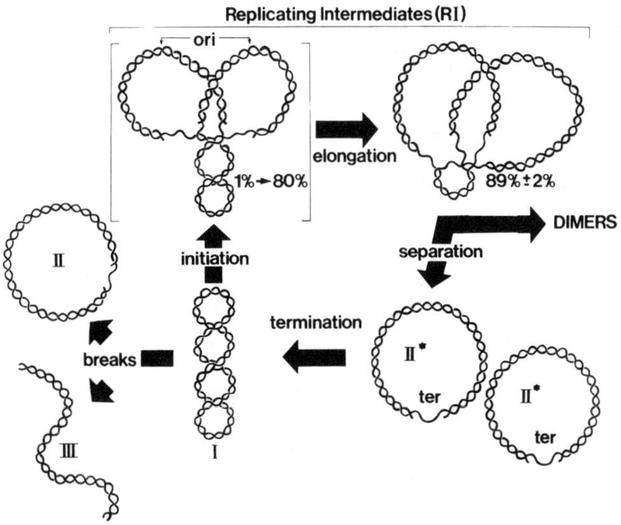

Fig. 2. SV40 DNA replication cycle during the period of maximum viral DNA synthesis. DNA forms are identified as I, II, II*, III and RI.

superhelical circular form of SV40 DNA (SV40(I) DNA). Replication then proceeds bidirectionally in the replicating intermediates (SV40 (RI) DNA) until the two forks arrive in a termination region (ter) approximately 180° from the origin and separation of sibling molecules takes place[14,22,26,47-50]. The nonreplicated portion of SV40(RI) DNA is superhelical and therefore must be covalently-closed. Presumably, a topoisomerase allows regular nicking and resealing to occur in front of the forks to permit DNA unwinding. In contrast to initiation, separation does not require a unique DNA site, although it may be promoted at many preferred DNA sites[47]. Replicating intermediates that are 90% completed (SV40(RI*) DNA) accumulate to a 2-3 fold excess over molecules at early stages in replication, suggesting a rate limiting step in separation of sibling molecules[36,46,47,49]. A similar rate limiting step in cellular DNA replication is implied by the accumulation of replicon size nascent DNA chains[9]. Separation of sibling molecules generates circular DNA monomers with their nascent DNA strand interrupted by a short (about 50 nucleotides) gap in the termination region[14,22,48,49,50]. When this gap is completed and the molecule sealed, replication terminates with the production of SV40(I)DNA. Since newly replicated DNA is rapidly assembled into nucleosomes, SV40 (I) DNA must contain superhelical twists when the chromosomal proteins are removed. Circular DNA monomers with apparently randomly placed interruptions in either strand (SV40(II) DNA) and linear monomers (SV40(III) DNA) may result from either damage incurred during DNA purification or intracellular nuclease activity as infected cells deteriorate. Catenated dimers represent an alternate pathway[44] in which SV40(RI*) DNA fail to separate properly[49] and then must be resolved by either recombination or a topoisomerase surveillance mechanism.

With the exception of initiation of DNA replication, the pathways of elongation, separation, and termination can all be faithfully continued in cell lysates[8], isolated nuclei[7], and soluble nuclear extracts[36,42,43] supplemented with cytosol. Nascent cellular[37] and viral[39] DNA is also assembled into nucleosomes under these conditions. These systems, together with intact virus infected cells, were used to analyze the events at replication forks.

Sequence of Events at SV40 DNA Replication Forks (Figure 3)

The sequence of events at SV40 replication forks appears to be typical of replication forks in both polyoma and mammalian chromosomes[9]. Discontinuous DNA synthesis occurs through a repeated initiation of short pieces of nascent DNA ("Okazaki fragments") about once every 135bp[1] with 80%[30] to more than 90%[3,5,19,21] originating from retrograde DNA templates. Therefore, DNA synthesis on forward arms of replication forks where the direction of synthesis is the same as the direction of fork movement is essentially a continuous process. The ratio of the number of Okazaki fragments to long nascent DNA chains reveals an average of about one Okazaki fragment per fork. Syn- IA thesis of Okazaki fragments is initiated on the 3'-termini of Synthesis of Okazaki fragments is initiated on the 3'-termini of uniquely-sized oligoribonucleotides (9±1 residues) of near-random base composition except for either an A or G ribonucleoside triphosphate at their 5'-ends[2,13,20,52]; essentially all SV40 RNA primers originate from retrograde DNA templates[19]. RNA primers are found on 30-50% of Okazaki fragments of all sizes, as well as on long nascent DNA[9]. Over 90% of Okazaki fragments are found in one of three sequential steps in their metabolism[1]: (i) About 20% are separated from longer nascent DNA chains by a phosphodiester bond interruption ("nick") that can be sealed with DNA ligase alone. (ii) About 30% are separated by a gap of single-stranded DNA and require both DNA polymerase and DNA ligase for joining. (iii) The remaining 50% face an RNA primer and require 5'-3' exonuclease, DNA polymerase, and DNA ligase activities to join to growing DNA strands. Excision of RNA primers occurs concurrently with synthesis and ligation of Okazaki fragments via a two-step mechanism that leaves 5'-terminal phosphates on nascent DNA[1,2]. Excision of the bulk of the RNA primer is independent of either concomitant DNA synthesis or joining of Okazaki fragments to longer DNA, while removal of the actual prN-p-dN$(pdN)_n$ linkage is accelerated 2-3 fold by DNA synthesis.

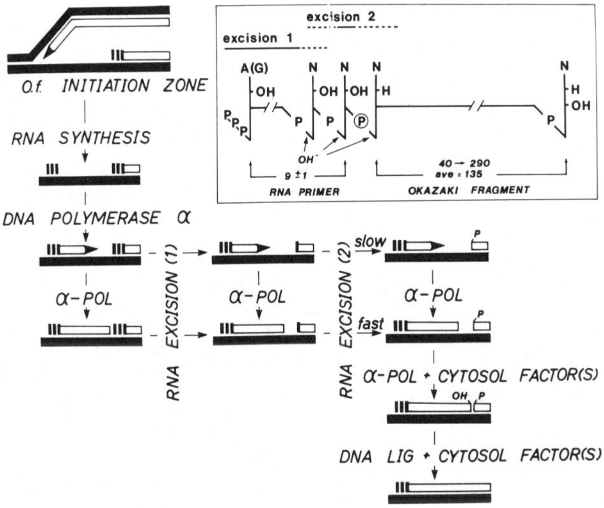

Fig. 3. Schematic representation of the metabolism of Okazaki fragments. Initiation of Okazaki fragments (O.f) occurs stochastically within an "initiation zone" via de novo synthesis of RNA primers (|||) complementary to the retrograde DNA template (▬). Within this zone, initiation events are promoted at preferred DNA sequences. The maturation of Okazaki fragments (▭▶) is represented on the vertical coordinate in four distinct steps: synthesis of RNA primers, elongation of DNA by DNA polymerase α, incorporation of the final deoxyribonucleotides (gap-filling), and joining of Okazaki fragments to growing DNA chains (ligation). Excision of RNA primers is represented on the horizontal coordinate in two distinct steps: excision of the bulk of the RNA which is independent of Okazaki fragment synthesis, and removal of the p-rN-p-dN-(pdN)n junction which is facilitated by concomitant DNA synthesis. Excision of all RNA primers is presumed to take place concurrently. A single example is illustrated for simplicity. Inset shows the structure of RNA primed Okazaki fragments with sizes given in nucleotides. N represents A, G, C, or T. Cleavage sites during alkaline hydrolysis are indicated by arrows. Excisions 1 and 2 refer to regions excised (solid bars) during replication; possible extensions of these regions are indicated by broken bars.

All DNA synthesis on both the forward and retrograde arms is normally carried out by DNA polymerase α[5,11,24,54]. All steps in DNA synthesis are inhibited by aphidicolin, a specific inhibitor of α-polymerase that has no effect on DNA ligase, but not by ddTTP which can selectively inhibit β- and γ- polymerases. Reconstitution experiments showed that neither β- nor γ- polymerase could substitute for α-polymerase in restoring the bulk of SV40 DNA replication in N-ethylmaleimide inactivated nuclear extracts. Completion of Okazaki fragments (gap-filling) is a unique step[1] requiring, in addition to α-polymerase, a protein cofactor(s) that is easily washed out of isolated nuclei and is found in cytosol from uninfected cells[5-7].

Two key observations support a stochastic model[1,4,5,6,9] for initiation of Okazaki fragments in which nascent DNA chains have a certain probability of being initiated at many sites within an approximately 290 nucleotide region of exposed retrograde DNA template ("Okazaki fragment initiation zone"). First, mature Okazaki fragments are not uniform in length, but vary from 40-290 nucleotides[1,5]. Therefore, since Okazaki fragments originate predominantly from retrograde DNA templates, they are not initiated at uniform intervals, but must be initiated at many distances from the 5'-ends of long nascent DNA strands. Second, all 16 possible RNA-DNA covalent linkages are found in replicating SV40 DNA at frequencies consistent with an essentially random distribution throughout the genome[2,5]. Therefore, since RNA primers are uniform in length, they must begin at a large number of different sites on the DNA template. In fact, the 5'-ends of nascent DNA chains have been located at many specific DNA sites throughout the genome[6,15,47], demonstrating that initiation of DNA synthesis does not occur randomly around the genome, but is promoted at many preferred DNA sequences. RNA primers are synthesized by an enzyme that is resistant to α-amanitin[9,19] and capable of partially substituting deoxyribo- for ribonucleotides[13,51].

Such a model explains the observation that the fraction of Okazaki fragments containing RNA at their 5'-ends is essentially the same for long fragments as for shorter ones[13,20]. Both "young" and "old" DNA chains are represented by nascent DNA of all sizes. Naturally, the shortest Okazaki fragments (10-30 nucleotides) contain a higher proportion of RNA primers[12,20] since they are enriched for newly initiated DNA chains. Assuming that the probability of removing an RNA primer increases with time after its synthesis, a fraction of the long nascent DNA strands would have RNA at their 5'-ends because some Okazaki fragments would be initiated close enough to allow ligation to occur before RNA primer excision. The question remained, however, as to what determines the size of an initiation zone; why are Okazaki fragments initiated an average of once every 135bp? Therefore, we turned our attention to the structure of replicating chromatin.

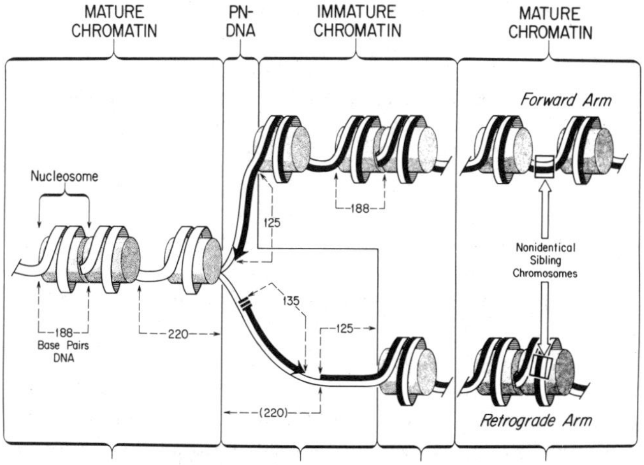

Fig. 4. Model for replication forks in SV40 chromosomes. Nascent DNA is represented by a black ribbon with arrowheads indicating the growing 3'-ends and three perpendicular bars indicating an RNA primer. An average of one Okazaki fragment per fork is represented on the retrograde arm. Numbers give the average distance in nucleotides. A nucleosome is a 110 A x 22 A cylinder consisting of $1\frac{3}{4}$ turns of duplex DNA (20 A in diameter) coiled around a histone octamer[65].

Chromatin Structure at Replication Forks (Figure 4)

Newly replicated SV40 DNA, like that of its host, is rapidly assembled into nucleosomes both in whole cells[9,17,23] and subcellular systems[3,37,39]. The structure of replication forks in native replicating SV40 chromosomes can be divided into at least four regions: (i) prereplicative mature chromatin, (ii) prenucleosomal DNA that encompasses the actual sites of DNA synthesis and includes Okazaki fragments, (iii) immature chromatin that encompasses regions of newly assembled nucleosomes that are more sensitive to nonspecific endonucleases than are regions of mature chromatin, and (iv) postreplicative mature chromatin consisting of newly assembled nucleosomes that cannot be distinguished from nucleosomes in nonreplicating, mature chromosomes. Nucleosomes of mature SV40 chromosomes are indistinguishable in sedimentation, histone composition (including H1) and nuclease sensitivity from those of their monkey cell host; each contains an average of 188 bp DNA coiled around an octamer of histones H2A, H2B, H3 and H4, with a "core" consisting of 146bp[38,39]. Improved electron microscopic techniques have revealed 24±0.2 nucleosomes per mature SV40 chromosomes[34] giving an average distance of

219 bp from one nucleosome to the next (5243 bp per genome[53]). Since these nucleosomes are arranged in a nearly random fashion around the genome[39,45], the regions of non-nucleosomal DNA vary from 0 to 62bp. These variable regions of internucleosomal DNA appear in intranuclear[38,57] as well as isolated SV40 chromosomes[39]. Comparison of replicating and mature SV40 chromosomes by analysis with micrococcal nuclease (MNase)[3,23], restriction endonucleases[45], and electron microscopy[58,59] demonstrated that the size and arrangement of nucleosomes on newly replicated chromatin was essentially the same as on mature chromatin. Nucleosomes were found directly in front and behind DNA replication forks as well as on both of the two sibling chromosomes (reviewed in[9]).

Despite the presence of nucleosomes on nascent cellular and viral DNA, comparison of the rates, extents, and products of MNase digestion revealed that newly replicated chromatin was more sensitive to MNase than nonreplicating chromatin in four ways: (i) a 2-6 fold increase in the initial rate of DNA digestion, (ii) a 25-50% increase in the extent of DNA digestion, (iii) a faster release of nucleosomal monomers, and (iv) a concurrent appearance of nascent DNA in a form significantly smaller than that associated with nucleosomes[3,9,23]. Some investigators have suggested that these nascent "subnucleosomal" DNA fragments originate from either nucleoprotein intermediates in nucleosome assembly or immature nucleosomes. However, in the case of SV40, these small DNA fragments actually originate from the nascent DNA found on both arms of replication forks before the first nucleosomes. Accordingly, this material, which includes Okazaki fragments, has been designated prenucleosomal DNA (PN-DNA; Figure.4).

PN-DNA is unambiguously recognized by its sensitivity to E. coli exonuclease III (Exo III) and the exonuclease coded by E. coli phage T7 gene 6 (Exo T7). Exo III degrades one strand of duplex DNA in the 3' to 5' direction and Exo T7 performs the same function in the 5' to 3' direction (Exo T7 can remove RNA primers). Since neither Exo III nor Exo T7 excises DNA from mature SV40 chromosomes or removes nascent DNA from nucleosomes at native replication forks[17], excision of nascent DNA from replicating chromosomes is limited to the region before the first nucleosome on each arm of the fork. The average distance from 3' and 5' ends of long nascent DNA strands to the first nucleosomes on either arm of a replication fork has been determined from exonuclease digestion of replicating SV40 chromosomes to be 125±20 nucleotides[17]. When replicating SV40 chromosomes were digested with MNase, both nucleosomal monomers and PN-DNA were rapidly released[3]. Predigestion of replicating chromosomes with either Exo III or Exo T7 decreased in part the amount of PN-DNA released by MNase; however, predigestion with both exonucleases resulted in complete removal of MNase sensitive PN-DNA[3]. In these experiments, the amount of radiolabeled DNA in nascent nucleosomal monomers relative to monomers released from mature SV40 chromosomes present in the same reaction remained constant. Therefore, PN-DNA originates from the sites of

DNA synthesis at replication forks and, apparently, from both arms of the forks. The second conclusion was confirmed by hybridization of PN-DNA to separated strands of SV40 DNA restriction fragments[3]. While isolated Okazaki fragments annealed predominantly, if not exclusively, to DNA strands representing the retrograde template, PN-DNA (like total SV40 DNA) annealed equally well to DNA fragments representing both forward and retrograde templates throughout the genome. Furthermore, PN-DNA represented at least twice as much nascent DNA as found in Okazaki fragments, and was observed even when Okazaki fragments represented an insignificant fraction of the labeled DNA in replicating chromosomes. These data demonstrate that PN-DNA does not simply represent a preferential release of Okazaki fragments as 5S duplex DNA.

PN-DNA was always recovered as a heterogeneously sized population of duplex DNA, resistant to single-stranded specific S1 endonuclease, but as sensitive to MNase as purified DNA[3]. Gel electrophoretic analysis of purified PN-DNA revealed a range of lengths from about 65 to 245bp, with a maximum at 120bp. Formaldehyde fixed PN-DNA had the same buoyant density in CsCl as purified DNA. On the other hand, nascent nucleosomal monomers from replicating chromosomes were resistant to MNase, had a buoyant density similar to chromatin, and sedimented at 11 S prior to deproteinization, but at 5-6 S afterwards. Therefore, isolated PN-DNA appears to be free of protein.

DNA replication proteins that were presumably associated with the PN-DNA region of replication forks inside the cell[9] were either lost when chromosomes were sedimented in 200mM NaCl[43], or not crosslinked to DNA by formaldehyde. In any event, it is not necessary to postulate the presence of a DNA-protein complex in order to explain the transient accumulation of PN-DNA during MNase digestions, because MNase digestion of purified SV40 DNA initially released fragments equivalent in size to PN-DNA[3]. Since PN-DNA originates from replication forks distributed throughout the genome, digestion of bare DNA at replication forks should lead to the same result. Random endonuclease cuts in the PN-DNA region should generate DNA fragments with an average size of about 100bp. The fraction of nascent DNA that is released as PN-DNA depends upon both the length of the radiolabeling period (i.e., the number of labeled nucleotides per fork) and the extent of digestion by MNase.

Synthesis of Okazaki fragments occurs in the PN-DNA region. Either Exo III or Exo T7 removed at least 80% of the radiolabel in Okazaki fragments[17], with a corresponding decrease in the fraction of PN-DNA observed[3]. Furthermore, 50-60% of the Okazaki fragments in either replicating chromosomes or purified replicating DNA could be released by single-strand specific endonucleases, the rest remained with the cleaved molecules because of single-stranded regions too small to be recognized under the conditions employed[16]. At least 90% of the Okazaki fragments released from formaldehyde-fixed chromosomes were not in nucleosomes[16]. The 10-20% of small nascent DNA chains that appear to be protected, together with the size variation of internucleosomal DNA suggest that the location of the first nucleosome may best be represented as a broad distribution about the mean of 125bp. Nevertheless, synthesis of most Okazaki fragments is initiated on nonnucleosomal DNA and completed prior to nucleosome assembly.

A region of immature chromatin is included in the model of eukaryotic replication forks to account for the observation that replicating SV40 chromosomes, predigested with Exo III and Exo T7 to remove PN-DNA still exhibit enhanced sensitivity to MNase. Furthermore, nascent nucleosomal dimers and trimers excised from replicating SV40 chromosomes are also more susceptible to MNase digestion than oligomers from mature chromosomes. The basis of chromatin maturation is not known, but may involve changes in histone composition and/or modification. The characteristics of PN-DNA and immature chromatin disappear with time as nascent chromatin completes its maturation process (reviewed in[9]. Chromatin maturation occurs even in the absence of concomitant DNA synthesis.

Because of the many advantages of subcellular systems in the study of biochemical mechanisms, we have characterized chromatin assembly during DNA replication in nuclei and nuclear extracts. The initial[3] and final[39] stages of SV40 chromatin in nuclear extracts supplemented with cytosol is essentially the same as mature chromatin formed in whole cells[37]. Neither mature SV40 chromosomes nor cellular chromatin prelabeled in intact cells underwent any changes during the in vitro incubations[37,39]. Our results with replicating SV40 chromosomes radiolabeled in nuclear extracts are in complete agreement with our results with chromosomes labeled in intact cells.

The accessibility of five specific DNA sequences to single-site restriction endonucleases was evaluated in both replicating and mature SV40 chromosomes[45]. Three restriction sites were in the region of the origin of replication and transcription, one was in the VP1 structural gene, and one was in the region for termination of replication (Figure 5). Viral chromosomes, isolated by three different methods, were incubated with saturating amounts of a single restriction enzyme and the DNA digestion products analyzed by electron microscopy and gel electrophoresis. Comparisons were made among mature chromosomes, and between the two arms of individual replication forks. Examples of unique phasing in which a particular site was either closed or open in all chromosomes were never observed, and the extents of cleavage were generally consistent with a near-random distribution of nucleosomes throughout the SV40 genome. Although variation in the accessibility of DNA sites near the origin of replication could be interpreted as "preferred" phasing in about 25% of replicating and mature SV40 chromosomes, the finding that two isoschizomers, Hpa II and Msp I, did not cut chromosomes to the same extent precluded an unambiguous interpretation of the extents of cleavage by individual restriction enzymes.

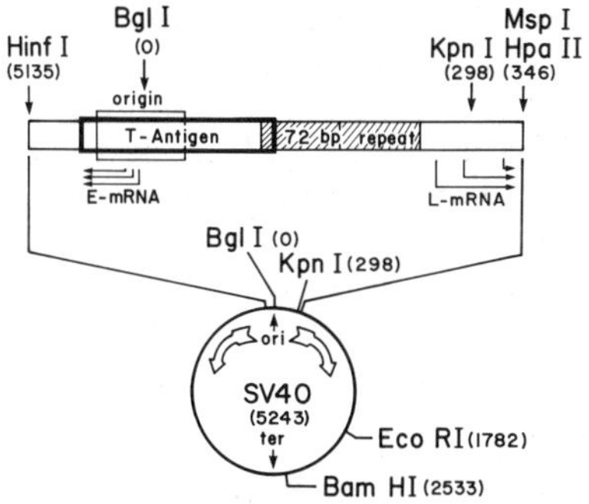

Fig. 5. General map of the SV40 genome showing the major initiation sites for early and late mRNA, region protected by T-antigen, regions for the origin and termination of replication, and a prominent repeated DNA sequence[50,53]. The numbers indicate base pairs. The single Hinf I site serves as a reference point. Restriction enzyme sites are numbered to the actual sites of cleavage.

The accessibility of specific DNA sequences in native SV40 chromosomes, including the origin region, was established within 400bp of replication forks during chromatin assembly[45]. This phase relationship remains essentially unchanged during the production of mature chromosomes and their re-entry into the replication pool. Each of the six restriction endonucleases studied cleaved the same fraction of its target sites in newly replicated chromatin as it did in mature chromatin. Similarly, the accessibility of the EcoRI and Bam HI sites in the unreplicated portion of replicating chromosomes was the same prior to its replication as it was after its replication and subsequent maturation into 70S chromosomes. Therefore, the accessibility of DNA sequences, including those near the origin, is not obviously related to replication.

Analysis of the accessibility of four specific DNA sequences in the newly synthesized portions of replicating SV40 chromosomes allowed a direct comparison of phasing in sibling molecules[45]. The fraction of replicating chromosomes in which only one of the two sibling molecules was cut was that expected if the structure of one was completely independent of the structure of the other. The fraction of replicating chromosomes containing an accessible restriction site in the newly replicated DNA was not only consistent with random phasing, but the probability of finding a restriction site that was either exposed or protected on both arms of a replication fork was also random. This suggests that chromatin assembly on one arm of a replication fork is not directed by assembly on the other arm, and is consistent with two different mechanisms of chromatin assembly occurring at a single replication fork. For example, it has been proposed for SV40 that all of the old histone octamers in front of the fork are retained on the forward arm while chromatin is assembled de novo on the retrograde arm with newly synthesized histones[9]. These two pathways of chromatin assembly may be related to the mechanisms of DNA replication since replication in SV40 is essentially continuous on the forward arm of a replication fork and discontinuous on the retrograde arm.

Hypothesis: Nucleosome Spacing Defines the Okazaki Fragment Initiation Zone (Figures 3 and 4)

If the rate limiting step in DNA replication is assumed to be "disassembly" of nucleosomes in front of replication forks to allow DNA unwinding, then the size of an Okazaki fragment initiation zone could be defined by prereplicative nucleosome spacing. As a fork advances from one nucleosome to the next, the forward arm will be duplex DNA, while the retrograde arm will be single-stranded DNA template with an average length of 220 nucleotides, the average center-to-center distance from one nucleosome to the next. However, these initiation zones will have a size distribution about the mean of ± 32 or ± 74 depending on whether core DNA alone (146bp) or core plus linker DNA (188bp) is considered because nucleosomes are irregularly, almost randomly, spaced. Therefore, using a stochastic model for initiation within this zone, one would expect a broad size range of Okazaki fragments with an upper limit of 250 or 290 nucleotides, an average size of about 110 nucleotides, and a heterogenous RNA primer sequence. Okazaki fragments are neither required nor excluded on forward arms. Since these expectations are borne out by the experimental results, the regularity of Okazaki fragment initiation and the heterogeneity in their lengths could be imposed on DNA replication by chromatin structure rather than DNA sequence[1,5,6,9]. In contrast, models that relate metabolism of Okazaki fragments to chromatin structure via initiation of DNA synthesis between prefork nucleosomes[18,20,32] would expect discontinuous DNA synthesis on both arms of replication forks, and maturation of Okazaki fragments to a relatively uniform size.

Involvement of Specific DNA Sequences in the Events at Native Replication Forks

Consideration of DNA sequence as an important parameter in DNA replication has generally been confined to the origin of replication where prokaryotic and eukaryotic organisms both begin DNA synthesis at a unique position on the chromosome[9]. However, the complexity of nucleic acid-protein interactions prompts one to anticipate that the rate of all such interactions will be significantly affected by DNA sequence. Therefore, it was surprising that neither completion of chromosome replication, which involves separation of sibling molecules and termination of DNA synthesis, nor repeated initiation of Okazaki fragments appeared to involve specific DNA sequences. Although the available data clearly show the absence of a unique DNA sequence in these operations, they do not rule out the use of "preferred" DNA sequences[9]. Therefore, we have begun to map the locations of nascent DNA chains on purified SV40(RI) DNA by specifically labeling either their 3' or 5'-ends and then measuring the distance from the labeled end to a uniquely defined restriction endonuclease site. The results demonstrate the presence of specific DNA sites where DNA replication is arrested in the termination region, and specific DNA sites within the origin of replication, as well as throughout the genome, that promote initiation of DNA synthesis.

(i) Preferred DNA Sequences that Arrest Replication Forks (Fig.6). SV40(RI*) DNA, 85-95% replicated, is 2-3 times more prevalent than an equivalent sample of SV40(RI) DNA at earlier stages in replication[46,49], suggesting that replication forks accumulate at specific DNA sites prior to separation of sibling molecules. To test this hypothesis, 3'-ends of nascent DNA chains were labeled by incubating purified SV40(RI) DNA from intact virus infected CV-1 cells with T4 DNA polymerase in the presence of one [α -^{32}P]dNTP and one unlabeled dNTP[47]. Control experiments showed that this method specifically labeled 3'-OH ends of DNA primers on a template without significantly altering their length. DNA chains greater than 2000 nucleotides were isolated by sedimentation in an alkaline sucrose gradient, and annealed with SV40(III) DNA, generated by cleavage at the origin of replication with Bgl I, to insure that all 3'-ends of DNA chains residing in the termination region exist as duplex DNA. This DNA was then digested with either Eco RI, Bcl I, Pst I or Bam HI restriction endonuclease to cut SV40 DNA near the termination site. The DNA products were denatured in glyoxal and fractionated by gel electrophoresis using SV40 DNA restriction fragments as size standards.

Several conclusions can be drawn from the locations of 3'-ends. First, nascent DNA chains accumulate at a minimum of two major and ten minor preferred DNA sites that are separated an average of 112bp over a distance of about 1320bp. Second, the two major DNA replication arrest sites are separated by about 470bp of unreplicated DNA centered at 2743bp or 52% of the genome from the origin of replication (defined by the Bgl I site). This region includes the normal termination site for DNA replication (Hind II + III B/G junction, Fig. 6 and refs.[22,48]). Therefore, most replication forks are arrested when replication is 91% completed, which accounts for the observed accumulation of SV40(RI*) DNA, presumably in preparation for separation of sibling molecules. Third, the arrangement of both major and minor DNA replication arrest sites suggests that progress of bidirectional

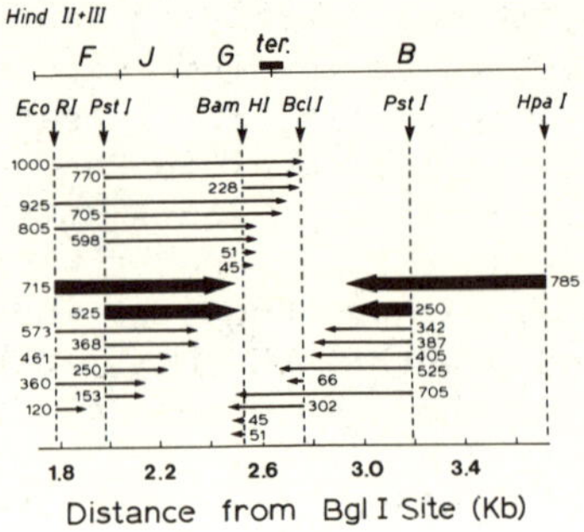

Fig. 6. Locations of 3'-ends of nascent DNA chains within the termination region of the SV40 genome. Thick arrows represent major accumulations of 3'-ends and thin arrows represent minor (about 5x less) accumulations. The average point of termination for SV40 replication forks (ter) was determined by the rate of appearance of newly replicated DNA throughout the genome and shown relative to Hind II + III DNA restriction sites. Numbers given the length of each DNA fragment in nucleotides (error is ± 3% of number shown). The distance in kilobases (Kb) is given from the Bgl I site (o) at the origin of replication based on a wild type SV40 strain containing 5243 bp (53) to the Eco RI (1782), Pst I (1992), Bam HI (2533), Bcl I (2770), Pst I (3238) and Hpa I (3735) sites of cleavage.

DNA replication is arrested when two forks are separated by about 500bp of unreplicated DNA. However, the distribution of these termination sites varied about the normal termination site by ± 450bp. This could account for the distribution of the short gap in the nascent strand of newly separated sibling molecules (SV40(II*) DNA) throughout a 730bp termination region[55]. Such variation may result from asynchronous arrival of replication forks. Electron microscopic analysis of replicating DNA revealed that in 78% of the molecules at any stage in replication, one fork had traveled up to 20% farther than its partner. Thus, replication forks may accumulate whenever they arrive at preferred DNA sites separated by about 500bp of unreplicated DNA (91% replication). Separation of sibling molecules may occur at this point regardless of which arrest sites are utilized.

<u>(ii) Preferred DNA Sequences that Initiate Okazaki Fragments</u>. To map the 5'-ends of nascent DNA, replicating SV40 DNA was purified from infected cells and the 5'-ends of nascent DNA chains labeled with ^{32}P-utilizing the polynucleotide kinase reaction[47]. Long [5'-^{32}P]DNA (12-16S) and Okazaki fragment [5'-^{32}P]DNA (2-6S) was isolated,

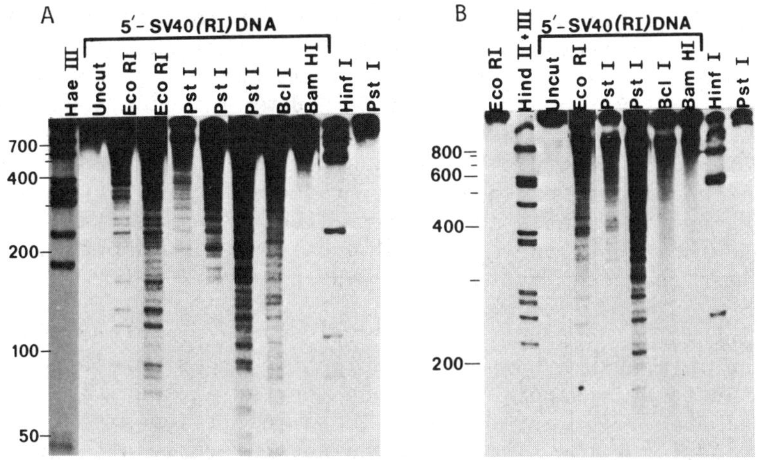

Fig. 7. 5'-ends of long nascent DNA strands are located at preferred DNA sites. The 5'-ends of nascent DNA from purified SV40(RI) DNA were labeled with ^{32}P. [5'-^{32}P]DNA longer than 2000 bases was then isolated, annealed with SV40 (III, Bgl I)DNA and digested with either Eco RI, Pst I, Bcl I or Bam HI. [5'-32P]DNA was then repurified, denatured in glyoxal and fractionated by electrophoresis in 5% polyacrylamide gels. Uniformly labeled SV40 DNA restriction fragments were included as length standards. Aliquots were allowed to migrate for different lengths of time to resolve the shorter (A) and longer (B) [5'-^{32}P]DNA chains. Numbers are in nucleotides. Two or three exposures are shown of Eco RI and Pst I lanes to accurately reproduce the data.

annealed to SV40 linear DNA and digested with a specific restriction endonuclease as previously done with [3'-^{32}P]DNA[47]. Before digestion, the 12-16S|^{32}P|DNA was longer than 2000 nucleotides, while after cleavage by either Eco RI, Pst I or Bcl I numerous discrete |^{32}P|DNA fragments from 40-600 nucleotides were observed in varying relative amounts (Fig. 7). The failure of Bam HI to release similar [^{32}P]DNA fragments and Bcl I to release fragments longer than 230 nucleotides confirmed that few, if any, replication forks had traveled through the termination site. Control experiments with randomly nicked molecules generated only smears of radioactivity when substituted for SV40(RI) DNA. For analysis of Okazaki fragments, DNA was first fractionated by gel electrophoresis[6] and only those fragments 100-200 nucleotides long were analyzed to insure that uncut DNA remained at the top of the gel.

At least 85% of these DNA fragments annealed to retrograde DNA templates so their orientation with respect to a given restriction site was unambiguous. As expected from the metabolism of Okazaki fragments (Fig. 3), the same results were obtained with either long DNA chains or Okazaki fragments. These data demonstrate the existence of preferred DNA sites that promote initiation of Okazaki fragments. Variation in the amounts of individual fragments shows that some sites are used more frequently than other sites. Assuming that these sites are found throughout the genome, the data predict about 110 to 130 major sites spaced an average of 44bp apart[47].

Using gel electrophoresis conditions that separate DNA chains differing in length by a single nucleotide, these preferred DNA sites were found to be clusters of 5'-ends located at specific nucleotides[6]. All four bases were represented. These initiation sites were also found by labeling the 3'-ends of the nascent chain after cleavage by a restriction enzyme that cut SV40 once at a single site, thus eliminating the possibility that some 5'-ends are blocked and cannot be labeled by polynucleotide kinase[6]. Similar results were obtained at the Msp I, Eco RI and Taq I sites, suggesting a common phenomenon throughout the genome. The 5'-terminal sequences on Okazaki fragments were the same as found on long nascent DNA chains, although a consensus sequence at 5'-ends was not evident. Since RNA primers are uniform in length, they too must begin at specific nucleotides. Therefore, nascent DNA chains appear to be initiated at many different, but specific DNA sequences.

(iii) Preferred DNA Sequences that Initiate DNA Synthesis within the Origin of Replication. The genetic locus required for initiation of SV40 DNA replication is a cis acting element consisting of 75bp of DNA[10] that is capable of autonomous replication when introduced into cells producing T-antigen[28]. Single base pair changes in this region alter the efficiency of replication[41]. Replication forks appear to originate within about 35bp of the Bgl I restriction site[26,47]. T-antigen, a product of SV40 gene A, is the only viral protein required for SV40 DNA replication[14,22,50], and binds specifically to three tandem DNA sites that span the ori region[56]. SV40 revertants of ori defective mutants were found to contain second site mutations in gene A[40]. These results show that initiation of DNA replication requires specific interactions between T-Ag and unique DNA sequences in the ori region, but they do not define where DNA synthesis is actually initiated.

In our previous experiments, the problem of determining the polarity of nascent DNA relative to the restriction site was solved by deduction using the facts that replication is bidirectional from a unique origin, Okazaki fragments originate predominantly from retrograde DNA templates, and only 5'-ends within 250 nucleotides of the restriction site are resolved by gel electrophoresis under conditions used for DNA sequencing. However, this approach cannot be used at

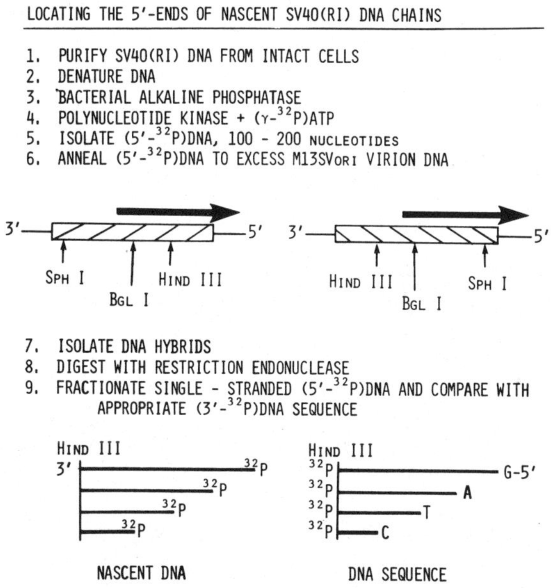

Fig. 8. Procedure for locating the 5'-ends of nascent DNA chains at the origin of SV40 DNA replication.

the origin of replication where small nascent DNA chains will be found on both sides of replication forks in early SV40(RI) DNA. Therefore, in order to unambiguously locate 5'-ends of nascent DNA at the origin as well as other genomic locations, unique sections of the SV40 genome were cloned into the DNA of E. Coli phage M13[15]. Since the virion DNA of this phage is a single-stranded circle of unique polarity, recombinants containing SV40 DNA in either orientation could be prepared in large quantities[29]. This permitted isolation of nascent SV40 DNA chains from specific genomic regions that annealed exclusively with SV40 DNA representing the template on only one arm of replication forks.

To locate the 5'-ends of nascent DNA in the ori region (Figure 8) SV40(RI) DNA labeled with [^{3}H]Thd in infected CV-1 cells, was purified[14,47], and the 5'-ends of nascent DNA chains were dephosphorylated with bacterial alkaline phosphatase before labeling with [γ-^{32}P]ATP in the presence of polynucleotide kinase[27]. Single-stranded [5'-^{32}P] DNA was fractionated by gel electrophoresis to obtain DNA chains 100-200 nucleotides long. This DNA was hybridized to M13 virion DNA containing a cloned ori region (either a 1346 nucleotide Sau 3A fragment containing one Hind III site or a 311 nucleotide Bst N1 fragment containing one Sph I site), and the DNA hybrids were digested with

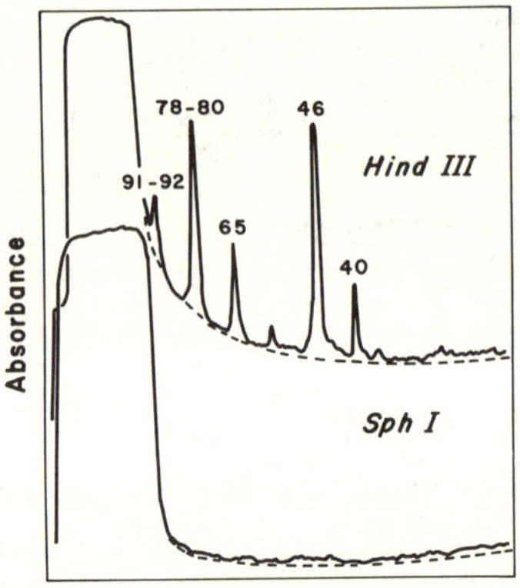

Fig. 9. The location of 5'-ends of nascent DNA from both ori strands. Nascent DNA was treated with RNase T2 prior to ^{32}P-labeling of their 5'-ends and then treated as described in Figure 8. Densitometer tracings were obtained from an autoradiogram of an 8% polyacrylamide-8 M urea gel. The broken line shows the tracing obtained when DNA was not cleaved by a restriction endonuclease.

either Hind III or Sph I, and the [5'-^{32}P]DNA strands fractionated by gel electrophoresis under conditions that resolve single nucleotide differences[27]. Concurrently, DNA fragments were prepared for sequencing that contained the same Hind III or Sph I site at one end, and that end contained a 3'-terminal ^{32}P-label. Application of the Maxam and Gilbert[27] chemical degradation technique to these fragments therefore provided the DNA sequence at the 5'-ends of those nascent DNA growing away from the origin and crossing the designated restriction site. Since the SV40 DNA sequence is known[50,53], the accuracy of our results was easily confirmed.

Before digestion by Hind III, of the[5'-^{32}P]DNA migrated as a broad band 100-200 nucleotides in size; after digestion a fraction of the labeled DNA was cut and migrated as shorter, discrete bands (Figure 9). Direct comparison with all possible 5'-ends of the same orientation (ie. DNA sequence) identified the 5'-terminal nucleotide on nascent DNA chains. Control experiments with 5'-labeled restriction DNA fragments substituted for nascent DNA chains confirmed the accuracy

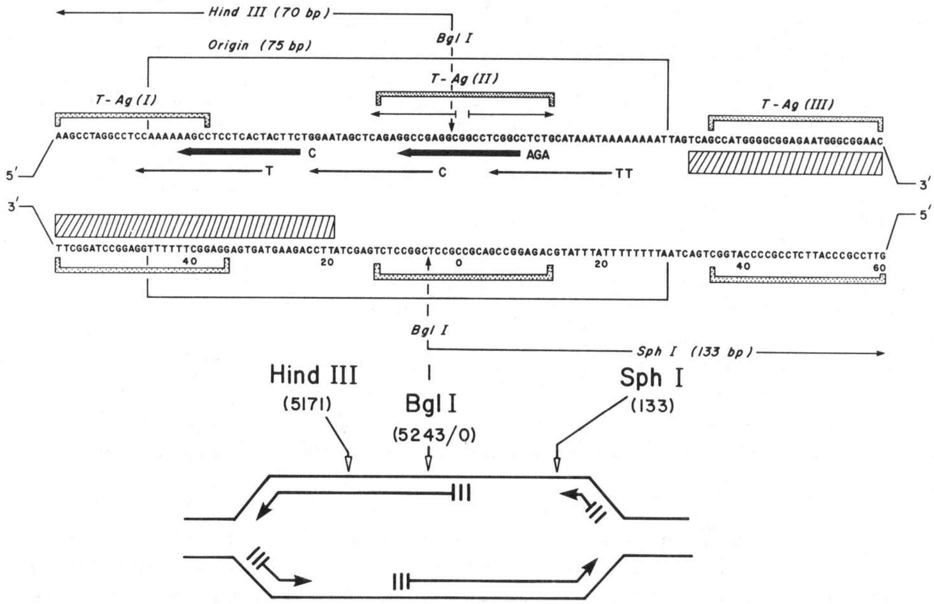

Fig. 10. Composite structure of a replication bubble at the origin of SV40 DNA. The three primary T-antigen binding sites, the sequences required for ori function, a 27bp palindrome and restriction endonuclease sites used in this study are indicated. The primary initiation sites for nascent DNA synthesis towards the Hind III site are shown by solid arrows beginning at their 5'-terminal nucleotides. Shaded areas indicate regions of the genome that were not yet examined. A lower resolution image of a single replication bubble is also shown with RNA primers indicated by three perpendicular lines.

of this procedure. Therefore, the 5'-ends of DNA chains with the same orientation as early mRNA (Figure 5) originated at two major and three minor locations within ori (Figure 10). All four nucleotides were represented at 5'-ends, although a compression effect[53] from the 27bp palindrome limited resolution of the 78-80 nucleotide band to AGA (Figure 10).

In contrast, Sph I digestion of the appropriate hybrid of opposite orientation did not release any labeled fragments shorter than 155 nucleotides (Figure 9). Control experiments showed that Sph I was capable of digesting the sample to completion under these conditions. Therefore, 5'-ends of nascent DNA with the same orientation as late mRNA were not found in an approximately 50 nucleotide segment of the ori region (Figure 10).

These data represent only 5'-ends of DNA and not RNA because SV40(RI) DNA was treated with high concentrations of RNase A during purification; the results were the same when this DNA was subsequently denatured and digested with RNase T2 prior to labeling 5'-ends. If the RNase A step was omitted, then each of the five DNA chains in Figure 10 could be shown to have RNA covalently attached to their 5'-ends in the following way. SV40(RI) DNA was denatured, treated with ATP and T4 polynucleotide kinase to phosphorylate all 5'-OH ends, and then incubated in KOH to hydrolyze RNA and generate new 5'-OH ends on any DNA chains that were covalently linked to RNA[67]. This DNA was then phosphorylated with $[\gamma\text{-}^{32}P]$ATP at the newly exposed 5'-OH ends, annealed to the appropriate M13SVori DNA, digested with Hind III and fractionated by gel electrophoresis. The same bands seen in Figure 9 were again found, but in different relative amounts.

These results demonstrate that nascent DNA chains are initiated at several specific nucleotides within the ori region, apparently using RNA primers. Each initiation site must represent a different population of DNA since the average size of an Okazaki fragment is twice as large as the ori region. DNA synthesis in the direction of early mRNA synthesis (towards the Hind III site) eventually becomes continuous, while synthesis in the opposite direction is already continuous in the region of ori we examined. If DNA synthesis is initiated sumultaneously on both sides of ori, then it must occur near the first T-Ag binding site. Alternatively, DNA synthesis could be initiated first in the direction of early mRNA synthesis, and then later in the opposite direction as the replication bubble opens. This form of staggered initiation occurs dramatically in mitochondrial DNA replication[9].

Specific Sequences in Native DNA Arrest the Progress of DNA Polymerase α

How does DNA polymerase α initiate, synthesize and finally complete DNA chains at replication forks? Studies of both prokaryotic and eukaryotic DNA replication have shown that part of the answer involves RNA primers, DNA polymerase accessory proteins, and the structure of chromatin at replication forks. In addition, part of the answer may also involve recognition of DNA sequence signals that promote either initiation or termination of DNA synthesis. Neither the exact nature of specific DNA sites in SV40 nor their influence, if any, on the activity of DNA polymerase α is known. However, it is clear that both RNA and DNA polymerases respond to specific sequences in their templates. For example, purified "primase" proteins from E. Coli phage T7 and phage T4 are able to initiate RNA synthesis at specific DNA sites and E. Coli RNA polymerase initiates transcription at specific nucleotides within the promoter DNA sequence and termin-

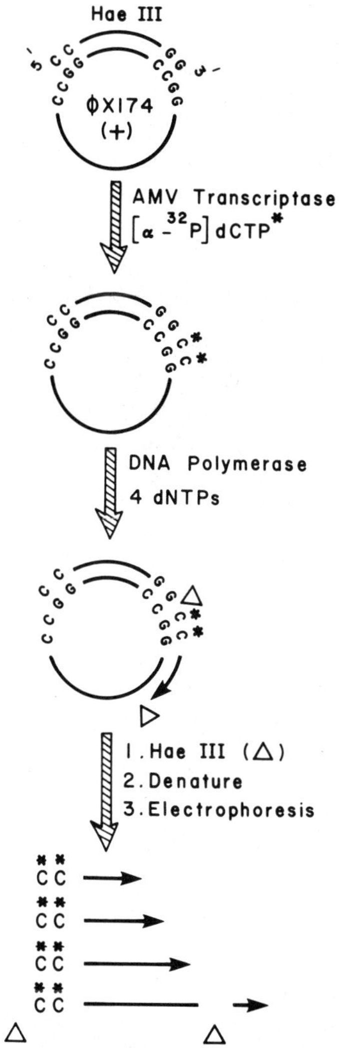

Figure 11. Outline of the procedure used to map the progress of DNA polymerase on a ϕX174 DNA template.

ates synthesis at specific DNA sequences[60]. RNA polymerase and Qβ replicase are two highly processive enzymes that pause at the distal side of hairpin structures[60,61]. Similarly, DNA synthesis by either E. coli DNA polymerase II[63], vaccinia virus DNA polymerase[62] or E. coli phage T4 DNA polymerase[64], but not by E. coli DNA polymerase I[62,63], is arrested at specific DNA sites. These sites are generally correlated with the presence of major palindromic (hairpin) sequences in the template, but the positions of 3'-ends of nascent DNA chains were not precisely located with respect to DNA sequence. DNA arrest (or pause) sites may represent a universal property of nucleic acid synthesizing enzymes, serving to increase the probability that a biological event will occur at a particular site by slowing down DNA synthesis.

As a first step in understanding how DNA polymerase α functions at native replication forks, we have developed a novel method to monitor the progress of DNA polymerase on a natural DNA template, nucleotide by nucleotide. Each nascent DNA chain contained the same amount of radiolabel regardless of its length and thereby provided a direct measurement of the number of molecules extended to a particular nucleotide. A primer-template of known DNA sequence was constructed by annealing an excess of single-stranded circular ϕX174 Hae III DNA restriction fragments (Figure 11). The 3'-end of this DNA primer was labeled by completing DNA synthesis at the cutting site with [α-^{32}P] dCTP in the presence of AMV reverse transcriptase. This primer-template was then incubated with CV-1 cell α-polymerase and all 4 unlabeled dNTPs at concentrations at least 10 fold higher than their apparent Km values.

It was quickly apparent that the rate of DNA synthesis on a natural DNA template varied dramatically with DNA sequence. Nascent [^{32}P]DNA chains accumulated at several specific sites along the template an average of 30-40 nucleotides apart (Figure 12). As expected, the fraction of primers used in the reaction increased with time of incubation and some of them were elongated to about 500 nucleotides. In order to identify the nucleotide at the 3'-ends of nascent DNA chains, the primer was removed at the end of the reaction by Hae III digestion so that newly synthesized chains less than 150 nucleotides could be resolved at the resolution of single nucleotide differences (Figure 11). In these experiments, α-polymerase could be followed only to the next Hae III site (194 nucleotides with the Hae III-4 primer). The result showed that DNA polymerase α was arrested at specific sequences as it advanced along the template (Figure 13). As the enzyme to primer-template ratio was increased, arrest sites close to the primer disappeared and new arrest sites were formed further downstream. Therefore, arrest sites are sequences at which α-polymerase dissociates from the template more frequently; increasing the enzyme concentration increases the probability of reinitiation at the same site. Strikingly similar results were found with DNA polymerase α from HeLa cells and calf thymus tissues, and phage T4 DNA polymerase. In contrast, DNA polymerase I from E. coli, an enzyme

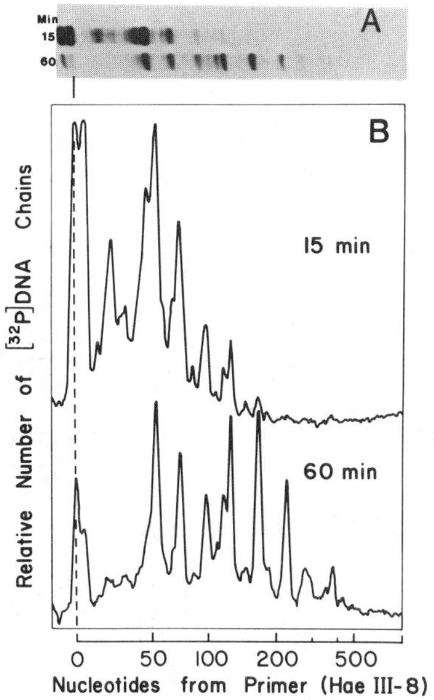

Fig. 12. Discontinuous elongation by DNA polymerase α. φX174 virion DNA annealed with the φX174 Hae III-8 DNA fragment was incubated with 0.5 units of CV-1 DNA polymerase for either 15 or 60 min in 50mM Tris-HCl (pH 8.0), 6mM $MgCl_2$, 20mM KCl, 5 μg/ml DNA, 100 μg/ml gelatin, 100 μM each of dATP, dCTP, dGTP and dTTP. The DNA products were denatured and then fractionated in a 5% polyacrylamide gel. SV40 DNA restriction fragments were included as size standards (not shown). (A) autoradiogram. (B) densitometer tracings of (A). The unextended Hae III-8[3'-^{32}P]DNA primer is indicated by a broken line. The scale on the horizontal axis indicates the number of nucleotides added to the primer.

not involved in replicating large amounts of DNA, failed to respond to DNA sequence signals. Direct comparison of the arrest sites with the DNA sequence of nascent DNA chains allowed identification of the 3'-terminal nucleotides on nascent DNA at arrest sites (Figure 13). Our results confirmed the reported DNA sequence for φX174[33].

In order to determine whether or not arrest sites resulted from interactions between DNA template regions widely separated on the φX174 DNA genome, the base template region of φX174 DNA used in Figure 13, was cloned into M13 DNA[29]. DNA synthesis was then initiated either on a short M13 DNA primer that left 99% of the M13 DNA single-stranded, or a long primer that converted all the M13 DNA into a duplex structure. In both cases, the arrest sites observed on the cloned φX174 DNA template were essentially identical with those observed in φXi74

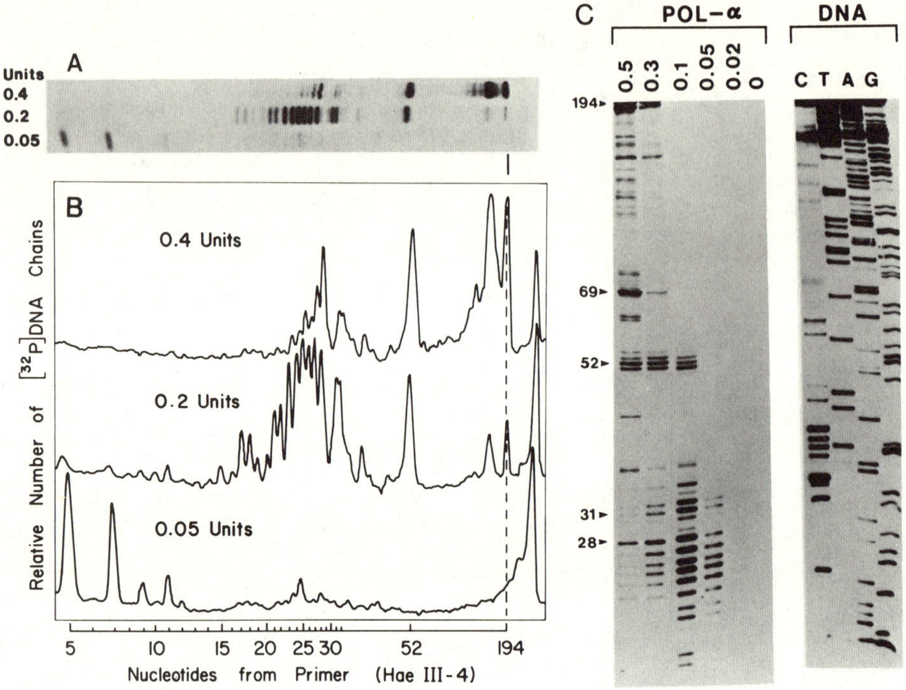

Fig. 13. High resolution analysis of discontinuous elongation by DNA polymerase α. ϕX174 virion DNA annealed with the ϕX1174 Hae III-4 DNA fragment was incubated with increasing amounts of CV-1 DNA polymerase α for 15 min under the conditions described in Figure 12. DNA products were then digested with Hae III to liberate nascent DNA chains containing 2 residues of $[^{32}P]$dCMP at their 5'-ends. The DNA products were denatured and then fractionated by electrophoresis in a 10% polyacrylamide gel for either 1.5 (A and B) or 2.5 hrs (C). (A) autoradiogram. (B) densitometer tracings of (A). (C) autoradiograms of the $[5'\text{-}^{32}P]$DNA chains synthesized by α-polymerase, and the $|5'\text{-}^{32}P|$DNA chains synthesized by *E. coli* DNA polymerase I in the presence of either d_2CTP, d_2TTP, d_2ATP or d_2GTP (indicated as C, T, A and G) to provide the sequence of nascent DNA[33]. The maximum length of the $[^{32}P]$DNA product (indicated by broken line in panel B) is limited by the distance to the next Hae III site (194 nucleotides). Numbers on the horizontal axis of (B) and vertical axis of (C) indicate nucleotides from the Hae III-4 DNA primer.

DNA alone. Therefore, the arrest of DNA polymerase at specific DNA sequences is a consequence of sequences between 24 bases upstream and 140 bases downstream from the arrest site. Arrest sites do not result from long range DNA-DNA interactions.

Out of 1200 nucleotides examined downstream from 11 Hae sites on the φX174 genome, 52 arrest sites were recognized where 3'-ends of nascent DNA had accumulated at least 3 fold over adjacent sequences. The distance between arrest sites varied from 1-140 nucleotides, and the sites themselves varied in intensity (3-10 fold), size (1-8 nucleotides with 63% consisting of 1-2 nucleotides) and composition (65% GC with 4-5 triplets containing 80% GC present 4-10 fold above their normal frequency in φX174 DNA and the absence of many AT rich triplets). Compositional analysis was carried out both with and without addition of the two adjacent nucleotides at each site. In both cases our conclusion was the same: specific GC rich sequences were preferred as arrest sites although a consensus sequence or family of sequences was not obvious. Analysis of the 30 nucleotides upstream and downstream of arrest sites revealed only a correlation with the presence of hairpin structures that could form a template obstacle in front of the polymerase (64% of arrest sites were within 8 nucleotides of the primer-proximal side of a hairpin, and regions free of arrest sites were free of hairpin structures). However, there was no clear relationship between the proximity of an arrest site to a hairpin and the intensity of the arrest site. Therefore, if one assumes that the rate of DNA synthesis normally varies with the sequence of the template, arrest sites may result from secondary structures that act as obstacles to the enzyme and thus amplify the normal variation in its progress. The actual stopping points will vary with the distance of specific preferred sequences from these obstacles, the stability of the obstacles, and the particular polymerase involved. If 5'-ends of DNA chains also act as barriers to an approaching polymerase, then the phenomenon of arrest sites may explain why α-polymerase alone is ineffective at completing synthesis of Okazaki fragments (ie. gap-filling,[54]).

Since accessory proteins can increase the fidelity, processivity, rate of synthesis, and specificity of initiation of replication as well as the initiation and termination of transcription, accessory proteins will likely modify the behaviour of DNA polymerase α. However, these proteins may act not only to suppress arrest sites in order to increase processivity, but they may also enhance certain arrest sites on native chromosomes in order to promote biological functions such as separation of sibling molecules, replication dependent recombination or transcription, and coordination of replication events on the two sides of the fork.

We have recognized two types of DNA polymerase α accessory proteins: proteins referred to as "cytosol factors" that are required to complete synthesis of Okazaki fragments on native replicating chromosomes (Figure 3), and proteins that stimulate initiation on

DNA primer-templates. The cytosol factors have been purified free of DNA polymerase and ligase activity, but have so far only been shown to affect endogenous α-polymerase activity in isolated nuclei. CV-1 α-polymerase and α-polymerase initiation protein(s) have been purified free of single and double-stranded endo- and exonucleases, RNA polymerase and ATPase activities. This initiation protein is a single-stranded DNA binding protein that appears analogous to the C1 factor described by Novak and Baril[66]. Our protein was purified from cell extracts as well as from a novel form of α-polymerase initially present in the cell extract ("holo"-enzyme). Initiation factor was then reconstituted with α-polymerase core. Either the initial "holo" enzyme or the purified reconstituted "holo"-enzyme stimulated incorporation of the first nucleotide on DNA primer-templates by 30-50 fold compared with α-polymerase core enzyme alone, and stimulation was specific for the homologous α-polymerase. Stimulation was most pronounced when the ratio of primer to template was low (eg. denatured cell DNA, ϕX174 primer-templates described in Figure 11, or parvovirus DNA). Although the initiation protein lowered the Km for 3'-OH primers, it did not affect the Km of dNTP substrates. In this way, the initiation protein increased the overall rate of DNA synthesis by 30-50 fold, but it had no effect on DNA arrest sites; arrest sites were still observed by lowering the "holo"-enzyme concentration. Using this novel form of DNA polymerase α, we are continuing our search for additional accessory proteins that will allow us to reconstitute native replication forks.

SUMMARY AND DISCUSSION

The combined studies on SV40, polyoma and cellular eukaryotic chromosomes suggest the following principles of chromosome replication:

(i) Initiation of DNA replication involves three steps: recognition of a preferred (specific) or unique DNA sequence, separation of DNA template strands, and, finally, initiation of DNA synthesis. The structural and enzymological requirements for initiation of the first nascent DNA chain in the origin region may be the same for all subsequent initiation events throughout the genome; only recognition and separation of the origin DNA templates may distinguish synthesis of the first Okazaki fragment from all others.

(ii) The use of many preferred DNA sequences, rather than a single required sequence, to promote initiation and arrest of DNA synthesis allows increased efficiency and control without also imposing the genetic burden of maintaining additional unique DNA sequence information. Preferred sites can be rearranged, deleted or inserted without impairing DNA replication. A unique (or highly preferred) DNA site as well as a specific protein that recognizes it (eg. T-antigen) is required for viral origins of replication to insure their preferential replication by the cell. Without a strong <u>ori</u>

region, the 1.7 microns of SV40 DNA would be lost in a sea of 10^6 microns of cell DNA. By analogy, cellular chromatin may have many classes of origins and recognition proteins to allow temporal and spacial control over replication. In contrast, unique termination sites are avoided because loss of such a site in eukaryotic cellular genomes would prevent separation of sibling chromosomes whereas loss of one origin in a tandem array would have little, if any, effect. However, preferred DNA sites that promote separation of sibling molecules would prevent interwinding of daughter chromatids (eg. catenated SV40 dimers) every 30-50 microns, a problem that would otherwise have to be resolved by a topoisomerase surveillance mechanism.

(iii) The rate-determining steps in replicon maturation are completion of Okazaki fragment synthesis (ie. gap-filling), separation of sibling chromatids, and termination of DNA replication (ie. completing the final gap-filling step in SV40 (II*) DNA). The separation can occur even when Okazaki fragments are present in the termination region[49]. In addition to the necessity of removing RNA primers, gap-filling is slow because DNA polymerase α has difficulty incorporating the final nucleotides.

(iv) The progress of DNA polymerase α, the enzyme normally responsible for all DNA synthesis at replication forks, is highly sensitive to the sequence and secondary structure of its template. This property may be involved in the arrest of replication forks on native chromosomes, or even selection of initiation sites. The activity of α-polymerase in situ will also depend on both accessory proteins and the chromatin structure of its template. For example, different accessory proteins dramatically stimulate the ability of α-polymerase to initiate DNA synthesis on a purified primer-template, and to complete the gap-filling step in the metabolism of Okazaki fragments on native chromosomes. However, if these factors affect the progress of α-polymerase equally at all nucleotides, DNA sequence signals such as arrest sites will still be read. Increasing the enzyme's processivity may only convert arrest sites into pause sites (the enzyme stops without leaving the template).

(v) DNA synthesis at native replication forks occurs on non-nucleosomal, but not necessarily bare, DNA.

(vi) The frequency of Okazaki fragment initiation is determined by chromatin structure which defines the size of an initiation zone by limiting the extent of DNA unwinding. In prokaryotes, where a nucleosome structure is absent, replication forks travel faster, and Okazaki fragments are initiated about once every 1500-2000bp even though DNA sequence signals for RNA primers appear at least once in 300bp.

(vii) Chromatin assembly rapidly occurs as soon as sufficient duplex DNA is available on either side of the fork. For SV40, this requires about 345bp, the distance from the 3' or 5'-ends of long nascent DNA chains to a nucleosome plus 220bp, the average distance from one nucleosome to the next. Prereplicative histone octamers do not dissociate, and in the presence of cycloheximide they reappear on the forward arm[9]. However, cycloheximide inhibition of protein synthesis also retards joining of Okazaki fragments as well as DNA synthesis[9]. Therefore, a preference for forward arms may simply reflect a preference for uninterrupted duplex DNA. Since the first nucleosome encountered on forward arms is a minimum of 125bp from the first prefork nucleosome, prereplicative nucleosomes apparently do not remain at the same DNA site during replication. Otherwise, the maximum distance should be 75bp (the average distance between two adjacent core particles).

(viii) Chromatin is not assembled at the same DNA sites on all molecules, although a choice of preferred DNA sites may exist. Even if chromatin assembly were tightly coupled to DNA replication, initiation of replication at a unique origin may not impose unique chromatin phasing with respect to DNA sequence because DNA synthesis can begin at several different sites within the ori region.

(ix) Chromatin assembly on one arm of a replication fork does not direct assembly on the other arm.

ACKNOWLEDGEMENTS

This work was supported by grants awarded to M.L.D. and P.M.W. from The National Institutes of Health and The National Science Foundation. R.T.H. and C.P. were supported by post-doctoral fellowships from the Damon Runyon-Walter Winchell Cancer Fund, and L.C.T. was a post-doctoral fellow of the American Cancer Society. M.E.C. and D.T.W. were supported by a National Research Service Award. We are indebted to Ann Kenneally for help in preparing this manuscript.

REFERENCES

1. S. Anderson and M. L. DePamphilis, J. Biol. Chem. 254, 11495, 1979,
2. S. Anderson, G. Kaufman and M. L. DePamphilis, Biochemistry 16, 4990, 1977.
3. M. E. Cusick, T. M. Herman, M. L. DePamphilis and P. M. Wassarman, Biochemistry, in press, 1981.
4. M. L. DePamphilis, Microbiology-1980 (ed. D. Schlessinger), p.259, 1980.
5. M. L. DePamphilis, S. Anderson, R. Bar-Shavit, E. Collins, H. Edenberg, T. Herman, B. Karas, G. Kaufman, H. Krokan, E. Shelton, R. Su, D. Tapper and P. M. Wassarman, Cold Spring Harbor Symp. Quant. Biol. 43, 679, 1979.

1. S. Anderson and M. L. DePamphilis, J. Biol Chem. 254, 11495, 1979.
2. S. Anderson, G. Kaufman and M. L. DePamphilis, Biochemistry 16, 4990, 1977.
3. M. E. Cusick, T. M. Herman, M. L. DePamphilis and P. M. Wassarman, Biochemistry, in press, 1981.
4. M. L. DePamphilis, Microbiology-1980 (ed. D. Schlessinger), p.259 1980.
5. M. L. DePamphilis, S. Anderson, R. Bar-Shavit, E. Collins, H. Edenberg, T. Herman, B. Karas, G. Kaufmann, H. Krokan, E. Shelton, R. Su, D. Tapper and P. M. Wassarman, Cold Spring Harbor Symp. Quant. Biol. 43, 679, 1979.
6. M. L. DePamphilis, S. Anderson, M. Cusick, R. Hay, T. Herman, H. Krokan, E. Shelton, L. Tack, D. Tapper, D. Weaver and P. M Wassarman, In "Mechanistic Studies of DNA Replication and Genetic Recombination", (B. Alberts, ed.), p.55, Acedemic Press, New York, 1980.
7. M. L. DePamphilis and P. Berg, J. Biol. Chem. 250, 4348, 1975.
8. M. L. DePamphilis, P. Beard and P. Berg, J. Biol. Chem.250, 4340, 1975.
9. M. L. DePamphilis and P. M. Wassarman, Ann. Rev. Biochem. 49, 627, 1980.
10. D. DiMaio and D. Nathans, J. Mol. Biol. 140, 129, 1980.
11. H. J. Edenberg, S. Anderson and M. L. DePamphilis, J. Biol. Chem. 253, 3273, 1978.
12. R. Eliasson and P. Reichard, J. Biol. Chem. 253, 7469, 1978.
13. R. Eliasson and P. Reichard, J. Mol. Biol. 129, 393, 1979.
14. G. C. Fareed and D. Davoli, Ann. Rev. Biochem. 46, 471, 1977.
15. R. T. Hay and M. L. DePamphilis, In "Structure and DNA-Protein Interactions of Replication Origins", (D. S. Ray, ed.), ICN-UCLA Symposia on Molecular and Cellular Biology, Vol. 21, Academic Press, 1981.
16. T. M. Herman, M. L. DePamphilis and P. M. Wassarman, Biochemistry 18, 4563.
17. T. M. Herman, M. L. DePamphilis and P. M. Wassarman, Biochemistry 20, 621, 1981.
18. D. R. Hewish, Nucl. Acids Res. 3, 69, 1976.
19. G. Kaufmann, J. Mol. Biol. 147, 25, 1981.
20. G. Kaufmann, S. Anderson and M. L. DePamphilis, J. Mol. Biol. 116, 549, 1977.
21. G. Kaufmann, R. Bar-Shavit and M. L. DePamphilis, Nucl. Acids Res. 5, 25, 1978.
22. T. J. Kelly Jr. and D. Nathans, Adv. in Virus Res. 21, 86, 1977.
23. K. H. Klempnauer, E. Fanning, B. Otto and R. Knippers, J. Mol. Biol. 136, 359, 1980.
24. H. Krokan, P. Schaffer and M. L. DePamphilis, Biochemistry, 18, 4431, 1979.
25. I. M. Leffak, R. Grainger and H. Weintraub, Cell 12, 837, 1977.
26. R. G. Martin and V. P. Setlow, Cell 20, 381, 1980.
27. A. M. Maxam and W. Gilbert, Proc. Nat. Acad. Sci. USA. 74, 560, 1977.

28. R. M. Myers and R. Tjian, Proc. Nat. Acad. Sci. USA. 77, 6491, 1980.
29. J. Messing, R. Crea and P. H. Seeburg, Nucl. Acids Res. 9, 309, 1981.
30. D. Perlman and J. A. Huberman, Cell 12, 1029, 1977.
31. C. P. Prior, C. R. Cantor, E. M. Johnson and V. G. Allfrey, Cell 20, 597, 1980.
32. B. H. Rosenberg, Biochem. Biophys. Commun. 72, 1384, 1976.
33. F. Sanger, A. R. Coulson, T. Friedman, G. M. Air, B. Barrell, N. Brown, J. Fiddes, C. Hutchinson, P. Slocombe and M. Smith, J. Mol. Biol. 125, 225, 1978.
34. S. Saragosti, G. Moyne and M. Yaniv, Cell 20, 65, 1980.
35. M. J. Seidman, A. J. Levine and H. Weintraub, Cell 18, 439, 1979.
36. M. J. Seidman and N. P. Salzman, J. Virol. 30, 600, 1979.
37. E. R. Shelton, J. Kang, P. M. Wassarman and M. L. DePamphilis, Nucl. Acids Res. 5, 349, 1978.
38. E. R. Shelton, P. M. Wassarman and M. L. DePamphilis, J. Mol. Biol. 125, 491, 1978.
39. E. R. Shelton, P. M. Wassarman and M. L. DePamphilis, J. Biol. Chem. 255, 771, 1980.
40. D. Shortle, R. F. Margolskee and D. Nathans, Proc. Nat. Acad. Sci. USA. 76, 6128, 1979.
41. D. Shortle and D. Nathans, J. Mol. Biol. 131, 801, 1979.
42. R. T. Su and M. L. DePamphilis, Proc. Nat. Acad. Sci. USA 73, 3466, 1976.
43. R. T. Su and M. L. DePamphilis, J. Virol 28, 53, 1978.
44. O. Sundin and A. Varshavsky, Cell 21, 103, 1980.
45. L. C. Tack, P. M. Wassarman and M. L. DePamphilis, J. Biol. Chem. in press, 1981.
46. D. P. Tapper and M. L. DePamphilis, J. Mol. Biol. 120, 401, 1978.
47. D. P. Tapper and M. L. DePamphilis, Cell 22, 97, 1980.
48. D. P. Tapper, S. Anderson and M. L. DePamphilis, Biochem.Biophys. Acta 565, 84, 1979.
49. D. P. Tapper, S. Anderson and M. L. DePamphilis, J. Virol., on press, 1981.
50. J. Tooze, DNA Tumor Viruses Cold Spring Harbor Press, 1980.
51. B. Y. Tseng and M. Goulian, J. Biol. Chem.255, 2062, 1980.
52. B. Y. Tseng, J. M. Erickson and M. Goulian, J. Mol. Biol. 129, 531, 1979.
53. H. Van Heuverswyn and W. Fiers, Eur. J. Biochem. 100, 51, 1979.
54. D. T. Weaver, H. Krokan and M. L. DePamphilis, J. Supramol. Str. suppl. 4, Abs. 895, p. 333, 1980.
55. M. C. Y. Chen, E. Birkenmeier and N. Salzman, J. Virol. 17, 614, 1976.
56. R. Tjian, Cold Spring Harbor Symp. Quant. Biol. 43, 679, 1979.
57. L. M. Hallick, H. A. Yokota, J. C. Bartholomew and J. E. Hearst, J. Virol. 27, 127, 1978.

58. M. M. Seidman, C. F. Garan and N. P. Salzman, Nucl. Acids Res. 5, 2877, 1978.
59. C. Cremisi, A. Chestier and M. Yaniv, Cold Spring Harbor Symp. Quant. Biol. 42, 409, 1978.
60. T. Platt, Cell 24, 10, 1981.
61. D. R. Mills, C. Dobkin and F. R. Kramer, Cell, 15, 541, 1978.
62. M. D. Challberg and P. T. Englund, J. Biol. Chem. 254, 7820, 1979.
63. A. L. Sherman and M. L. Gefter, J. Mol. Biol. 103, 61, 1976.
64. C-C. Huang, J. E. Hearst and B. Alberts, J. Biol. Chem. 256, 4087, 1981.
65. A. Klug, D. Rhodes, J. Smith, J. J. Finch and J. O. Thomas, Nature 287, 509, 1980.
66. B. Novak and E. F. Baril, Nucl. Acids Res. 5, 221, 1978.
67. R. Okazaki, S. Hirose, T. Okazaki and Y. Kurosowa, Biochem. Biophys. Res. Comm. 62, 1018, 1975.

RECOGNITION OF ORIGIN OF REPLICATION BY INITIATION FACTORS IN ESCHERICHIA COLI

Masamichi Kohiyama, Annick Jacq and Claude Reiss

Institut de Recherche en Biologie Moléculaire
Centre National de la Recherche Scientifique
Université Paris VII, 2, Place Jussieu
75 005 Paris, France

HISTORICAL ASPECTS OF INITIATION

The notion of initiation of DNA replication was suggested for the first time by Replicon hypothesis of Jacob, Brenner and Cuzin, in which replication is clearly separated from initiation[1].

This idea was based on two facts; firstly it had been shown that *E. coli* can pursue replication but not reinitiation in the absence of protein synthesis[2]. Secondly, a temperature-sensitive mutant of *E. coli* had just been isolated which pursues DNA synthesis for only a half cycle of DNA replication when placed at a nonpermissive temperature[3]. The only machinery of DNA synthesis known at that time was DNA polymerase, isolated by A. Kornberg[4]. The temperature sensitivity of this enzyme was examined immediately after isolation of this mutant afterwards designated as *dna* A[5]. The negative answer provoked some confusion in the understanding of temperature sensitivity. Later, however, it was clearly shown that the machinery of DNA synthesis in a *dna* A mutant was normal, even at high temperature. In fact, when the mutant was irradiated by γ-rays, producing nicks in the chromosomes, it resumed DNA replication even at non-permissive temperature[6]. Isolation of the mutant lacking DNA polymerase of Kornberg[7] made it possible to isolate the DNA polymerase III coded by the dna E gene, as the principle replicase of *E. coli*[8].

Thus, the problem of initiation was clearly defined and separated from that of chain elongation. The first question raised concerned the kind of enzymatic reactions involved in initiation. Two answers were obtained in different systems: in E. coli RNA synthesis, because of rifampycin inhibition of initiation[9], and, in ϕX174 RF replication, site specific nicking carried out by cis A protein[10]. Site specificity is one of the most fundamental and complicated aspects of initiation. Although the replicon hypothesis stipulated a genetically defined point called the origin of replication, from which chromosome replication starts, it was several years before such an entity, consisting of the isolation of a cis acting mutant in DNA synthesis of lambda phage, was genetically demonstrated[11]. The necessity of such a region for DNA replication was clearly shown in a plasmid by the in vitro recombination technique[12]. Thus the replication origin (ori) can be defined as a locus required for DNA replication in order to maintain one replicon in an autonomous state. Such a DNA fragment from E. coli has been found by two groups using different techniques; Hirota's group looked for a fragment ori-C, which enables a DNA carrying an ampicillin resistant gene to replicate autonomously in recombination techniques[13], whereas von Meyenberg's group isolated a lambda transducing phage carrying ori-C, from which they derived a minichromosome containing ori-C without any viral DNA[14]. In E. coli, their ori-C is a unique site without which E. coli cannot be viable unless provided with the F factor integrated in the chromosome[15].

The initiation should comprise a series of reactions performed at ori-C by several proteins. In fact, twelve genetically different mutants of E. coli have been isolated which show the expected phenotype of initiation deficiency, i.e., delayed arrest of DNA synthesis. None of the gene product has yet been isolated due to the lack of an in vitro assay system of initiation.

Some of the proteins may interact directly with ori-C. Analysis of such interactions is complicated by the fact that ori-C segments apparently complex with membrane proteins[16]. Participation of the membrane in the cell cycle, especially equipartition of daughter chromosomes, was first suggested in the replicon hypothesis. Among the twelve gene products possibly participating in initiation, some may be hydrophobic proteins. This possibility is high in the case of the dna P gene product because it concerns phenethyl alcohol resistance[17].

The first step of initiation should be recognition of ori-C by some proteins. As mentioned earlier, there are indications that RNA polymerase participates in initiation. This possibility is highly probable for the following reasons: In the plasmid col E_1, in vitro initiation of DNA replication is shown to be dependent on RNA polymerase[18]. In E. coli, the initiation is sensitive to rifampycin[9]. Finally, the temperature sensitive initiation mutant dna A can be restored phenotypically by introduction of a special class of rifampycin resistant mutation of the β subunit of RNA polymerase[19].

If RNA polymerase participates in initiation, three possiblities can be envisaged for its mode of participation: Firstly, there may exist a protein which facilitates the interaction between RNA polymerase and ori-C by its capacity to recognize ori-C. Secondly, although RNA polymerase can recognize some promotors in ori-C, for initiation it requires some proteins which switch RNA synthesis to DNA synthesis. And finally, RNA polymerase plays a role in transcriptional activation[20] which separates complementary strands, thus making it possible for dna G protein to synthesize some primers[21].

In any case, initiation of DNA replication involves a fundamental problem in biology, ie. the mechanism of the recognition of promotor sites by RNA polymerase, which has not yet been clearly elucidated. If RNA polymerase simply recognizes a nucleotide sequence, all promotors in E. coli should have a common sequence. However, even though the importance of Pribnow's box has been discussed[22], its base sequence is not constant for all promotors. Therefore, some factors other than a linear base sequence would have to be introduced for recognition. The palyndrome structure or a cruciform of DNA[23] may influence the recognition if they really exist in vivo. However, these hypotheses neglect one possibility: that a base sequence can be easily read with a single-stranded DNA rather than a double helix. Assuming this, sequence recognition may be accompanied by a partial melting of double stranded DNA. Therefore, the stability of a site should be taken into account during recognition. In fact, several researchers have plotted A-T base pair frequency in a DNA according to its base sequence without giving any interpretation. The stability of DNA depends not only on AT or GC pairs, but also on their neighbour sequences. Thus, the same profile of stability can be obtained for different base sequences[24]. The computer program for stability of a DNA based on its base sequence has been described by Gabarro et al[24], who established the program according to the model of Azbel[25]. The virtue of this approach was demonstrated in several cases where the homologous sequences of Pribnow's box (two sites) are always localized in two segments with low stability, separated by one high stability segment[26]. The point mutations in the two low stability sites of Pribnow's box increase the stability, which may interfere with RNA polymerase action[26].

INITIATOR PROTEINS

1. dna A

The existence of initiator proteins has been shown by genetic studies; twelve independently located mutations can affect DNA synthesis in such a way that only initiation is inhibited.

One class of mutant, dna A, has been very well analyzed by many groups. Thus, the location of the gene is precisely demonstrated by cloning techniques (81.8 min) and the gene product at the denatured state was first estimated as 54 K daltons[27]. However, the molecular weight of 48-49 K daltons[28,29] was very recently obtained in experiments in which the identity of dna A protein was assured by isolation of an amber mutant. Another discrepancy between these studies is that, according to the latter, the dna A gene product is a weak basic protein, while according to the former, it is an acidic protein.

What is the biochemical function of dna A protein? The following experiment[30] indicates that it is involved in the step in which RNA polymerase action is required; some dna A-ts mutants arresting DNA synthesis at non-permissive temperature can resume the synthesis when placed back at the permissive temperature. When rifampicyn is added at the moment of shift to permissive temperature, DNA synthesis is not restored. On the contrary, chloramphenicol cannot inhibit this resumption. Therefore, dna A protein functions at the moment of RNA polymerase action. Concomitant with this conclusion, some dna A-ts mutants are phenotypically suppressed by certain rpoB mutations[19]. What then is the precise action of dna A protein? Some indirect experiments have shown that it may mimic the action of the rho factor[31]; several dna A-ts mutants synthesize tryptophane synthetase more efficiently at high temperature than does a wild type strain. This acceleration is observed neither at permissive temperature nor in a dna A-ts having a deletion at the attenuator region of trp operon. The activation is also observed in thr operon[32]. Therefore, a plausible model is that the dna A protein arrests RNA synthesis at some specific site, such as the attenuator site, so that DNA synthesis can take over using RNA of the correct size and position. However, the real situation is more complicated. In the phe A operon, the inactivation of dna A protein decreases biosynthesis of prephenate dehydratase, the opposite effect to that in the trp operon[32].

The complexity of dna A protein resides not only in its action but also in its molecular state. In fact, this temperature sensitive mutation can be phenotypically suppressed by many different gene mutations[33,34]. This type of genetic analysis can indicate the possible complex of dna A protein with other proteins, and also the sites on which the protein reacts. In terms of co-functioning with other factors, it has already been mentioned that the β subunit (rpo B) of RNA polymerase can be envisaged as a partner of dna A protein. Because some mutations of rpo B phenotypically reverse the temperature sensitive mutation of rho[35] one can also envisage that these three proteins cooperate with each other. However, it has not yet been observed that some rho mutations reverse the phenotype of dna A-ts.

The suppression of dna A-ts by rpo B mutation is not a general rule; some rifampicin resistant mutants cannot reverse the temperature sensitivity. Inversely, some rpo B mutations which are not rifampicin resistant can reverse the temperature sensitive growth. One of the particular rpo B dna A suppressor mutations can provoke a stable DNA replication[37] in the presence of chloramphenicol[38]. Thus the cooperative interaction between the β subunit of RNA polymerase and dna A protein have been both genetically and functionally established.

Very recently, a suppressor mutation for dna Z-ts[39] was found in the region where dna A is located[36]. Although the precise location of this suppressor mutation is not at present shown to be in dna A, it may suggest the possibility of the interaction between dna A protein and holopolymerase III[40]. Another result obtained from the study of λ dv replication in the dna A mutant also indicated the same type of hypothesis, that is, an interaction between dna A protein and replication factors.

The basis for this interpretation is that λdv susp mutant could not replicate in dna A-ts at non permissive temperature[41]. Because P protein interacts with dna B protein[42], the triangle interaction between the dna A protein and the P-dna B complex may well be the explanation for this observation. Thus, these observations support the idea of binding of the dna A protein with replication factors, which can also be expressed as a kind of primosome, as found in ϕX174 DNA initiation[43]. This idea coincides well with the fact that there are at least six differently located suppressor mutations for dna A-ts[31].

Among the dna A-ts suppressor mutations, it might be possible to find the mutant of the dna A protein acting site, in particular ori-C. Only one mutation has been found near ori-C without a precise location[31].

2. Other initiation factors

One of the temperature sensitive mutants in which the initiator of replication is clearly shown to be defective is that of one of the dna C mutants[44]. Some other mutations in the same gene shut off elongation[45].

Thus, the dna C protein participates not only in initiation but also in elongation. This bivalency is found even in dna B protein; most of the mutations affect elongation except for dna B 252, which modifies only the initiation. These facts are understandable because the dna B protein forms a complex with the dna C protein[46]. As a same type of initiation factor, the DNA gyrase may also be cited[47]. One of the gyr B mutants shows the same type of DNA replication kinetics as an initiation mutant[48]. However, it is possible that DNA gyrase is required for termination of replication (separation of two daughter chromosomes) and not for initiation.

The final example among well known mutants is the phenetylalcohol and temperature-sensitive initiation mutant called dna P[17]. There are two reasons why the dna P gene product is a membrane protein; the gene controls the resistance to phenetylalcohol, and, at non-permissive temperature, the mutant accumulates less potassium ions than at permissive temperature. This interpretation may be too simple, because phenetylalcohol does not inhibit K^+ accumulation in the wild type.

ORIGIN RECOGNITION

The first problem in origin recognition is in reading the base sequence of ori-C. Since ori-C cannot code for a protein[49], it probably serves as a recognition site for initiation proteins. How, then, do these proteins recognize the sites in ori-C? As an analogy to reading the triplets for protein synthesis, the base sequence of ori-C can be interpreted by certain rules. Thus, the idea of iterons instead of triplets was first presented in lambda phage[50] and then in R6K plasmid[51] and in E. coli[52]. In the case of the lambda phage origin, a block of polynucleotides composed of 18 base pairs was found to iterate four times with a few nucleotides as spacers. However, each iteron is not exactly identical to the others, though the differences are limited to a few bases. These four iterons may form a clover structure which could be involved during origin recognition by the O-protein. In the same type of decoding, we can cite the search for palindromes. The possible existence of a cruciform found in plasmid[23] indicates some importance for the palindromic structure in site recognition, without knowing its mechanism. Admittedly, the modification of the DNA structure at ori-C is important in recognition; it is essential to take into consideration the Z form DNA, a left handed double helix found with oligo d(G-C): d(G-C)[53]. It is logically admitted that during DNA replication, positive supertwists are accumulated in front of replication forks. One way to relieve this structural constraint is to induce left handed DNA in which positive supertwists could be formed. Although the idea is very attractive, the question arises of whether or not a left handed DNA can be obtained in a natural DNA. The form V reported on a plasmid DNA may support its possible existence in a natural DNA[54]. Another interesting observation is that the methylation of cytosine facilitates the formation of a Z form-like structure in poly d(G-mC)[55].

With these observations as keys for decoding the base sequence, let us consider the replication origin of E. coli.

The origin of E. coli replication, defined as a minimum segment which can support an autonomous replication of non-autonomously replicating genes, represents a segment of DNA containing around 240 base pairs[52]. The earliest Okazaki pieces have also been found in this segment, although they stretch in only one direction[56]. The question can be raised as to whether or not this minimum segment covers other functions of ori C, such as membrane attachment or bidirectionality of replication. In fact, there is an indication which suggests the importance of right stretch (between base pair 284 to 489) of the minimum ori for bidirectionality[57]. Within the minimum ori, there are three iterons of a base sequence. Some bases in iterons could be deleted without the loss of ori function for autonomous replication[52]. Therefore, the iterons may not play an important role in ori-C. Another feature of ori-C, noticed in the base sequence, is that the frequency of adenine methylation sites (GATC) is ten times higher than average. Because no MboI sites have been found in ori-C, it is evident that they are methylated. The same is true for cytosine methylase site. One of the interesting interpretations of this fact is that ori-C may be easily transformed to left handed DNA because of these methylations. If methylation of bases is important for ori function, its absence may disturb cell growth. The adenine methylase mutant (dam) was isolated several years ago, but grows normally[58]. However, the fact that the double mutants of dam, either with pol A or rec (A,B,C) are lethal has not yet been clearly explained[59].

Thus, the direct reading of the ori-C base sequence is not mature apparently due to lack of keys in the upper considerations. Therefore, some other method of decoding should be envisaged. As previously described, the RNA polymerase plays an important role in initiation and can be used as the key to understanding the mechanism of ori-C functioning. Thus the promotors for RNA polymerase have been sought in ori-C. In vitro transcription experiments on the minichromosome pCM 959 indicate two main starting points: One located on the left side (167) of the Hind III site stretching to the left, and the other (303) situated on the right side of the Hind III site growing to the right site of the Hind III site growing to the right[60]. The first RNA can be used as a primer for the synthesis of Okazaki pieces, starting either from 106 or 71 with the same polarity as the RNA[56]. These switching points may be created by the action of dna A 'rotein. The replication in the right direction may be initiated by the other RNA, thus leaving a no man's land corresponding to about 150 base pairs between the two promoters. If this scheme is possible what happens on each complementary strand of RNA synthesis? Our recent studies suggest that each complementary strand is involved in membrane attachment. We hypothesized several years ago that ori-C binding to membranes is mediated by a membrane protein having an affinity for DNA. In order to isolate such a hypothetical protein, we began to obtain many possible DNA binding proteins from E. coli membrane fractions. Among seven proteins obtained, we have found one which demonstrates specific affinity for ori-C[61]. This affinity has been shown to be due to the recognition of ori-C DNA by our protein, called B'. By testing many restriction fragments of the mini-chromosome pCM959, two recognition sites have been demonstrated: One is located around Bam H1 (92) and the other between Xho-I-(417) and Pst-1(489). Because the recognition of these sites by B' proteins is only possible with single strand DNA, the polarity of each site has been determined. The first site, around Bam H1 (92), should be on the 3'OH→5P strand, the complementary strand of the template of right side RNA synthesis[62]. These facts suggest the involvement of B' protein in daughter chromosome segregation, as well as in the initiation of replication, because the first recognition site is in the neighbourhood of the first Okazaki pieces' synthesis. However, its intrinsic role can only be determined by isolation of mutants for B' protein.

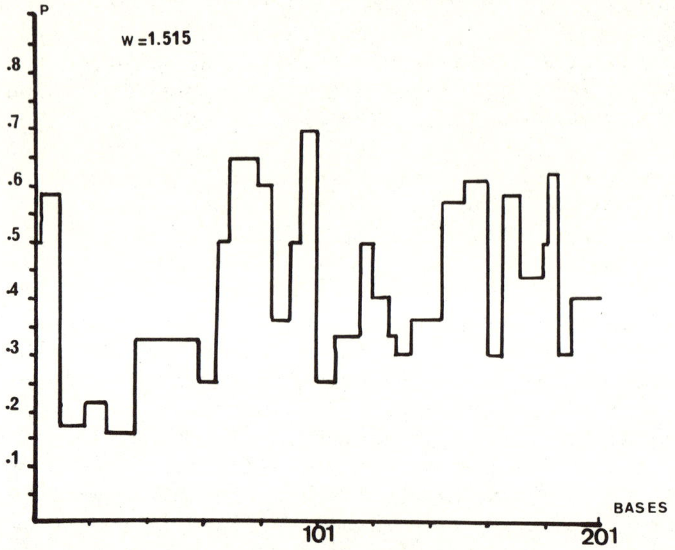

Fig. 1. Stability profile of ori-C sequence extending from base pair 1 to base pair 200[49]. Ordinate: relative stability $p=(T_m-T_{AT})/T_{GC}-T_{AT})$[24].

Finally, it should be taken into consideration that the base sequence decoding in double-stranded DNA might be accompanied by DNA melting. Therefore, it is important to measure the melting curves of ori-C DNA and whether or not low or high melting regions have something to do with palyndromes, iterons or initiation signals. As described earlier, the stability of a DNA whose base sequence is known can be calculated according to the equations of Azbel.

Figure 1 shows the stability of ori-C DNA calculated by this method. However, one of the particularities of ori-C DNA is the fact that it has many 6-methyladenines. The replacement of adenine by 6-methyladenine in poly d(A-T) had been shown to drastically decrease the melting temperature[63]. Taking the neighbour sequence of 6-methyladenine (G-6mA T C) into account, the relative stability of 6-methyladenine T is then introduced into the calculation. The stability curve of ori-C thus obtained is shown in Figure 2. The introduction of methyl adenine effectively lowers the stability, especially in the region between 101 and 161. Interestingly, from this unstable region, the first Okazaki piece as well as RNA can both be synthesized in the left-hand direction. Several sharp drops in this region offer more promotor sites for RNA synthesis.

A preliminary result from the experiment of the B' protein protecting region against DNase digestion shows that, in fact, B' protein may cover the region between 105 and 144 of the nontemplate strand of the first Okazaki piece synthesis. This coincidence may enhance

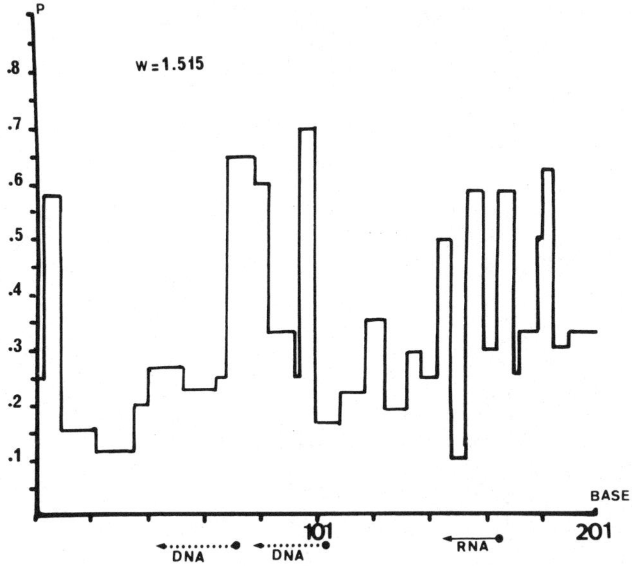

Fig. 2. Stability profile of ori-C sequence when the adenine bases in the adenine methylated sites have been methylated.

o.... Possible start of an RNA primer synthesis, according to Okazaki et al[56].

o—— Possible start of an RNA primer synthesis, according to the localization of the RNA polymerase promoters within ori-C[60].

the importance of this region, as well as the importance of the stability of DNA. Thus it is very important to obtain the experimental value for the stability of ori-C DNA.

REFERENCES

1. F. Jacob, S. Brenner and F. Cuzin, Cold Spring Harbor Symp. Quant. Biol. 28, 329, 1963.
2. O. Maaløe, Cold Spring Harbor Symp. Quant. Biol.26, 45, 1961.
3. M. Kohiyama, H. Lamfrom, S. Brenner and F. Jacob, C. R. Acad. Sci. 265, 1820, 1963.
4. I. R. Lehman, M. J. Bessman, E. S. Simms and A. Kornberg, J. Biol. Chem. 239, 222, 1958.
5. Y. Hirota, A. Reyter and F. Jacob, Cold Spring Harbor Symp. Quant. Biol. 33, 677, 1968.
6. M. Kohiyama, Cold Spring Harbor Symp. Quant. Biol. 33, 317, 1968.
7. P. De Lucia and J. Cairns, Nature 224, 1164, 1969.
8. M. Gefter, Y. Hirota, T. Kornberg, J. Wechsler and C. Barnoux, Proc. Natl. Acad. Sci. USA, 68, 3150, 1971.

38. T. Kogoma, T. A. Torrey and T. Atlung, J. Supramolec. Struct. and cell Biochem., suppl. 5, 340, 1981.
39. C. C. Filip, J. S. Allen, R. A. Gustafson, R. G. Allen and J. R. Walker, J. Bacteriol., 119, 443, 1974.
40. W. Wickner and A. Kornberg, J. Biol. Chem., 249, 6244, 1974.
41. G. Kellenberger-Gujas and A. J. Podhasjska, Mol. Gen. Genet., 126, 17.
42. R. D'Ari, A. Jaffe-Brachet, D. Touati-Schartz and M. Yarmolinsky, J. Mol. Biol., 94, 341, 1975.
43. R. L. Low, K. Arai and A. Kornberg, Proc. Natl. Acad. Sci., 78, 1436, 1981.
44. W. H. Schubach, J. D. Whitmer and C. I. Davern, J. Mol. Biol., 74, 205, 1973.
45. P. Carl, Mol. Gen. Genet., 109, 107, 1970.
46. S. Wickner and J. Hurwitz, Proc. Natl. Acad. Sci. USA., 72, 921, 1975.
47. M. Gellert, K. Mizobuchi, M. H. O'Dea and H. A. Naoh, Proc. Natl. Acad. Sci. USA., 73, 3872, 1976.
48. E. Orr, F. Neil, I. Fairweather, B. Holland and R. Pritchard, Mol. Gen. Genet., 177, 103, 1979.
49. K. Sugimoto, A. Oka, H. Sugisaki, M. Takanami, A. Nishimura, Y. Yasuda and Y. Hirota, Proc. Natl. Acad. Sci. USA., 76, 757, 1979.
50. D. Moore, K. Denniston-Thompson, K. Kruger, M. Furth, B. Williams, D. Daniels and F. Blattner, Cold Spring Harbor Symp. Quant. Biol. 43, 155, 1978.
51. D. Stalker, R. Kotler and D. Helinski, Proc. Natl. Acad. Sci. USA., 76, 1150.
52. A. Oka, K. Sugimoto, M. Takanami and Y. Hirota, Mol. Gen. Genet. 178, 9, 1980.
53. A. M. Wang, G. J. Quigley, F. J. Kolpak, J. L. Crawford, J. H. van Boom, G. van der Marel and A. Rich, Nature, 282, 680, 1979.
54. U. H. Stettler, H. Weber, T. Koller and C. Weissman, J. Mol. Biol. 131, 21, 1979.
55. M. Behe and G. Felsenfeld, Proc. Natl. Acad. Sci. USA., 78, 1619, 1981.
56. T. Okazaki, S. Hirose, A. Fujiyama and Y. Kohara, Mechanistic studies of DNA replication and genetic recombination. ICN-UCLA Symp. Molec. Cell. Biol., Bruce Alberts ed. (Academic press, New York), 19, 429, 1980.
57. M. Meijer and W. Messer, J. Bacteriol., 143, 1049, 1980.
58. M. G. Marinus and N. R. Morris, J. Bacteriol., 114, 1143, 1973.
59. M. G. Marinus and N. R. Morris, J. Mol. Biol., 85, 309, 1974.
60. H. Lother, H. Buhk, G. Morelli, B. Heimann, T. Charkraborty and W. Messer, Structure and DNA-protein interactions of replication origins. ICN-UCLA Symp. Molec. Cell.Biol., Bruce Alberts ed. in press.
61. A. Jacq and M. Kohiyama, Eur. J. Biochem., 105, 25, 1980.

62. A. Jacq, H. Lother, W. Messer and M. Kohiyama, Mechanistic Studies of DNA replication and Genetic Recombination. ICN-UCLA Symp. Molec. Cell. Biol., Bruce Alberts ed. (Academic press, New York), 19, 189, 1980.
63. J. D. Engel and P. H. von Hippel, J. Biol. Chem., 253, 927, 1978.

THE TERMINUS OF CHROMOSOME REPLICATION OF *E. coli* PHENOTYPIC SUPPRESSION OF A *dnaA* MUTATION BY PLASMID INTEGRATION NEAR *terC*

Jacqueline Louarn, Philippe Legrand, Josette Patte
Jean-Michel Louarn

Centre de Recherche de Biochimie et de Genetique
Cellulaires du C.N.R.S.
118 route de Narbonne
31062 Toulouse Cedex, France

The existence of a fixed termination region for chromosome replication in *E. coli* has been proposed[1,2,3,4]. In particular, we have shown that when a *dnaAts* mutation is phenotypically suppressed by an integrated plasmid (Integrative Suppression[5]), the replication forks initiated from the plasmid always meet in the *rac* (min 30)-*man* (min 35.5) region, irrespective of the plasmid insertion site on the chromosome. The terminus of replication, *terC*, was thus described primarily as a locus inhibiting replication fork movement in either direction. In addition, the termination step might be involved in regulatory operations of the cell cycle, as previously proposed[6,7], but this possibility remains poorly documented. In the course of our previous analyses, as well as in other studies on integrative suppression by plasmid R100 derivatives[8], integrative suppression by plasmid integration in a large region surrounding *terC* (grossly between 15 min and 45 min on the genetic map of Bachmann et al[9] was never observed. If the restriction in the distribution of integration sites along the chromosome is related to *terC* functions, its analysis could constitute a way to investigate the role of the terminus.

Using a system of site-specified suppressive integration, we have been able to show that plasmid integration near *terC* confers a conditional phenotype Sin (for suppressive integration) on a *dnaAts* mutant: cell viability at high temperature is observed only in poor growth media, but not in rich media such as that used in previous work for the selection of integratively suppressed Hfr's. These Rms (for rich medium sensitive) bacteria present a second constraint for integrative suppression: for integration sites closest to *terC*, only one of the two possible plasmid orientations is stable. Chromosome replication in Rms bacteria was analysed, and indications that the stop signal at terC can be transgressed by replication forks have been obtained. Derivatives of Rms bacteria, resulting either from plasmid transposition or from chromosomal mutation, have been selected for growth ability in rich medium. A preliminary characterization of one of these mutations is presented here.

SITE-SPECIFIED INTEGRATIVE SUPPRESSION. THE EXPERIMENTAL DESIGN.

We have postulated that if the chromosome and the plasmid share a common sequence of sufficient size, homologous recombination could greatly favor plasmid integration within the chromosomal homologous sequence. Prophage Mu insertions proved to be valid in this respect. When a cointegrate pAR132::Mu such as pLN1 or pLN2 (see figure 1; pAR132 is an isolate of the RTF component of plasmid R100-1) was introduced into a dna46 mutant[13] harboring a prophage Mu integrated in the malA locus (min 74.7), the resulting strain displayed a colony-forming ability at 42° close to 7.10^{-3}, more than 100 times higher than that observed using the parental plasmid pAR132. These tr derivatives were all proficient for chromosome transfer, and transfer data indicate that : i) the origin of transfer is located in the vicinity of the chromosomal Mu insertion; ii) the direction of transfer is, in most cases, the unique direction predictable from the respective orientations of the Mu prophages, assuming that a homologous genetic exchange between the plasmid and the chromosome is responsible for chromosome mobilization. Direct evidence for the absence of autonomous plasmid in these strains was obtained by sedimentation analysis of an alkaline lysate, as described by Freifelder[14]. Thus, integrative suppression in this case is preferentially obtained through a homologous recombination between the Mu sequences.

Integration by homologous recombination is reversible : after two successive subclonings at 30° of the Sin Hfr's thus formed, all the resulting colonies tested contained few if any tr bacteria : in most cases the bacteria had returned to the R^+ ts state; in 5% of the colonies, the bacteria had lost the plasmid, as inferred from the loss of the tetracycline resistance encoded by this plasmid. This indicates that the integrated state is not maintained once the selective pressure for temperature resistance is removed, and clearly supports the idea that plasmid integration is responsible for the tr character.

PLASMID INTEGRATION NEAR terC : CONDITIONAL PHENOTYPE Sin

When the chromosomal Mu insertion is located in the terminus region (prophages Mu inserted for instance in gal, min 17, trp, min 27, man, min 35.5 and ptsM, min 40; see figure 1), integration of pLN1 or pLN2 within the resident prophage is never detected when the

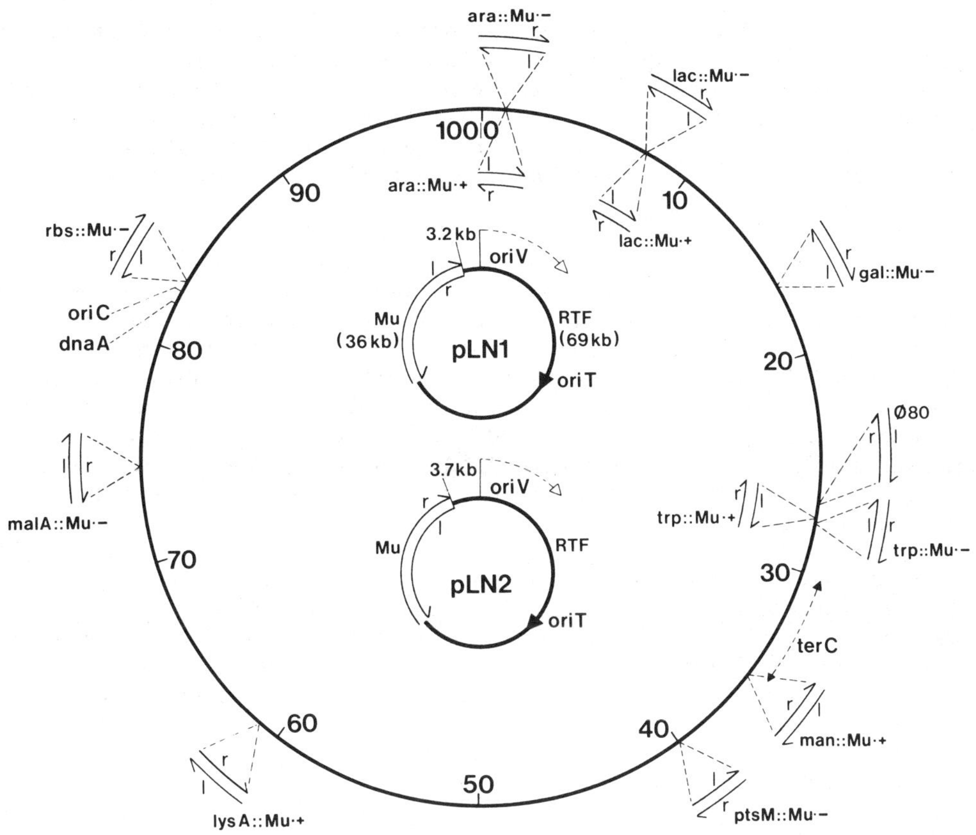

Fig. 1. Prophage insertions and plasmid structure.
The figure shows the location on the E. coli chromosome of the different Mu prophages used as specific sites for pLN1 and pLN2 integration, and of the prophage ϕ80 used as physical marker. Prophage Mu orientations were established by chromosome mobilization, using pLN1 and pLN2 as mobilizer plasmid[2]; + or - orientations are given according to Campbell et al[10]. Moreover, the trp:: Mu and man:: Mu insertions have been located and orientated on the 500 kb long physical map of the terminus region, recently constructed in our laboratory by J.P. Bouché (to be published). Prophage ϕ80 orientation is from[11]. Arrowheads on the prophage strands represent 3' ends. The schematic structure of pLN1 and pLN2 is redrawn from[2,12] and unpublished data of P. Legrand. With respect to the origin of replication, oriV, and the orientation of unidirectional replication from this origin, prophage insertion sites and orientations are given; the position of the origin of transfer, oriT, and the direction of conjugational transfer from this origin are also shown.

search for Sin Hfr's is carried out using a rich growth medium such as L Agar[15] : Sin Hfr's appear with a low frequency on this medium (about that observed when the parental plasmid pAR132 is used as a suppressor agent), and display sites of plasmid integration scattered over the chromosome. Clearly, integrative recombination of the plasmid within these chromosomal prophages cannot lead to a viable Sin phenotype under these conditions.

The inhibition of integrative suppression near terC is removed when the selection of tr clones is carried out in minimal medium, such as M9 medium[15] containing 1% glycerol or 1% aspartate as the carbon source. On this medium, a high proportion of tr clones can be detected (4 to 8.10^{-3}) which, using all the criteria presented above, appear to contain Sin Hfr bacteria formed by plasmid integration within the chromosomal prophage. As expected, the tr bacteria obtained on M9 medium are rich medium-sensitive (and thus designated as Rms bacteria below), and plate at a low efficiency (10^{-2} to 10^{-4}) on LA plates incubated at 42°.

DNA synthesis and cell mass increase in a typical Rms Sin Hfr (formed by pLN2 integration within a trp:: Mu insertion) and in a medium-resistant Sin Hfr (formed by pLN2 integration within a lysA : : Mu insertion) were compared under different growth conditions. No gross difference was detected between the two strains incubated at 42° in M9-aspartate (the generation time T was close to 300 min.), M9-glycerol (T close to 180 min) or M9-glucose (T close to 150 min). As illustrated in figure 2, upon transfer from M9-glucose to LB medium and incubation at 42°, both strains displayed a rapid increase in the rate of cell mass synthesis with a doubling time ranging from 30 to 50 min. In the medium-resistant strain, the rate of DNA synthesis remained unchanged for 1 hour, then increased to a doubling time of about 60 min, which is the generation time of this Hfr cultivated in LB medium under steady state conditions. In the Rms strain, on the contrary, the nutritional shift-up had no effect on the rate of DNA synthesis, even after a 4 hour incubation period. The unbalanced growth of Rms bacteria is accompanied by intensive filamentation, and eventually by cell death, the colony forming ability dropping to 2% of the initial value within 3 or 4 hours. When the Rms Sin Hfr was lysogenic for bacteriophage λ, some phage induction was also observed, since the amount of free phages per OD unit of the culture increased 5 to 10 fold after a 4 hour incubation in LB medium. Lethal growth might thus be accompanied by some induction of the SOS repair system. The following sequence of events, which remains to be proven, can be proposed : filamentation and death could be consequences of SOS induction (as observed in tif mutants[16]), due to the imbalance between DNA synthesis and cell mass synthesis triggered by the nutritional shift-up. If this is true, the primary defect in Rms bacteria should be their inability to initiate replication rounds at a higher rate than the one attained in M9-glucose medium at 42°.

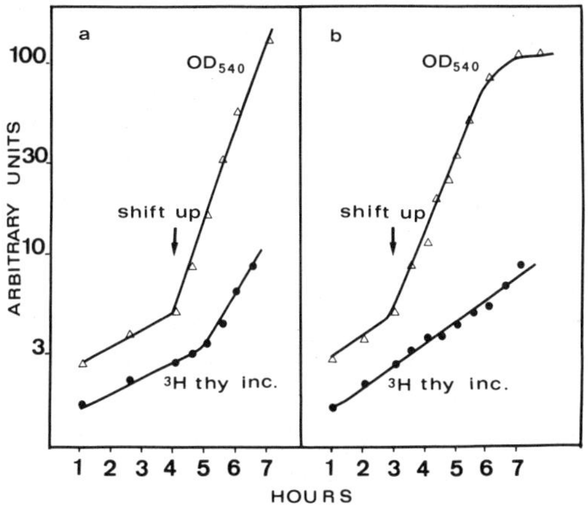

Fig. 2. Unbalanced growth in Rms Hfr's after a nutritional shift-up. Temperature-resistant clones obtained by insertion of pLN2 in a lysA::Mu insertion (a) and in a man::Mu insertion (b) were grown in M9-glucose at 42° in the presence of [^{3}H]-thymine (4μCi/10μg/ml). At an O.D.$_{540}$ nm close to 0.5, the cells were diluted 5-fold in prewarmed LB medium (8μCi/20 μg/ml of [^{3}H]-thymine), and incubated at 42°. At intervals, samples were removed to determine the TCA-precipitable radioactivity and the optical density at 540nm.

PLASMID INTEGRATION NEAR terC : THE ORIENTATION CONSTRAINT

Integration of pLN1 or pLN2 in sites relatively remote from terC, such as ptsM: :Mu and gal: :Mu, results in Rms Sin Hfr's which contain the suppressor RTF component in either orientation within the chromosome. The situation is different when plasmid integration in sites closer to terC (man: :Mu and trp: :Mu) is specified : whatever the plasmid used, all the stable Sin Hfr's formed by integration within these chromosomal prophages promote a late conjugational transfer of terC. When the opposite orientation of the integrated plasmid is the result of a direct recombination between the two prophages, an additional inversion event occurring between the invertible G regions[17] of the two Mu prophages is implied (data not shown). Since, in this situation, a mixture of "direct" and "inverted" Hfr's is frequently detected in recently isolated tr clones, we believe that a selective advantage is attached to a particular orientation of the suppressor plasmid integrated near terC.

The phenomenon of orientation preference for suppressive integration of plasmid R100 derivatives is not restricted to the termination region. Using the site-specified integration system, we have also observed that among Sin Hfr's able to grow in rich medium, those formed by plasmid integration midway between oriC and terC on both oriC-terC arms display the highest viability when they promote chromosomal transfer in the direction terC first to oriC late (Figure 3). It is puzzling that this orientation is opposite that observed for the closest integration sites to terC.

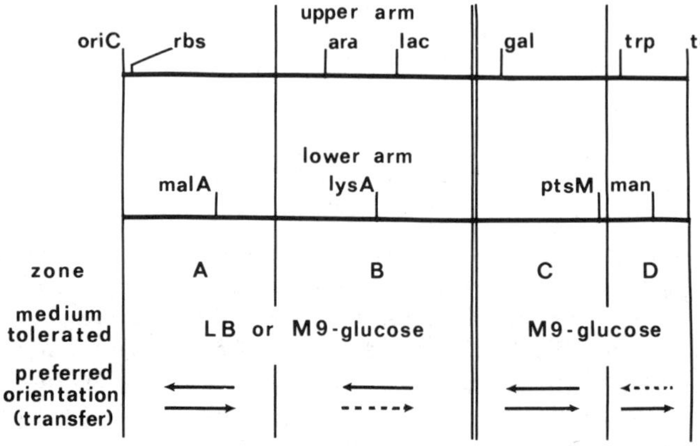

Fig. 3. Chromosome zoning with respect to integrative suppression by pLN1 and pLN2.
We have determined the medium tolerance and the orientation stability of Sin Hfr's resulting from the integration of pLN1 or pLN2 within the various Mu chromosomal prophages indicated on the map. The symmetrical disposition of the different zones on both oriC-terC arms is given only tentatively, also taking into account previous results[1]. Plasmid orientations are indicated by arrows, which refer to directions of conjugational transfer in Sin Hfr's. Preferential orientations are indicated by continuous arrows; disadvantaged ones by broken arrows.

The fact that the zones where a certain orientation of the plasmid is preferred are symmetrically distributed around the oriC-terC axis, indicates that plasmid orientation plays a role in Sin Hfr viability through plasmid-originated asymmetric event(s) probably related to chromosome replication. It is worth noting here that the major mode of replication of the autonomous plasmid is the unidirectional one[18,19,12]; reported in Figure 1; bidirectional replication has been detected only for the integrated state[1,20], or for a minority of replicating molecules examined under synchronous conditions[19]. The mode of chromosome replication under plasmid control thus appears as a possible essential factor controlling medium restriction and orientation preference. Some aspects of this replication mode are presented in the following section.

BEHAVIOR OF PLASMID-INITIATED FORKS REACHING terC

Although shown to be bidirectional[1,20] and to terminate at terC[1], chromosome replication under the control of the R100 replication machinery has not yet been precisely described. We report here studies on replication in Rms Sin Hfr's, which were designed to demonstrate possible differences in the recognition of terC by opposite replication forks.

The mode of chromosome replication in Rms Sin Hfr's has been investigated by determining the direction of replication and relative frequency of 3 markers: the two Mu prophages involved in the integration process of the suppressor plasmid, and a ϕ80imm$^{\lambda}$ prophage integrated at attϕ80 (min 27).

The utilization of plasmid origin and the bidirectionality of replication was demonstrated using the trp::Mu strain family. When replication is initiated at oriC, Okazaki fragments made during the replication of the prophage ϕ80 are about 80% r strand DNA; when replication is shifted under the control of pLN1 or pLN2 integrated within trp::Mu, the opposite strand preference is observed among the ϕ80-specific Okazaki fragments (Table 1 and Figure 4). Since prophage ϕ80 is located between trp and oriC (at 0.2 min from trp; figure 1), the inversion in the direction of replication of this prophage is a clear indication of the use of the plasmid origin in the Sin Hfr's. In the same Hfr's, the genomic ratio Mu/ϕ80 (measured by hybridization using pulse-labeled DNA to take into account replicating bacteria only) was found to be close to 2 (Figure 4). Considering the close vicinity of attϕ80 and trp, this result is consistent with an identical number of copies per replicating cells of the sequence flanking the integrated plasmid, that is with an initiation process occurring roughly synchronously on both sides of the suppressor plasmid origin. The observation that Mu-specific Okazaki fragments extracted from the Sin Hfr due to pLN2 integration within trp::Mu display almost no polarity is also consistent with bidirectional replication from the integrated plasmid (Table 1).

The behavior of replication forks reaching terC from an origin located near terC was analyzed in the ptsM::Mu and man::Mu strain families. These strains are characterized by the presence of terC on the short chromosomal arm separating the prophage ϕ80 from the Mu insertion; the ϕ80 prophage direction of replication should therefore depend upon the efficiency of the replication fork blockage at terC. Results obtained are reported in Table 1 and in Figure 5. In none of these experiments are the ϕ80-specific Okazaki fragments highly polarized, as if this prophage has a tendency to be replicated in either direction in these Sin Hfr's. A clear indication that the terminal block is not absolute was obtained: when replication is under the control of pLN2 integrated within man::Mu, the direction of replication of the prophage ϕ80 is mainly opposite that observed

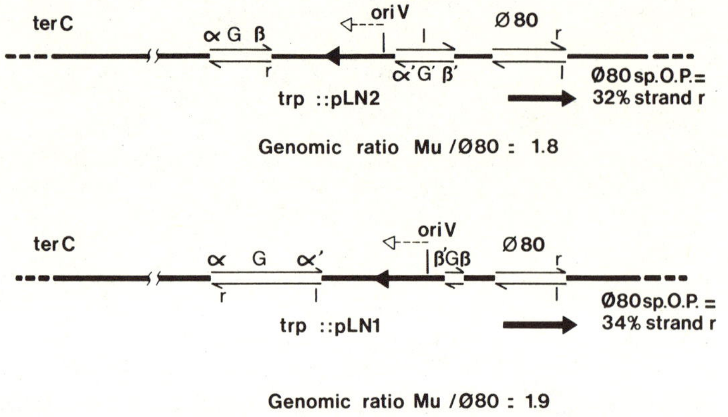

Fig. 4. Bidirectional replication in Rms Sin Hfr's. The figure shows the organization of the Mu, RTF and ϕ80 DNA's in a "direct" Hfr derived from LN521 (upper part) and in an "inverted" one (lower part) derived from LN811. The direction of replication of the ϕ80 prophage is given, deduced from data of table 2. The genomic ratio Mu/ϕ80 was established as follows: cultures of the Hfr strains were performed in M9-glucose medium at 42°, as explained in figure 2 legend, then pulse-labeled at 42° for 2 min after a 2 min period of thymine starvation (500μCi [^{3}H] thymidine/50 OD_{540nm}units) ; a culture of the F^{-} strain LN783 was pulse-labeled similarly to serve as control ; DNA extraction and hybridization to filters loaded with Mu or ϕ80 DNA were performed as described in Louarn et al[1] ; the genomic ratio Mu/ϕ80 in the Hfr strains was calculated by dividing the ratio: radioactivity fixed to Mu DNA filter / radioactivity fixed to ϕ80 DNA filter, by the same ratio obtained using LN783 pulse-labeled DNA; in this latter strain the two prophages Mu and ϕ80 should be exactly at the same number of copy per cell, considering their proximity on the genetic map.

when replication is initiated at oriC. Some terminus transgression is also observed in the direct Hfr formed by pLN1 integration within ptsM::Mu. However, since in the direct Hfr formed by pLN2 integration within the same Mu insertion, the prophage ϕ80 is replicated mainly in the same direction as that from oriC, plasmid orientation appears to be involved in terminus recognition (these experiments were repeated several times, with very reproducible results). As shown in Figure 5, the fork initiated in the same direction as in the unidirectional replication of the autonomous plasmid seems to display the greatest ability to transgress terC. It is tempting to correlate

Table 1. Strand preferences of Okazaki fragments made during prophage Mu or prophage $\phi 80$ replication in various strains.

Strain	Mu insertion	plasmid	Ratio : strand r/strands r+1 among Okazaki fragments specific of: Mu prophage	$\Phi 80$ prophage
LN783	trp::Mu	none	0.31	0.74
LN521	"	pLN2	0.25	0.82
Hfr LN521 direct	"	"	0.46	0.32
Hfr LN811 inverted	"	pLN1	0.50	0.34
LN638	ptsM::Mu	none	0.72	0.73
Hfr LN534 direct	"	pLN1	0.51	0.50
Hfr LN535 direct	"	pLN2	0.44	0.61
LN774	man::Mu	none	0.70	0.79
LN776	"	pLN2	0.71	0.77
Hfr LN776 direct	"	pLN2	0.52	0.40

Directions of replication were established by determining the strand preference of short Okazaki fragments made during the replication of the different prophages. Unidirectional replication of a given prophage always results in a 70 to 80% strand preference in our hands[1]. The culture of the Rms clones was achieved in M9-glucose supplemented with thymine, leucine and isoleucine (in the course of the present work, it was noticed that leu Sin Hfr's were pseudoauxotrophs for isoleucine, for unclear reasons). At $O.D_{540nm}$ close to 0.5, the cells were shifted into M9 medium without thymine for 2 min (to deplete thymine nucleotide pools) before labeling with [^{3}H]-thymidine for 15 sec. Arrest of incorporation, cell lysis, purification of short (< 10 S) Okazaki fragments, and DNA-DNA hybridization to isolated phage strands were performed as previously described[1,2].

this observation with the requirement for a certain plasmid orientation: the best viability for suppressive integration by the RTF machinery in the vicinity of terC is obtained when the fork reaching this site first is the one most able to move through it.

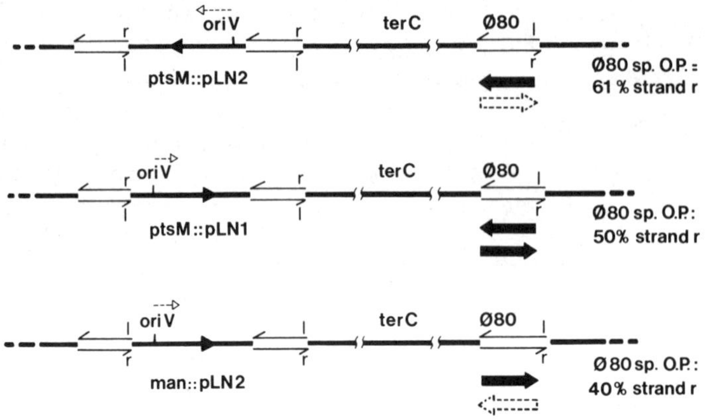

Fig. 5. Replication of the terC region in Rms Sin Hfr's. The figure shows the organization of Mu, RTF and ϕ80 DNAs in Rms Hfrs derived from LN534, LN776, and the proposed directions of replication of the ϕ80 prophage, as deduced from the strand preferences of Okazaki fragments specific to this prophage, in the different Hfr's (table 2).

THE rms-1 MUTATION

In a culture of Rms Sin Hfr bacteria, derivatives able to grow on rich medium at 42° are relatively abundant (up to 10^{-2}). We have begun the study of 90 clones able to grow on rich medium, which were independently derived from the Rms Sin Hfr obtained by pLN2 integration within man::Mu. The Rmr (rich medium resistance) phenotype of the 90 clones was associated: i) with plasmid transposition to a site closer to oriC in 19 "transposed" Hfr's; ii) with a loss of transfer ability in 15 clones; iii) with the maintenence of the original transfer properties in 56 clones; among these, 16 were able to grow on minimal as well as on rich medium at 30° or 42°, and 40 displayed a Rms phenotype at 30°. At present, only the genetic control of this latter but predominant phenotype has been investigated. Further characterization of a clone belonging to the major class provided the following information.

Maintenance of the Sin phenotype

Taking advantage of the fusaric acid sensitivity of cells harboring a Tn10 transposon[21], Tc^s derivatives of the Rmr clone have been obtained. This was done at 30° on M9 medium. The loss of the plasmid, presumably obtained by the selection procedure, is accompanied by a return to a ts state, indicating that the ability to grow at 42°

is under the control of the integrated plasmid in the original Rmr bacteria. One Tc^s derivative, named LN850, was further examined. Its Rms phenotype at 30° is preserved, so that LN850 cells appear viable only in M9 medium at 30°.

Involvement of a chromosomal mutation

Due to the relative instability of the Hfr state in site-specified plasmid integrations, the plasmid present in the Rmr clone parent of LN850 could be introduced into other dnaA46 strains. Once transferred, the plasmid behaves in all respects identically to pLN2. On the other hand, the reintroduction of pLN2 into LN850 cells was possible, indicating that LN850 is indeed plasmid-free. Temperature-resistant derivatives from LN850/pLN2 could be obtained at the high level characteristic of site-specified integrations, and the resulting Sin Hfr's transfer their chromosome from man and are able to grow on rich medium at 42°. In addition, the parent plasmid of pLN2, pAR132 (figure 1), has been introduced in LN850, and temperature resistant derivatives have been selected on rich medium. These clones are, as expected, Hfr's but they display no restriction for plasmid integration in the terminus region, contrary to the original dnaA46/pAR132. These data clearly indicate that a chromosomal mutation is responsible for the medium tolerance of Sin Hfr derivatives of LN850, even when the plasmid integration sites are close to terC.

The identity test

The frequent occurrence of the mutant type exemplified by LN850 makes it likely that the mutation responsible for the medium tolerance at 42° of its Sin Hfr derivatives is also responsible for its Rms phenotype at 30°. To test this possibility, several spontaneous mutants of LN850 able to grow on LA medium at 30° have been isolated and rendered R^+ by introduction of pLN2. Subsequently, Sin Hfr's were derived, which turned out to result from pLN2 integration within man::Mu and to harbor a Rms phenotype at 42°. These observations sustain the contention that strain LN850, in addition to the dnaA46 mutation, harbors a second mutation, provisionally named rms-1, which is characterized by the following properties : i) its presence corrects for the Rms phenotype of terminally integrated Sin Hfr's; ii) its presence induces a Rms phenotype which is expressed when chromosome replication is initiated at oriC, but not from a plasmid origin.

Mapping of the rms-1 mutation

Conjugations using a set of donor Hfr's and a streptomycin-resistant derivative of LN850 as recipient, followed by the selection of ex-conjugants able to grow on L Agar, have been performed. The rms$^+$ allele is frequently transferred by Hfr PK3 (approximate origin position 78 min; clockwise transfer), but not by Hfr KL25 (approximate origin position 84 min; clockwise transfer). About 90% of the rms$^+$ ex-conjugants also acquired the dnaA$^+$ allele, indicating that the rms and dnaA genes are linked but separable. Attempts to co-transfer the rms-1 mutation with a pyrE mutation (min 81.3) and a rbs mutation (min 83.8) by P1 transduction have failed. At present, the most likely map position of the rms gene is thus the 78-80 min region. Work is in progress to refine this preliminary mapping.

Growth of rms-1 bacteria

In the original Rmr Hfr derivative of the Rms Sin Hfr formed by pLN2 integration within man::Mu insertion, incubation at 42° in LB medium results in a generation time of 50 min. Clearly, the inhibitory effect on plasmid replication initiation caused by its vicinity to terC is suppressed in the rms-1 mutant.

DNA cell mass synthesis have been compared in the female strain LN850 and its rms$^+$ ancestor LN770 (figure 6). In M9-glucose, the two strains grow with identical growth rates. A 1,5 higher DNA/mass ratio is, however, observed in LN850 (it is worth recalling that this ratio is low in dnaA46 bacteria, about half that of dnaA$^+$ cells: 1). Upon transfer to LB medium and incubation at 30°, DNA synthesis was found to rapidly equilibrate at a faster but identical rate in the two strains (doubling time 45 min). The rate of cell mass synthesis also increases in these two strains but, whereas DNA and mass synthesis are rapidly balanced in the parent LN770, the rms^{-1} mutant bacteria display a rate of mass synthesis which is clearly lower (doubling time 60 min after a 1 hour incubation) than that of DNA synthesis. A 6 to 7 hour incubation of LN850 bacteria in LB medium (with serial dilution to keep the OD540nm between 0.2 and 0.5) results in a doubling of the generation time. No gross effect on cell viability measured on M9-glucose plates, was detected in the experiments, indicating that in rich medium rms^{-1} bacteria are not killed but divide for a number of generations insufficient to result in a visible colony. Under non-permissive conditions, these bacteria are characterized by an uncoupling of DNA and cell mass synthesis resulting in an increase in the DNA content per cell (figure 7).

DISCUSSION

Taking advantage of a plasmid integration system programmable for a number of defined sites on the chromosome, we have determined the conditions required for viable replication of the chromosome under the control of a plasmid installed in the vicinity of the replication terminus.

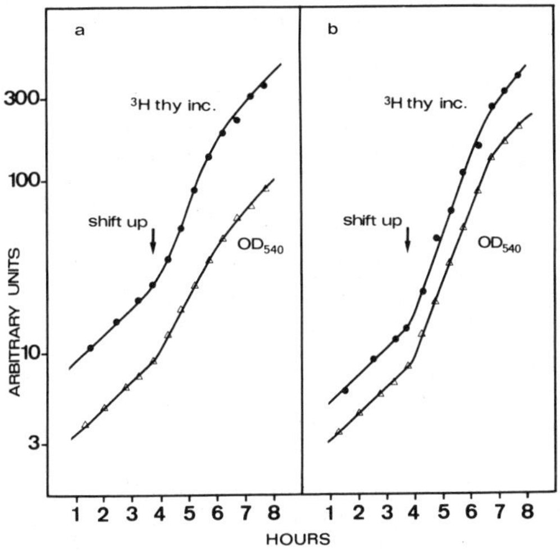

Fig. 6. Growth of rms-1 bacteria after a nutritional shift-up. Growth of the F^- dnaA46 rms-1 LN850 strain was compared to that of its direct ancestor LN770 (F^- dnaA46 man::Mu). The two strains were grown in M9-glucose medium in the presence of [^{3}H] thymidine (4μCi/10μg/ml) up to OD_{540} nm close to 0.5, then diluted in pre-warmed LB medium (8μCi/20μg/ml of [^{3}H] thymine) and incubated at 42°. At intervals, samples were removed to determine the TCA-precipitable radioactivity and the optical density at 540nm.

(a) : LN850

(b) : LN770

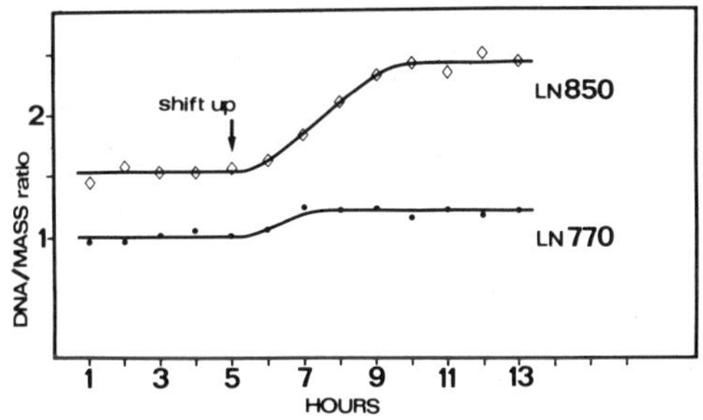

Fig. 7. Evolution of the DNA/cell mass ratio in LN850 cells after a nutritional shift-up.

Data of figure 6 were used to calculate the DNA/mass ratio of LN850 cells as compared to that displayed by LN770 cells. Arbitrarily, the DNA/mass ratio of LN770 in M9-glucose was taken as 1.

Our main observation is that integration of the R100 replication machinery near terC leads to cells displaying a conditional Sin phenotype, viable at 42° in minimal medium only. Rms Sin Hfr's, when transferred into rich medium, seem to die due to under-initiation of chromosome replication, perhaps through induction of the SOS system. Our proposal is that proximity to terC of the effective origin of replication may exert an inhibitory effect on the frequency of initiation which overcomes the "normal" control of this frequency, presumed to be exerted by the cell mass to origin ratio[22]. Among the possible models, the frequency of initiation might be dependent upon the duration of the replication cycle, in addition to other "normal" controls. For instance, some factor required for initiation at the plasmid origin could also be engaged in the replication complex, and reach the initiation concentration only when liberated from the complex at the termination step.

The second constraint observed in Sin Hfr's formed by plasmid integration near terC is that of a certain plasmid orientation. We also observed that plasmid-initiated forks seem to recognize terC with different efficiencies. Since in the favored orientation near terC, the fork travelling along the short plasmid-terC arm is the one displaying the highest ability to move through terC, the capacity of terminus transgression might confer a selective advantage. It should tend to decrease the replication time (i.e. the time elapsing between an initiation event and the subsequent completion of chromosome replication), and thus favor initiation at the plasmid origin, according to the model presented above. An improved description of the mechanism of chromosome replication under plasmid control is needed to clarify the role of plasmid orientation for Sin Hfr viability.

Whatever the precise mechanisms, the observation of the constraints restricting suppressive integration near terC, strengthens the contention that this site is important in the cell cycle. The role of terC, in addition to its function as a point of arrest, absolute or not, for replication fork movement, is actually not defined by these experiments. However, one point deserves mention : since transgression of the terminus (i.e. unidirectional replication of terC) is compatible with cell viability, it seems unlikely that fork function at terC is required for the occurrence of some subsequent step of the cell cycle.

Our last comment concerns the rms-1 mutation. So far, the effects recognized for this mutation are : i) a suppression in Sin Hfr's of the inhibition of replication initiation at the plasmid origin when the plasmid is integrated near terC; ii) in rms-1 bacteria incubated in rich medium, an uncoupling of the DNA synthesis to cell mass synthesis when replication is initiated at oriC; this uncoupling is apparently due to a lower rate of cell mass increase than in rms^+ bacteria. There are many possible explanations for the medium effect on cell mass synthesis in F^- rms-1 bacteria. One might suggest, for instance, mutations provoking non-regulated expression of genes controlling non-compensable metabolic steps. Since Rms Sin Hfr's grown in a rich medium die due to under-initiation, a mutation-induced lower rate of cell mass synthesis might tend to restore cell viability under these conditions. In this view, however, it is hard to explain why such "corrected" Hfr's display generation times remarkably short for Sin Hfr's. This and the uncoupling effect observed in F^- rms-1 bacteria plead in favor of a specific action in the cell cycle regulatory network for the rms gene. To our knowledge, the phenotype attached to the rms-1 mutation is unique in that it represents an example of bacteria displaying a normal response of DNA synthesis to culture condition changes, associated with an altered response of cell mass synthesis. Another example of an uncoupling between DNA and cell mass synthesis is offered by the partial revertant of dnaA46, dnaAcos, but this mutant is characterized by an over-initiation at the non-permissive temperature of 30°, associated with an apparently normal rate of mass increase[23]. The recA$^+$-dependent stable replication is also a system in which replication continues when protein synthesis is limited[24]. However, mutants which are constitutive for stable replication are not reported to be medium-sensitive[25].

Clearly, no precise function can be attributed at present to the rms gene, and more work is needed to assess the specificity of the rms function in the cell cycle, particularly with respect to initiation and/or termination of chromosome replication. Our hope is that the selection procedure introduced by the search for derivatives of Rms Sin Hfr's that are able to grow in rich medium will allow the study of a new set of cell cycle mutants.

REFERENCES

1. J. Louarn, J. Patte and J. M. Louarn, Evidence for a fixed termination site of chromosome replication in Escherichia coli K12, J. Mol. Biol. 115, 295-314, 1977.
2. J. Louarn, J. Patte and J. M. Louarn, Map position of the replication terminus on the Escherichia coli chromosome. Mol. Gen. Genet. 172, 7-11, 1979.
3. P. Kuempel, S. Duerr and N. Seeley, The terminus of the chromosome in Escherichia coli inhibits replication forks. Proc. Natl. Acad. Sci. USA. 74, 3927-3931, 1977.
4. P. Kuempel, S. Duerr and P. Maglothin, Chromosome replication in an Escherichia coli dnaA mutant integratively suppressed by prophage P2. J. Bacteriol. 134, 902-912, 1978.
5. Y. Nishimura, L. Caro, C. M. Berg and Y. Hirota, Chromosome replication in Escherichia coli. IV. Control of chromosome replication and cell division by an integrated episome. J. Mol. Biol. 55, 441-456, 1971.
6. D. J. Clark, The regulation of DNA replication and cell division in E. coli B/r. Cold Spring Harbor Symp. Quant. Biol. 33, 823-838, 1968.

7. N. C. Jones and W. D. Donachie, Chromosome replication, transcription and control of cell division in Escherichia coli. Nature New Biology 243, 100-103, 1973.
8. A. Nishimura, Y. Nishimura and L. Caro, Isolation of Hfr strains from R^+ and $ColV2^+$ strains of Escherichia coli and derivation of an R' lac factor. J. Bacteriol. 116, 1107-1112, 1973.
9. B. J. Bachmann and K. Brooks Low, Linkage map of Escherichia coli K12, Edition 6. Microbiol. Rev. 44, 1-56, 1980.
10. A. Campbell, D. Berg, D. Botstein, E. Lederberg, R. Novick, P. Starlinger and W. Szybalski, Nomenclature of transposable elements in procaryotes. DNA insertion elements, plasmids and episomes, 15-22, 1977.
11. N. Franklin, The N operon of lambda : Extent and regulation as observed in fusions to the tryptophan operon of Escherichia coli. In the bacteriophage lambda (A. D. Hershey, ed.). Cold Spring Harbor Press, New York, 621-638, 1971.
12. P. Legrand, J. P. Bouche and J. M. Louarn, Direction of deoxyribonucleic acid transfer and replication in a derivative of plasmid R100-1. J. Bacteriol. 140, 1105-1108, 1979.
13. Y. Hirota, A. Ryter and F. Jacob, Thermosensitive mutants of E. coli affected in the process of DNA synthesis and cell division. Cold Spring Harbor Symp. Quant. Biol. 33, 677-693, 1968.
14. D. Freifelder, A. Folkmanis and I. Krischner, Studies on Escherichia coli sex factors : Evidence that covalent circles exist within cells and the general problem of isolation of covalent circles. J. Bacteriol. 105, 722-727, 1971.
15. J. Miller, Experiments in molecular genetics. Cold Spring Harbor, New York : Cold Spring Harbor Laboratory, 1972.
16. J. George, M. Castellazzi and G. Buttin, Prophage induction and cell division in E. coli. III. Mutations sfiA and sfiB restore division in tif and lon strains, and permit the expression of mutator properties of tif. Molec. Gen. Genet. 140, 309-332, 1975.
17. D. Kamp, R. Kahmann, D. Zipser, T. R. Broker and L. T. Chow, Inversion of the G DNA segment of phage Mu controls phage infectivity. Nature 271, 577-580, 1978.
18. E. Ohtsubo, J. Feingold, H. Ohtsubo, D. Mickel and W. Bauer, Undirectional replication of three small plasmids derived from R factor R12 in Escherichia coli. Plasmid 1, 8-18, 1977.
19. L. Silver, M. Chandler, E. Boy de la Tour and L. Caro, Origin and direction of replication of the drug resistance plasmid R100-1 and of a resistance transfer factor derivative in synchronized cultures. J. Bacteriol. 131, 929-942, 1977.
20. M. Chandler, L. Silver and L. Caro, Suppression of an Escherichia coli dnaA mutation by the integrated R factor R100-1. III Origin of chromosome replication during exponential growth. J. Bacteriol. 131, 421-430, 1977.
21. B. R. Bochner, H. C. Huang, G. L. Schieven and B. N. Ames, Positive selection for loss of tetracycline resistance. J. Bact. 143, 926-933, 1980.

22. R. H. Pritchard, Control of DNA replication in bacteria. In DNA synthesis: present and future: I. Molineux and M. Kohiyama ed., Plenum publ., New York, 1-26, 1978.
23. G. Kellenberger-Gujer, A. J. Podhajska and L. Caro, A cold sensitive dnaA mutant of E. coli which overinitiates chromosome replication at low temperature. Molec. Gen. Genet. 162, 9-22, 1978.
24. K. G. Lark and C. A. Lark, recA^{+}-dependent DNA replication in the absence of protein synthesis : characteristics of a dominant lethal mutation, dnaT, and requirement for recA^{+} function. Cold Spring Harbor Symp. Quant. Biol. 43, 537-549, 1979.
25. T. Kogoma, A novel Escherichia coli mutant capable of DNA replication in the absence of protein synthesis. J. Mol. Biol. 121, 55-69, 1978.

A MUTATION THAT BLOCKS INITIATION OF DNA SYNTHESIS IN HAMSTER CELLS

Roger Hand, Eric Eilen and Claudio Basilico

McGill Cancer Centre and Department of Medicine
McGill University, Montreal and
Department of Pathology, New York University
School of Medicine, New York

ABSTRACT

The DNA negative mutant of Syrian hamster cells, ts BN-2 stops synthesizing DNA within 2 h of shift to the nonpermissive temperature of 39.5°C. This occurs in exponentially growing cells and synchronized cells put at restrictive temperature at the start of S phase. Although the amount of DNA synthesized is reduced, the fraction of cells in S phase is similar to wild-type BHK-21 cells. DNA fiber autoradiography shows that in the mutant at restrictive temperature the rate of DNA chain elongation is not reduced. Initiation is decreased as shown by an increase in the intervals between active replication origins and a decrease in the relative frequency of initiation events immediately after shift to nonpermissive temperature.

The regulation of DNA synthesis is an important part of the control of cell growth. Little is known about this regulation in higher organisms - humans and other mammals. The use of conditional lethal mutants in procaryotes has proven extremely useful in the analysis of DNA replication in those organisms[1]. A similar approach with mammalian cells would seem to hold equal promise. There are several temperature-sensitive mutants of established mammalian cell lines that are DNA negative. The most extensively characterized is the mouse L cell mutant A-1 isolated by Thompson et al.[2]. The mutation here appears to block the recompaction of replicated chromatin[3]. Sheinin and her coworkers have shown the effects of this on S phase DNA and chromatin synthesis[4-7] and the details are presented in her article elsewhere in this volume. Other DNA negative mutations, such as the C-1 mutation from L cells mutant[8] and ts 2 from mouse 3T3 cells[9], have been analyzed in part and still others, such as ts 20, and ts 22, from mouse 3T3 cells[10] and ts 13A and ts 15C from Chinese hamster ovary cells[11] have been isolated but the biochemical lesions have not been characterized.

One temperature-sensitive mutant, BN-2, derived from the Syrian hamster line BK 21[12], has been studied in vivo[13]. At nonpermissive temperature, the cells are stopped in the growth cycle immediately before entry into the S phase or shortly after the G1-S interface has been passed. The cells also show premature chromosome condensation of the type characteristic of S phase cells. The maturation of nascent DNA chains proceeds normally. These findings suggest that the mutation blocks the initiation of new chain sythesis.

Initiation of synthesis may be the key step in the regulation of DNA synthesis[14]. The circular bacterial genome has a single initiation point. Once initiated, the new DNA chains are elongated via fork-like growing points that move in both directions from the origin until they meet at a point halfway around the genome. In mammalian cells, the situation is more complex, since the mammalian cell contains 1000 times more DNA than the bacterial cell.

The mammalian genome is made up of many replication units[15,16]. Each unit is about 50μm long (about 150 kb) and a typical mammalian cell may have 50,000 of them. Synthesis is initiated at the center of each unit and is completed in about 1h. Once synthesis starts, the nascent chains are elongated in opposite directions by displacement of the replication forks towards the outlying termini. Initiations are staggered through the 8h DNA synthesis phase and clusters of units are activated at intervals. Replication can therefore be considered as a coordinated series of initiation and chain elongation events.

DNA fiber autoradiography[17,18] is useful as a method for resolving the coupled processes of initiation and elongation by providing measurements of each. In this procedure, active replication units can be labeled with [^{3}H]thymidine, and their locations identified along the extended fibers of the genome by autoradiography and light microscopy. The cells have been gently lysed on a microscope slide and the released DNA fibers spread using the same technique as one would use to prepare a blood smear. With practice, one can get well spread preparations of individual fibers. With appropriate labeling protocols, these can be used to analyze initiation and chain elongation events.

The purpose of this work was to look directly at both steps in DNA synthesis - initiation and elongation - in _ts_ BN-2 to determine which is blocked by the mutation. This article reviews data published previously[19].

When exponentially growing cells are shifted to 39°C, DNA synthesis, as measured by [^{3}H]thymidine incorporation, declines rapidly,

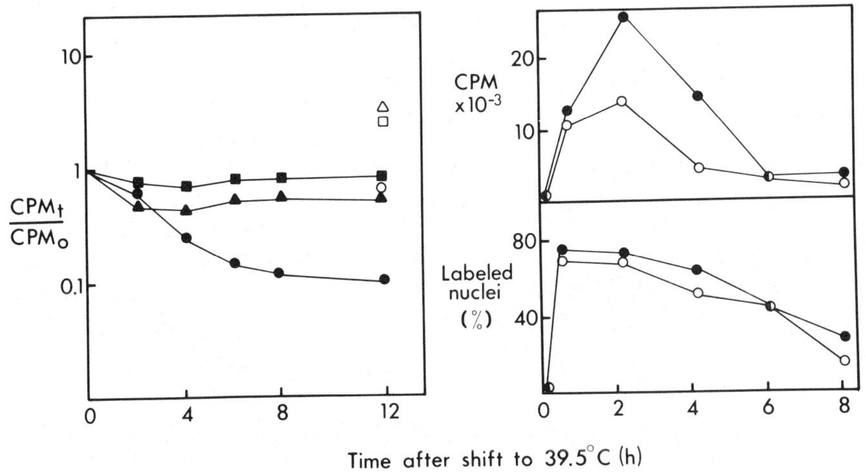

Fig. 1. Mutant phenotype of ts BN-2 in exponentially growing cells and in synchronized cells. Left panel, exponentially growing cells at the nonpermissive temperature of 39.5°C (closed symbols) or permissive temperature of 33.5°C (open symbols) were labeled with [^{3}H]thymidine to measure DNA synthesis (●,O), with [^{3}H]uridine to measure RNA synthesis (▲,△) or with [^{3}H]amino acids to measure protein synthesis (■,□). Acid precipitable radioactivity was measured by liquid scintillation counting. This figure is drawn from data presented by Nishimoto et al.[13]. Right panel, wild-type BHK cells (●) and ts BN-2 cells (O) at 33.5°C synchronized according to the method of Dooley and Ozer[23] - isoleucine deprivation followed by replenishment in the presence of hydroxy urea - were released into the S phase by replacing the medium with hydroxyurea with fresh medium from which hydroxyurea was omitted. The cells were placed at 39.5°C at the time of medium replacement. They were pulse-labeled with [^{3}H] thymidine for 15 min at intervals after the temperature shift. They were assayed by scintillation counting for ^{3}H incorporation into acid-precipitable material (top) or by autoradiography for the fraction of labeled nuclei (bottom).

reaching about 25% of the initial value by 4 to 6h after shiftup, and continuing to decrease therafter (Figure 1, left). Protein and RNA synthesis do not decline through 12h. At the permissive temperature, macromolecular synthesis is normal and the cells have a generation time of 16h.

The mutant phenotype in synchronized cells is shown in Figure 1, right. Mutant and wild-type cells were blocked at the start of S-phase at the permissive temperature by isoleucine deprivation followed by hydroxyurea treatment. The cells were released from the block at the restrictive temperature. The upper panel shows that the mutant cells incorporate [^{3}H]thymidine into DNA as well as wild-type cells for the first part of S-phase at the restrictive temperature. From 2 to 8h, DNA synthesis in the mutant cells decreases. This is not caused by a failure of cells to enter S-phase, since, as shown in the lower panel, the fraction of cells synthesizing DNA is similar in wild-type and mutant cells throughout the 8h of the experiment.

At restrictive temperature, DNA synthesized in the mutant as short fragments in synchronized S phase cells is incorporated into large molecules. After a short pulse with [^{3}H]thymidine, short pieces of DNA are synthesized in mutant cells at permissive and restrictive temperatures. As measured by velocity centrifugation in alkaline sucrose gradients, these newly synthesized molecules are incorporated in large chains - greater than 50 S - after a 3h chase at both temperatures (data not shown).

The results of the centrifugation experiments suggest that chain elongation is normal in the mutant, but a more precise estimate can be provided by DNA fiber autoradiography. For these studies, we used a step-down protocol to label the DNA. Cells were pulsed with [^{3}H] thymidine of high and low specific activities for consecutive 30-minute periods and prepared for fiber autoradiography as described previously[20]. This produces regions along the DNA autoradiograms of high and low grain density. Individual replication units with bidirectional replication show two types of grain patterns. These are illustrated in Figure 2, in which the upper panel is DNA from wild-type cells, and the lower, DNA from mutant cells. Those units that begin replication before the radioactive pulse have a gap separating the two stretches of heavy and light grain density. One is shown at X. Those units that begin replication during the radioactive pulse of high specific activity have a central heavy density track with flanking light density tracks. One is shown at Y. These patterns may be used to measure DNA synthesis. For instance, the rate of DNA replication fork movement may be determined by measuring the length of the heavy tracks on units like the one marked by the arrowheads in Fig. 2b at X, where the beginning of the pulse of high specific activity is marked by the appearance of grains and the end by the transition from heavy to light grain density. The distance between active origins can be measured using patterns such as seen at the arrowheads in Figure 2a,

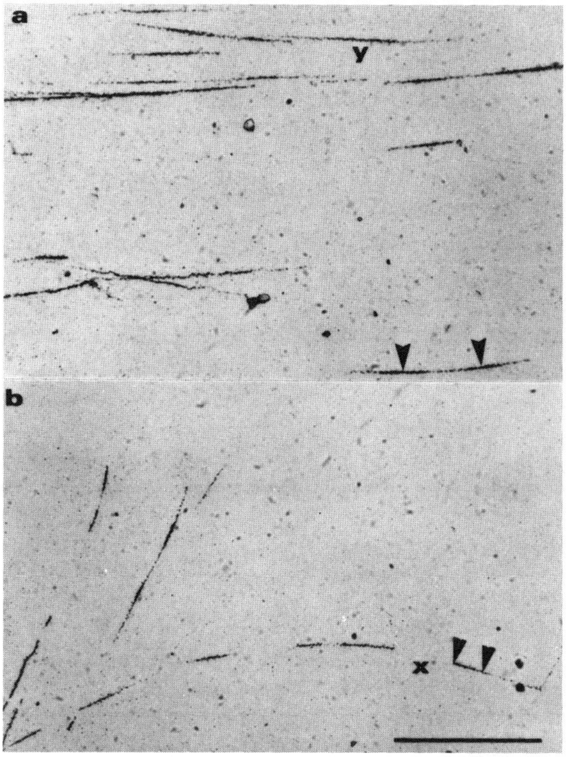

Fig. 2. DNA fiber autoradiography of wild-type BHK 21 cells and mutant ts BN-2 cells. Exponentially growing cells were shifted to 39.5°C and labeled for 30 min with [^{3}H]thymidine of high specific activity (60 Ci/mmole) and then for an additional 30 min with [^{3}H]thymidine of low specific activity (4.5 Ci/mmole). They were prepared for fiber autoradiography, and, after exposure and development, analyzed by light microscopy. a) Wild-type cells. b) Mutant BN-2 cells. X shows a unit that began replication before the pulse and Y shows a unit that began replication during the pulse. In a, the arrowheads indicate the interval between two active origins. In b, the arrowheads show a track with heavy grain density from a unit that began synthesis before the pulse. The autoradiogram could be used to measure rate of chain elongation. The bar at lower right is 100μm. Magnification, 250.

in which the heavy grain density marks the origin on individual units.

In these experiments, groups of cell culture dishes containing wild-type or mutant cells were shifted to the restrictive temperature while others were maintained at the permissive temperature throughout the experiment. Cells were subjected to the labeling protocol at time 0 and at several other times up to 8h after shift to restrictive temperature.

The results of measurements of the rate of replication fork displacement are shown in Figure 3. The rate in wild-type cells at restrictive temperature varies from 0.74 to 0.86μm/min. The mutant cells have a slightly higher rate at all time points, with an average increase of about 30% above the values in wild-type cells. The rates are significantly faster in mutant cells at 0, 3 and 8h.

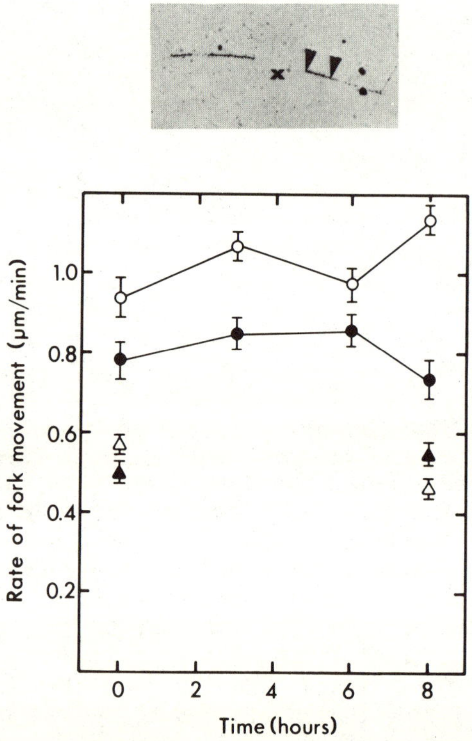

Fig. 3. Rate of replication fork movement. Exponentially growing cells were shifted to the restrictive temperature of 39.5°C at time 0. At intervals, samples of cells were labeled with [^{3}H]thymidine and processed for autoradiography as described in the legend to Fig. 2. Fifty autoradiograms of the type shown at the top of the figure were measured from each sample to determine the rate. Error bars show the standard errors of the arithmetic means. (●), BHK-21, 39.5°C; (O), BN-2, 39.5°C; (▲) BHK-21, 33.5°C; (Δ), BN-2, 33.5°C.

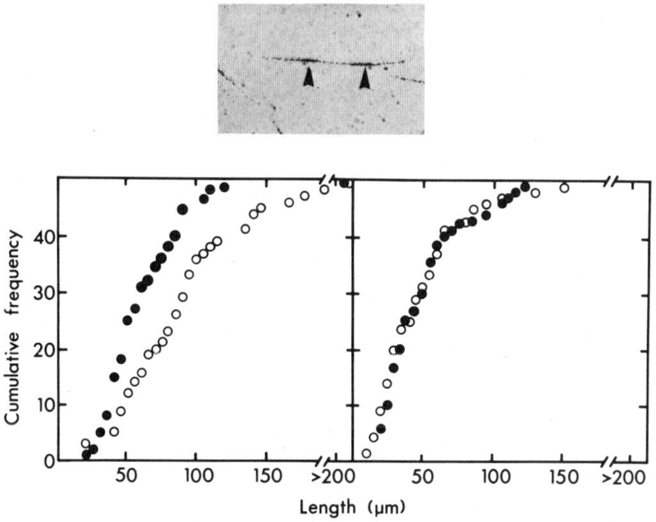

Fig. 4. Interval between initiation sites. The cumulative frequency distributions of these intervals were determined by measuring the distances between replication origins on the type of autoradiograms shown at the top of the figure. (●), BHK-21 cells; (O), BN-2 cells. In the left panel, the measurements are from cells that were shifted to 39.5°C at the start of the pulse with [^{3}H]thymidine. In the right panel, the measurements are from cells kept at the permissive temperature of 33.5°C.

The rate of elongation is lower in both mutant and wild-type cells at permissive temperature. We have observed this dependence of DNA chain elongation on temperature in Chinese hamster ovary cells as well[22].

The interval between initiation sites was determined by measuring the center-to-center distances between adjacent sites. The cumulative frequency distributions are presented in Figure 4. At restrictive temperature, the intervals are larger in the mutant than in the wild-type immediately after temperature shift. At the permissive temperature, mutants and wild-type cells show similar distributions.

For statistical analysis, we calculated the geometric means of these distributions along with those from samples kept at 39.5°C for longer times. Although origins are at non-random intervals along chromosomes, the exact type of non-random distribution is unknown. The geometric mean provides a measure of central tendency that lends itself more readily to statistical evaluation than does the median. In addition, the geometric mean is clearly more reliable than the arithmetic mean since the distributions are not normal in shape but are skewed toward larger sizes. The results are presented in Table 1.

Table 1. Intervals between initiation sites

Time at 39.5°C(h)	Cell line	Interval: Geometric mean(μm)	Interval: 95% confidence limits	
0	WT	56.4	52.7 to 60.3	*
	BN-2	79.3	72.8 to 86.3	
3	WT	63.3	58.6 to 68.2	*
	BN-2	88.7	81.5 to 96.6	
6	WT	70.0	63.7 to 76.9	*
	BN-2	83.5	76.7 to 91.0	
8	WT	61.7	56.7 to 65.9	*
	BN-2	98.1	92.7 to 103.9	

* $P < 0.01$, Student's t-test

The interval between initiation sites at restrictive temperatures in mutant cells is increased at all times and this is statistically significant at 0, 3 and 8h. This increase is most likely caused by decreased initiation. The potential origins between active origins are not used at the non-permissive temperature.

As another measure of initiation, we compared the relative frequency of patterns showing initiation before (X, Figure 2b) or during (Y, Figure 2a) the pulse. With a constant pulse time, the percentage of each pattern is fixed. If some factor reducing initiation becomes effective at the beginning of the pulse, the percentage of patterns showing initiation during the pulse(Y) will be reduced. As shown in Table 2, the percentage of these patterns in wild-type cells increases as the cells are shifted from the permissive temperature. This increase probably reflects an increase in initiation associated with the switch to higher temperature. The increase is less marked in the mutant cell line, again indicating reduced initiation.

The results presented here demonstrate that ts BN-2 has a defect in replication that can be attributed to decreased initiation but not to retarded chain elongation. In the mutant cells at restrictive temperature, DNA chain elongation is normal since, by sucrose gradient analysis, full-sized DNA molecules are synthesized after a 3h chase with unlabeled thymidine and, by fiber autoradiography, replication fork movement is not retarded. Initiation is defective, since there is an increased distance between initiation sites, suggesting the

Table 2. Relative Frequency of Initiation

Cell line	Temperature(^{o}C)	Initiation(%) Before pulse	Initiation(%) During pulse	Number scored
WT	33.5	59	41	425
	39.5	39	61	493
BN-2	33.5	60	40	353
	39.5	49	51	383

dropout of sites that normally would be active, and there are fewer initiations - or fewer patterns showing initiations - immediately after shift to higher temperature.

The faster rate of chain elongation in the mutant requires explanation. Earlier studies showed that units in which the origins were more widely separated had faster rates of chain elongation[21]. With larger intervals between origins, there may be less constraint to unwinding of the DNA double helix. With a long stretch of DNA unwound ahead of the growing point, the complex of replication proteins at the fork may be able to move processively along the template at a faster rate.

There are other possible explanations. If initiation is inhibited, then there will be relatively more molecules of replication factors such as enzymes available to form functional replication complexes. If we postulate that, in normal chain elongation, the process slows as certain cellular factors are consumed, then a greater availability of these factors in the mutant under restrictive conditions (because fewer active replication complexes are needed) might allow for faster rates of elongation.

Nishimoto et al.[13] showed that the BN-2 mutation stops some of the cells in late G-1 and stops others after they have passed into S phase. Our findings of an initiation defect are consistent with their findings. The protein or proteins required to initiate synthesis on individual replication units would be used on those active at the beginning of S as well as on those that begin synthesis during the S phase. In this view, the biochemical events required to begin S phase would be the same as those required to maintain it. Identification of the BN-2 mutant protein and its wild-type counterpart should allow this hypothesis to be tested experimentally.

ACKNOWLEDGEMENTS

Colette Oblin and Pat Santanello provided technical assistance. This work was supported by research grants from the Medical Research Council of Canada, National Cancer Institute of Canada, and the United States Public Health Service and a contract from the United States Public Health Service.

REFERENCES

1. S. H. Wickner, DNA replication proteins of Escherichia coli. Annu. Rev. Biochem. 47, 1163-1192, 1978.
2. L. H. Thompson, R. Mankovitz, R. M. Baker, J. E. Till, L. Siminovitch and G. F. Whitmcre, Isolation of temperature-sensitive mutants of L cells. Proc. Natl. Acad. Sci. USA. 66, 377-384, 1970.
3. G. Setterfield, R. Sheinin, I. Dardick, G. Kiss and M. Dubsky, Structure of interphase nuclei in relation to the cell cycle. Chromatin organization in mouse L cells temperature sensitive for DNA replication. J. Cell Biol. 77, 246-263, 1978.
4. R. Sheinin, Preliminary characterization of the temperature-sensitive defect in a mutant mouse L cell. Cell 7, 49-57, 1976.
5. R. Sheinin and R. Guttman, Semi-conservative and non-conservative replication of DNA in temperature-sensitive mouse L cells. Biochim. Biophys. Acta. 479, 105-118, 1977.
6. R. Sheinin and P. N. Lewis, DNA and Histone synthesis in mouse cells which exhibit temperature-sensitive DNA synthesis. Somat. Cell Genet. 6, 225-239, 1980.
7. R. Sheinin, P. Darragh and M. Dubsky, Some properties of chromatin synthesized by mouse L cells temperature-sensitive for DNA replication. J. Biol. Chem.253, 922-926, 1978a.
8. S. A. Guttman and R. Sheinin, Properties of ts C1 mouse L cells which exhibit temperature-sensitive DNA synthesis. Exp. Cell Res. 123, 191-205, 1979.
9. M. L. Slater and H. L. Ozer, Temperature-sensitive mutants of Balb/3T3 cells: Description of a mutant affected in cellular and polyomavirus DNA synthesis. Cell 7, 289-295, 1976.
10. K. K. Jha, M. Siniscalco and H. L. Ozer, Temperature-sensitive mutants of Balb/3T3 cells. III. Hybrids between ts 2 and other mouse mutant cells affected in DNA synthesis and correction of ts 2 defect by human X chromosome. Somat. Cell Genet. 6, 603-614, 1980.
11. P. R. Srinivasan, R. S. Gupta and L. Siminovitch, Studies on temperature-sensitive mutants of Chinese hamster ovary cells affected in DNA synthesis. Somat. Cell Genet. 6, 567-582, 1980.

12. T. Nishimoto and C. Basilico, Analysis of a method for selecting temperature-sensitive mutants of BHK cells. Somat. Cell Genet. 4, 323-340, 1978.
13. T. Nishimoto, E. Eilen and C. Basilico, Premature chromosome condensation in a ts DNA-mutant of BHK cells. Cell 15, 475-483, 1978.
14. R. Hand, Eukaryotic DNA: organization of the genome for replication. Cell 15, 317-325, 1978.
15. H. J. Edenberg and J. A. Huberman, Eukaryotic chromosome replication. Annu. Rev. Genet. 9, 245-284, 1975.
16. R. Sheinin, J. Humbert and R. C. Pearlman, Some aspects of eukaryotic DNA replication. Annu. Rev. Biochem. 47, 277-316, 1978.
17. J. Cairns, The Chromosome of Escherichia coli. Cold Spring Harbor Symp. Quant. Biol. 28, 43-46, 1963.
18. J. A. Huberman and A. D. Riggs, On the mechanism of DNA replication in mammalian chromosomes. J. Mol. Biol. 32, 327-341, 1968.
19. E. Eilen, R. Hand and C. Basilico, Decreased initiation of DNA synthesis in a temperature-sensitive mutant of hamster cells. J. Cell Physiol. 105, 259-267, 1980.
20. R. Hand and I. Tamm, DNA replication: rate and direction of chain growth in mammalian cells. J. Cell Biol. 58, 410-418, 1973.
21. R. Hand, Regulation of DNA replication on subchromosomal units of mammalian cells. J. Cell Biol. 64, 89-97, 1975.
22. R. Hand and J. R. Gautschi, Replication of mammalian DNA in vitro: evidence for initiation from fiber autoradiography. J. Cell Biol. 82, -85-493, 1979.
23. D. C. Dooley and H. L. Ozer, Replication kinetics of three sequence families in syncronized L cells. J. Cell Physiol. 90, 337-350, 1976.

POLYOMA AND CELL CHROMATIN REPLICATION STUDIED IN MOUSE CELLS WHICH EXHIBIT TEMPERATURE-SENSITIVE DNA SYNTHESIS BECAUSE THEY ARE S^{ts} or $G_1{}^{ts}$

Rose Sheinin, Richard Colwill
and Peter R. Ganz

Department of Microbiology and Parasitology
Faculty of Medicine, University of Toronto
Canada M5S 1A1

ABSTRACT

Mammalian cells exhibiting temperature sensitive (ts) DNA synthesis have been exploited for two major purposes. The first was to elucidate specific biochemical reactions of cell cycle progression, the proteins which function therein and their genetic determinants. The second objective was to explore the mechanisms of interaction of polyoma virus DNA replication with those for the host genome; a specific goal being the identification of host proteins obligatorily-linked to viral DNA synthesis and integration into the cellular DNA. As a result of comparative biochemical and biophysical studies of cells which are dna^{ts}/S^{ts} or $G_1{}^{ts}$, a modified model of chromatin replication has been formulated. We have demonstrated that the ts protein of the ts^{C1} dna^{ts}/S^{ts} or $G_1{}^{ts}$, a modified model of chromatin of polyoma virus DNA. The gene product of the ts A1S9 dna^{ts}/S^{ts} mouse L-cell, which normally participates in conversion of newly-replicated single-stranded DNA of $\not> 5 \times 10^6$ daltons to chromosomal DNA of $\simeq 10^9$ molecular weight, plays a role in establishing or maintaining chromatin conformation essential for DNA replication and determines the novobiocin sensitivity of the cells.

INTRODUCTION

The biochemical-genetic exploration of the physiology of living things was initiated with eukaryotic organisms. In 1941[1] Beadle and Tatum set about identifying enzymatic pathways of intermediary metabolism in Neurospora crassa. This experimental approach has been most successfully applied to prokaryotes. The mechanisms of DNA replication of bacteria and their viruses have been revealed in their very great complexity by biochemical-genetic analysis over a period of more than two decades[2-6]. This work has resulted in the de novo, in vitro replication of physiologically active DNA[7-10]; this under circumstances in which many of the enzyme/proteins which participate in the process remain unidentified, unisolated and uncharacterized.

These major successes have led once again to the application of the tools and concepts, which emerged from the biochemical-genetic studies with prokaryotes, to lower[11-13] and higher eukaryotes[14-16] to probe the cell duplication cycle and the key process of chromatin replication. This has necessitated the isolation and at least partial characterization of mutant eukaryotic cells[11-17].

Because a mutation in most genes which determine chromatin - DNA replication is likely to be lethal, it is desirable to work with those which are conditionally lethal; in particular, those which are temperature-sensitive (ts (cf[11,12,15,17])). A very large number of ts mammalian cells have now been isolated (cf[15,17]). Many have been shown to arrest at a specific point in the cell duplication cycle, upon expression of the ts defect (cf[15]). These will, of course eventually exhibit ts DNA synthesis. However, very few ts mammalian

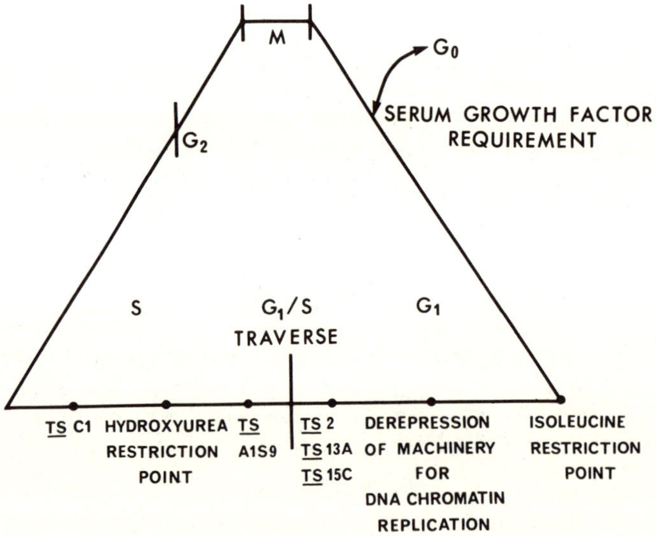

Fig. 1. Location of ts mutations on a temporal map of mammalian cell cycle progression.

cells have been shown to carry a mutation in a gene which directly controls DNA replication or chromatin replication during the DNA-synthetic phase of the cell cycle.

THE MAMMALIAN CELL DUPLICATION CYCLE

The orderly, sequential and coordinately-regulated process by which a cell is replicated has been operationally divided into several stages[18-20]. As is indicated in Figure 1, a newly formed cell enters a pre-DNA-synthetic period, termed gap-1 or G_1. It then progresses through G_1, and enters the S-phase in which the chromatin DNA and protein is duplicated. In the subsequent gap-2, or G_2 period the chromatin is modified biochemically and physically in preparation for chromosome condensation and segregation, which occur during mitosis. This M phase is terminated as the cell moves through cytokinesis producing the progeny daughter cells.

CELL CYCLE PROGRESSION AS IT AFFECTS CHROMATIN REPLICATION IN MAMMALIAN CELLS

The G_1 Phase

The major event in G_1 of particular relevance to a consideration of chromatin replication occurs fairly late in G_1. It is the derepression of formation of the machinery of DNA replication. As shown by our own work[15,16,21] and that of others[22-24], this results in the formation of all of the enzymes of DNA synthesis in a pattern reminiscent of the derepression of classical operons[25]. When cells which are growing normally in an intact organism or in culture stop dividing, they usually arrest in G_1 prior to this derepression event. In some instances growth is halted in a pseudo - G_1 state, termed G_0, outside the active cell duplication cycle. When such quiescent cells are reactivated, they progress into and through G_1 to the derepression event. Once cells move beyond this point (termed "start" in yeast[11,12]) they are usually destined to traverse the G_1/S interface and proceed through S_1, G_2, M and cytokinesis[20].

Little more is known about the biochemistry or genetic determinants of this trigger mechanism. It is recognized that major structural reorganization of the chromatin takes place during G_1. The chromatin is in its most condensed interphase state early in G_1 and it takes on an increasingly less compact conformation as the cell moves to late G1[26,27]. It has been suggested that this physical rearrangement occurs in order to accommodate the machinery for DNA replication, made in preparation for entry into S phase.

S Phase

S phase is, by definition, that period in which the DNA is synthesized[18]. It also embraces the formation of the chromatin histones[28], high mobility group proteins[29], other structural[30] and non-structural proteins and their deposition onto the newly-replicated DNA, giving rise to the primary nucleosomal subunit organization[31] and subsequent folding and modelling to produce the duplicated nucleofilaments[32]. The biochemical events which initiate the G_1/S traverse and which result in termination of S phase by G_1 entry remain unidentified.

Progression Through G_2 and M

Although little (if any) de novo synthesis of chromatin components occurs after S phase, there is significant modification - both structural and chemical - of those already made. This[33] results in the compacted chromatids fashioned late during G_2[34]. At the same time, modulation of the cytoskeleton occurs[35] perhaps in preparation for subsequent chromosome mobilization, movement and segregation during mitosis.

ENZYME/PROTEINS OF CHROMATIN REPLICATION

There is now a great body of knowledge about the enzymes and proteins which participate in the replication of DNA in mammalian cells (cf[16,36-43]). Multiple DNA polymerases, ligating enzymes, topoisomerases, DNA-binding proteins (DBP's) and DNA-dependent ATP-ases have already been identified. Optimal DNA synthesis has been shown to depend upon the interaction of specific single-strand DPB's and/or ATPases with template DNA, making the latter more accessible for replication by the DNA polymerases. Information concerning the synthesis of histones and non-histone chromosomal proteins which is coupled to DNA replication during S phase is abundant[44-49].

Having made these statement, it is important to recognize the major gaps in our knowledge. Thus, it is entirely likely that not all the enzyme/proteins of DNA replication have been identified. Nor has it been possible to assign to those already known, a specific function within the framework of cell cycle progression and/or the total process of chromosome duplication and segregation. Little is known about the derepression of chromatin replication, or about the determinants of regulation of coupling of synthesis of chromatin proteins to DNA synthesis. Such regulation proceeds at the level of transcription and translation[44-46,49].

BIOCHEMICAL-GENETIC STUDIES OF CHROMATIN REPLICATION USING *ts* MOUSE CELLS

Mutant Cell Availability

Any biochemical-genetic study depends upon the availability of mutant cells of choice and of *in vitro* systems for assaying the gene product involved. In the present context, this means cells which are *ts* in DNA replication and *in vitro* systems which yield effective DNA and/or chromatin synthesis.

As noted earlier, very few ts mammalian cells have been isolated which are in fact dna^{ts} or S^{ts} (cf[15]). The available evidence suggests that two mouse L-cells[14,48,50], two BHK cells[51,52] and perhaps one CHO cell[53] may have these properties. On the other hand, there are a number of *ts* cells which exhibit temperature inactivation because they are G_1^{ts} (cf[15,54,55]). The non-permissive temperature arrest points of some of these *ts* mutants are indicated in Figure 1.

IN VITRO SYSTEMS FOR THE STUDY OF DNA SYNTHESIS

To dissect the process of DNA replication in *ts* mammalian cells we have developed *in vitro* systems using wild-type cells and two dna^{ts}/S^{ts} isolates. These utilize cell-free homogenates, isolated nuclei and cytosol fractions[56-58].

With these tools of biochemical genetics, we have proceeded towards three goals:

(1) Identification, isolation and characterization of the particular gene products encoded in the *ts* loci of interest.
(2) Utilization of late G_1^{ts} and dna^{ts}/S^{ts} mouse cell mutants to probe the mechanism of replication of polyoma (Py) virus DNA.
(3) Elucidation of the inter-relationships between chromatin structure, function and replication using late G_1^{ts} and dna^{ts}/S^{ts} mutants.

STUDIES WITH ts A1S9 CELLS

We have now demonstrated by six different methods[15,47,56-63] that *ts* A1S9 mouse L-cells, selected on the basis of ts growth[64], are temperature-sensitive in DNA replication. As expected for dna^{ts} cells, they arrest in S phase upon temperature inactivation, and are therefore also S^{ts}. We have shown that the primary defect in *ts* A1S9 cells is restricted to the nucleus[15,56,61,62]. The ts protein normally participates in the conversion of single-stranded DNA of 5×10^6 molecular weight, newly-replicated in the semi-conservative mode[60] to chromosomal DNA of $\simeq 10^9$ daltons[62,63].

In vivo studies suggest that the major known enzymes and proteins which effect polydeoxyribonucleotide chain synthesis are made and are functional in ts A1S9 cells inactivated at the non-permissive temperature (npt) of 38.5°C. Such cells make Okazaki fragments and convert them to larger molecular weight DNA at control rates, as revealed by pulse-chase experiments[62]. This results in the continuous accumulation of the 5×10^6 molecular weight DNA for at least 6 hours at the npt[62]. DNA fibre radioautography[57] has revealed that movement of the DNA replication fork along the template of active replication units (R.U's) is unimpeded as the ts A1S9 defect is expressed.

Two different kinds of in vitro experiments have been performed in pursuit of this problem; one involving direct assay of specific enzymes, the other using a more complex in vitro system based on isolated nuclei which express the ts phenotype[56]. Crude extracts and partially-purified enzyme preparations of temperature-inactivated ts A1S9 cells have been assayed for DNA polymerase (M. Gold, unpublished) and polynucleotide ligase activity (Robertson and Sheinin, unpublished). No evidence for temperature-sensitivity of the enzymes was obtained.

It has been established that intact nuclei isolated from temperature-inactivated ts A1S9 cells are unable to carry out DNA synthesis comparable to that supported by nuclei from control ts A1S9 or wild-type (WT-4) cells[56]. Nor can they be stimulated into activity by fully-functional nuclei or cytosol from control cells. Of greater significance is the finding that cytosol from temperature-inactivated ts A1S9 cells, when added to nuclei from control cells, can effectively reconstitute whole cell homogenate activity. In contrast, reconstitution was not observed with nuclei of heat-inactivated ts A1S9 cells. These nuclei are also refractory to activation by nuclear proteins extracted from control and temperature-inactivated nuclei; all of which enhance the DNA-synthetic activity of control nuclei and effect DNA replication from purified polyoma (Py) virus or calf thymus DNA[56,57]. If the inactive nuclei are disrupted by sonic vibration, the DNA freed of structural constraints is able to serve as template for endogenous DNA synthesis[57].

These in vivo and in vitro experiments have been interpreted as indicating that the ts A1S9 protein is not likely to be DNA polymerase-α or the major polynucleotide ligase. This conclusion derives support from our demonstration that temperature-inactivated ts A1S9 cells support full multiplication of infectious Py virus[58,63], which is dependent upon these (and other cellular) enzymes for replication of the viral DNA[21,65]. Since temperature-inactivated ts A1S9 cells carry out repair replication[15,60] it seems unlikely that the ts A1S9 locus codes for the DNA polymerase-β, which may be the major repair enzyme of mammalian cells[66]. The DNA polymerase-γ is also an improbable candidate, since temperature-inactivated ts A1S9 cells support full and normal replication of mitochondrial DNA[61], which utilizes the γ-enzyme[67]. The poly (ADP-ribose) polymerase, an enzyme known to modify chromatin protein structure, has also been ruled out[68].

These various observations suggest that inactivation of the ts A1S9 protein renders the DNA inaccessible or unavailable for replication. This hypothesis is compatible with DNA fibre radioautographic analyses[57] which reveal decreased frequency of initiation of R.U's, as increased distance between initiation sites along the DNA extracted from ts A1S9 cells undergoing temperature inactivation. It also accords with earlier radioautographic and biochemical studies[15,64] on the pattern of entry into and exit from S phase, during temperature inactivation of ts A1S9 cells.

One possible explanation which would accommodate all of the observations postulates that thermal denaturation of the ts A1S9 protein gives rise to an abnormal conformation of the chromatin-bound DNA. It has been demonstrated in prokaryotic[5,69] and eukaryotic organisms[70,71] that unless the template DNA has the correct supercoiled conformation, initiation of replication does not occur. The results of our recent studies[57,72] suggest that the structure of the chromatin-bound DNA of temperature-inactivated ts A1S9 cells is indeed unlike that of control cells. Such DNA when bound to the intercalating agent ethidium bromide, sediments in a pattern which suggests abnormal supercoiling.

It had been demonstrated with prokaryotes[5,69,73] that novobiocin interacts with an ATP-dependent subunit of the enzyme DNA gyrase. As a result, the normal process for conferring a correct conformation on replicating DNA is interrupted, leading to cessation of DNA formation. Inhibition of DNA synthesis by novobiocin was reported for yeast[71] setting the stage for our studies on the novobiocin sensitivity of ts A1S9 cells. These cells are about twice as sensitive to novobiocin as are wild-type cells, other dna^{ts} cells, and other ts mammalian cells[72]. It is especially interesting to find that a ts A1S9 revertant, designated ts^+ AR (Sparkuhl and Sheinin, unpublished), exhibits wild-type sensitivity to novobiocin, suggesting that the nov^R property may be genetically-determined by the ts A1S9 locus. Our most recent experiments indicate that ts A1S9 cells may be deficient in a novobiocin-binding protein of the chromatin, readily detected in wild-type cells (Colwill and Sheinin, unpublished). These various observations suggest that expression of the ts A1S9 defect results in a more highly-constrained conformation of the nuclear DNA than occurs normally. In this configuration the ts A1S9 DNA may be unable to function as template for replication, thereby explaining the ts phenotype.

We have exploited ts A1S9 cells which replicate Py DNA with no impairment, under conditions which restrict normal cellular DNA synthesis, to re-examine the pattern of viral DNA synthesis[58]. The usual Py DNA replicating intermediates (R.I's) were detected, as was newly-made mature form I Py DNA. However, in addition, very large molecular weight, linear molecules ranging in size from 6 to 20 unit genome lengths were observed. The usual R.I's were synthesized early in the course of viral infection in mouse L-cells; whereas the latter forms accumulated late during the infectious cycle.

Restriction enzyme analysis of the large molecular weight Py DNA forms showed them to be arranged in a "head to tail" organization, with junctions near the site on the viral DNA for cleavage by the Eco RI endonuclease. The synthesis of all forms of Py DNA was unaffected by expression of the ts A1S9 defect. Clearly the ts A1S9 gene product is not required for viral DNA replication. Although Py infection does not correct the ts A1S9 defect, it does result in the usual derepression of synthesis of the cellular DNA. Thus, there is as much as a 20-fold stimulation of synthesis of aberrant cellular DNA in Py-infected ts A1S9 cells incubated at the npt[58,63].

STUDIES WITH ts C1 MOUSE L-CELLS

The ts C1 mouse L-cell, selected on the basis of ts growth[74] has also been shown to be dna^{ts}/S^{ts}[15,16,50,61,76]. Radioautographic, biochemical and biophysical studies have established that the ts C1 gene product acts throughout S phase. Its temperature inactivation results in arrest of the cells in S phase beyond the much earlier arrest point of ts A1S9 cells (Fig. 1).

The available evidence indicates that the ts C1 protein is required continuously through S phase for the semi-conservative replication of the DNA[15,60,61]. DNA fibre radioautographic analysis reveals that initiation is not the major defect of temperature-inactivated ts C1 cells[76]. It points rather to a defect in completion of polydeoxyribonucleotide chain elongation on R.U's.

The ts protein of ts C1 mouse L-cells has a half-life of 3-4 hours at 38.5°C[50]. It is absolutely required for nuclear DNA replication, but not for mitochondrial DNA synthesis[61]. The ts C1 gene product is essential for Py DNA replication, apparently at a termination stage in polydeoxyribonucleotide chain elongation[77].

Recently an in vitro system has been established for the study of DNA replication in nuclei isolated from control ts C1 cells, and from Py-infected cells[77]. In both cases, rapid temperature inactivation has been demonstrated in vitro, again confirming the conclusion that the ts C1 gene product participates in Py DNA replication.

The ts C1 protein remains unidentified. It does however seem unlikely that it is the DNA polymerase-α, DNA polymerase-β, DNA polymerase-γ (cf[15]) or poly (ADP) -ribose polymerase[68].

G_1^{ts} MAMMALIAN CELLS

Cells which are ts in a function of the non-S phase segment of the cell cycle may, upon arrest at the npt, exhibit ts DNA synthesis. This is, in fact, the case with three mutant cell types with which we have worked; the ts 2 BalB/C-3T3 mouse fibroblast[78] and the ts 13A and 15C derivatives of Chinese hamster ovary (CHO) cells[79]. These ts mutants have been classified as G_1^{ts} on the basis of the following experiments (unpublished findings; see also[15,54]).

Cells starved of isoleucine arrest in the G_1 phase, as noted in Figure 1. When they are released from such amino acid starvation at the permissive temperature (pt) they move synchronously through one, and perhaps two duplication cycles. If they are released at the npt, they are unable to make this progression, and indeed remain fixed in G_1. Appropriate upshift and downshift manipulations subsequent to restoration of isoleucine, suggest that the execution point for phenotypic expression of the BalB/C 3T3-ts 2, CHO-ts 13A and CHO-ts 15C loci may be placed late in G_1 (see Fig. 1).

The ts 2, ts 13A and ts 15C cells, like other mammalian cells, are subject to early S phase arrest by treatment with hydroxyurea (Fig. 1). Cells treated with this anti-metabolite at the pt, cycle synchronously through S phase, mitosis and beyond, upon drug removal and continued incubation at the pt. If however, they are released from hydroxyurea arrest at the npt, they progress through S phase, divide once and then arrest, presumably in the G_1 phase.

That these cells are not affected in an S phase function is more dramatically illustrated by studies in which cells were initially treated with hydroxyurea at the npt. The cells survive this treatment in the sense that upon removal of hydroxyurea, they progress through the ongoing S phase, whether they are incubated at the pt or the npt. Whereas these cells continue to divide at the pt, they are unable to do so at the npt, presumably due to the cumulative secondary effects of temperature inactivation (to be discussed below).

The final set of experiments in this series exploited the observation that when mitotic cells are fused with cells in G_1, S and G_2 phase, they undergo a characteristic pattern of premature chromosome condensation[80]. Using these criteria we have confirmed the G_1^{ts} assignment for ts 2 mouse fibroblasts and ts 13A and ts 15C CHO cells[55].

PLEIOTROPIC EXPRESSIONS OF DEFECTS IN MAMMALIAN CELLS TEMPERATURE-SENSITIVE IN DNA SYNTHESIS

The foregoing studies have established that we have in hand two mouse L-cells which are dna^{ts}/S^{ts}, one 3T3 mouse fibroblast which is G_1^{ts} and two CHO cells which are G_1^{ts}. These findings are in accord with complementation analyses which reveal that the *ts* A1S9, *ts* C1, *ts* 2, *ts* 13A and *ts* 15C loci are genetically distinct (cf[15,53,81]). We have therefore proceeded to examine a number of secondary expressions of each *ts* gene product, with particular emphasis on structural-functional inter-relationships of chromatin constituents.

Patterns of DNA Replication

When dna^{ts}/S^{ts} ts A1S9 and *ts C1* cells and G_1^{ts} *ts* 2, CHO-*ts* 13A and CHO-*ts* 15C cells are incubated at the npt, the first phase of expression of the *ts* phenotype with respect to DNA synthesis is manifest in suppression of normal semi-conservative replication[15,54,60]. This phase, equivalent to one cell generation period, is followed by a second in which normal semi-conservative DNA synthesis is maintained at 1-5% of control levels for a second generation time equivalent. Thereafter, semi-conservative DNA synthesis is totally replaced by repair replication.

Chromatin-DNA and -Protein Synthesis

Our studies of the coupling of chromatin protein synthesis to DNA replication in the five *ts* mammalian cells of interest, have yielded intriguing differences between the dna^{ts}/S^{ts} and G_1^{ts} cells. In *ts* A1S9 and *ts* C1 mouse cells, temperature inactivation of DNA replication was followed by cessation of synthesis of all histones[15,47,48] and a majority of NHCPs (preliminary findings). This was not the case with *ts* 2, *ts* 13A and *ts* 15C cells. In fact, synthesis of chromatin proteins continued at control rates until very late after temperature upshift. Even in the second and third stages of temperature-inactivation of DNA replication, histone synthesis was seen to proceed at 70% of control levels. On the basis of these observations and those obtained with CHO-LI G_1^{ts} cells[82,83], we have tentatively concluded that cells which are truly dna^{ts} will exhibit coupled inhibition of the replication of chromatin-DNA, histones and the NHCPs, upon expression of the *ts* defect. This need not be the case with G_1^{ts} cells which are mutant in a protein which acts to move cells across the G_1/S traverse.

Chromatin Structural Organization

Quantitative morphometric analysis coupled with transmission electron microscopy has been used to probe the relationship between chromatin structure and replication in the *ts* cells of interest here. We found that with the two dna^{ts}/S^{ts} mouse L-cells, temperature inactivation of DNA synthesis was followed by marked disaggregation of the nucleoplasmic heterochromatin, which began late during the first generation interval post-temperature shift (pts)[15,75]. This phenomenon was not seen in the G_1<u>ts</u> BalB/C-3T3 <u>ts</u> 2 mouse fibroblasts[54], nor in the CHO-<u>ts</u> 13A and CHO-<u>ts</u> 15C cells (unpublished observations). It did, however, occur in WT-4 cells directly blocked in DNA replication by anti-metabolites, but not in cells indirectly prevented from replicating their DNA by cycloheximide treatment or arrested in G_1 by isoleucine starvation[84].

On the basis of these studies, and a larger body of evidence summarized elsewhere[15], we have developed a model for the regulation of chromatin replication in eukaryotic cells which suggests the following: Depression of the formation of the machinery of chromatin replication proceeds late in the G_1 phase and results in the unleashing of several separate, but co-ordinately regulated, biochemical pathways. These include: (i) structural reorganization of the condensed chromatin in preparation for replication; (ii) synthesis of histones and other chromosomal proteins; (iii) replication of DNA and (iv) modification of histones and NHCPs and their deposition onto newly-replicated DNA.

Although activated late in G_1, these biosynthetic and morphogenetic pathways are put into effect during S. Protein modification proceeds throughout S and the subsequent G_2 period to prepare the chromatin for chromosome condensation and segregation which occurs in M. Our studies with the <u>ts</u> mutant cells indicate that expression of mutation in G_1 beyond the point of derepression of the enzyme/proteins of chromatin duplication, can result in uncoupling of DNA replication from the first two processes noted above.

SUMMARY AND FUTURE PROSPECTS

It is clear from our own work, and that of many others (cf[14,15,17,22]) that the tools and concepts of biochemical genetics can be successfully applied to the unravelling of the rather complex processes of chromatin replication and cell cycle progression in higher eukaryotic cells. In addition, they can be used to dissect out the biochemical reactions contributed by the host cell to the replication of viruses. Clearly, solution of the total problem by this experimental approach depends ultimately on having large numbers of cells which are mutant in each cistron of the gene complement involved. In addition, it is essential to be able to assay for each function encoded in each mutant locus.

Unfortunately for our present purposes, few dna<u>ts</u>/S<u>ts</u> mammalian cells have as yet been isolated. G_1<u>ts</u> cells are more readily obtained[17,85-87]. However, too few physiological or biochemical diagnostic landmarks of progression through each phase of the cell cycle have been established to permit full exploitation of even these.

We anticipate that more extensive studies with the ts cells described herein will provide us with important information, concepts and tools to permit more rapid advances in the biochemical genetic study of eukaryotic chromatin replication.

ACKNOWLEDGEMENTS

The continuing financial support of the Medical Research Council of Canada and the National Cancer Institute of Canada (N.C.I.C.) is gratefully acknowledged. Richard Colwill is a Research Student of the N.C.I.C. and Peter R. Ganz is a Fellow of The Leukaemia Society of America.

REFERENCES

1. G. Beadle and W. Tatum, Proc. Natl. Acad. Sci. USA 27:499-506 (1941).
2. A. Kornberg, this Volume (1981).
3. B. M. Alberts, J. Barry, P. Badinger, R. L. Burke, U. Hibner, C. C. Liu and R. Sheridan, in:"Mechanistic Studies of DNA Replication and Genetic Recombination - INC.UCLA Symposia on Molecular and Cellular Biology," XIX, B. Alberts and C. M. Fox, eds., Academic Press, New York, pp. 449-473 (1980).
4. M. Kohiyama, this Volume (1981).
5. K. N. Kreuzer and N. R. Cozzarelli, J. Bact. 140:424-436 (1979).
6. T. F. Meyer and K. Geider, J. Biol. Chem. 254:12436-12441 (1979).
7. A. Kornberg, CRC Crit. Rev. Biochem. 7:23-43 (1979).
8. J. Hurwitz, ibid 45-74 (1979).
9. C. C. Richardson, L. J. Romano, R. Kolodner, J. E. LeClerc, F. Tamanoi, M. J. Engler, F. B. Dean and D. S. Richardson, Cold Spring Harbor Symp. Quant. Biol. 43:427-440 (1978).
10. S. H. Wickner, Ann. Rev. Biochem. 47:1163-1191 (1978).
11. L. H. Hartwell, J. Cell Biol. 77:627-637 (1978).
12. J. R. Pringle, J. Cell. Physiol. 95:393-406 (1978).
13. D. T. Stinchcomb, K. Struhl and R. W. Davis, Nature 282:39-45 (1979).
14. R. Baserga, J. Cell Physiol. 95:377-382 (1978).
15. R. Sheinin, in:"Nuclear and Cytoplasmic Interactions in the Cell Cycle, G. L. Whitson, ed., Academic Press, New York, pp. 105-166 (1980).
16. R. Sheinin, J. Humbert and R. E. Pearlman, Ann. Rev. Biochem. 47: 277-316 (1978).
17. C. Basilico, Adv. Cancer Res. 24:223-266 (1977).
18. S. Howard and S. R. Pelc, Heredity (Suppl.) 6:261-273 (1953).
19. D. M. Prescott, Adv. Genet. 18:99-177 (1976).
20. A. B. Pardee, R. Dubrow, J. L. Hamlin and R. F. Kletzien, Ann. Rev. Biochem. 47:715-750 (1978).

21. R. Sheinin, in:"The Molecular Biology of Viruses," J. S. Colter and W. Parynchych, eds., Academic Press, New York, pp. 627-643 (1967).
22. R. Baserga, "Multiplication and Division in Mammalian Cells," Marcel Dekker, New York (1976).
23. S. Kit, Molec. Cell Biochem. 11:161-182 (1976).
24. R. Baserga, N. Engl. J. Med. 304:453-459 (1981).
25. F. Jacob, S. Brenner and F. Cuzin, Cold Spring Harbor Symp. Quant. Biol. 28:329-348 (1963).
26. C. Nicolini and R. Baserga, Chem. Biol. Interactions 11:101-116 (1975).
27. M. Bustin, FEBS Lett. 70:1-10 (1976).
28. V. E. Groppi and P. Coffino, Cell 21:195-204 (1980).
29. J. M. Walker, E. Brown, G. H. Goodwin, C. Stearn and E. W. Johns, FEBS Lett. 113:253-257 (1980).
30. J. Busch, N. R. Ballal, M. R. S. Rao, Y. C. Choi and L. I. Rothblum, in:"The Cell Nucleus, Vol 5, Pt. B," H. Busch, ed., Academic Press, New York, pp. 415-468 (1980).
31. R. Kornberg, Ann. Rev. Biochem. 46:931-954 (1977).
32. E. M. Bradbury, Differentiation 13:37-40 (1979).
33. I. L. Goldknopf, F. Rosenbaum, R. Sterner, G. Vidali, V. G. Allfrey and H. Busch, Biochem. Biophys. Res. Commun. 90: 269-277 (1979).
34. F. Back, Intl. Rev. Cytol. 45:25-64 (1976).
35. E. E. Dirksen, D. M. Prestcott and C. F. Fox, ICN-UCLA Symp. Molec Cell Biol., Vol. XII - Cell Reproduction (1979).
36. M. L. DePamphilis and P. M. Wasserman, Ann. Rev. Biochem. 49: 627-666 (1980).
37. A. Falaschi, F. Cobianchi and S. Riva, Trends Bioch. Sci. 5:154-157 (1980).
38. J. J. Champoux, Ann. Rev. Biochem. 47:449-479 (1978).
39. R. M. Benbow, C. B. Breaux, H. Joenje, M. R. Krauss, R. W. Lennox, E. M. Nelson, N. S. Wang and S. H. White, Cold Spring Harbor Symp. Quant. Biol. 43:597-602 (1978).
40. P. A. Fisher and D. Korn, in:"Mechanistic Studies of DNA Replication and Genetic Recombination. INC-UCLA Symposia on Molecular and Cellular Biology," XIX, B. Alberts and C. F. Fox, eds., Academic Press, New York, pp. 6551664 (1980).
41. M. Mechali, J. Abadiedebat and A-M. de Recondo, J. Biol. Chem. 255:2114:2122 (1980).
42. T-S. Hsieh and D. Brutlag, Cell 21:115-125 (1980).
43. M. I. Baldi, P. Bennedeti and G. P. Tocchini-Valentini, Cell 20: 461-467 (1980).
44. T. W. Borun, in:"Cell Cycle and Cell Differentiation," Vol. 7, J. Reinert and H. Holtzer, eds., Springer-Verlag, New York, pp. 249-290 (1975).
45. S. C. R. Elgin and H. Weintraub, Ann. Rev. Biochem. 74:725-77 (1975).
46. L. H. Kedes, Ann. Rev. Biochem. 48:837-870 (1979).
47. R. Sheinin, P. Darragh and M. Dubsky, J. Biol. Chem. 253:922-926 (1978).

48. R. Sheinin and P. N. Lewis, Somat. Cell Genet. 6:227-241 (1980).
49. S. J. Hochhauser, J. L. Stein and G. S. Stein, Intl. Rev. Cytol. 71:96-243 (1981).
50. S. A. Guttman and R. Sheinin, Exptl. Cell Res. 123:191-205 (1979).
51. T. Nishimoto, E. Eilen and C. Basilico, Cell 15:475-483 (1978).
52. T. Nishimoto, T. Takahashi and C. Basilico, Somat. Cell Genet. 6: 465-476 (1980).
53. A. McCracken, Somatic Cell Genetic, in press (1982).
54. R. Sheinin, manuscript in preparation.
55. L. Naismith and R. Sheinin, manuscript in preparation.
56. J. Humbert and R. Sheinin, Canad. J. Biochem. 56:444-452 (1978).
57. R. Colwill, C. Schwartz, R. Hand and R. Sheinin, in press (1982).
58. R. Ganz and R. Sheinin, in press (1982).
59. R. Sheinin, P. Darragh and M. Dubsky, J. Biol. Chem. 253:922-926 (1978).
60. R. Sheinin and A. Guttman, Biochim. Biophys. Acta 479:105-118 (1977).
61. R. Sheinin, P. Darragh and M. Dubsky, Canad. J. Biochem. 55:543-547 (1977).
62. R. Sheinin, Cell 7:49-57 (1976a).
63. R. Sheinin, J. Virol. 17:697-704 (1976b).
64. L. H. Thompson, R. Mankovitz, R. M. Baker, J. E. Till, L. Siminovitch and G. F. Whitmore, Proc. Natl. Acad. Sci. USA 66:377:384 (1970).
65. S. Kit, Adv. Virus. Res. 11:173-221 (1968).
66. U. Hubscher, C. C. Kuenzle and S. Spadari, Proc. Natl. Acad. Sci. USA 76:2316-2320 (1979).
67. A. I. Scovassi, R. Wicker and U. Bertazzoni, Europ. J. Biochem. 100:491-496 (1979).
68. P. Savard, G. Poirier and R. Sheinin, Biochem. Biophys. Acta 653: 271-275 (1981).
69. K. Mizuuchi, L. M. Fisher, M. H. O'Dea and M. Gellert, Proc. Natl. Acad. Sci. USA 77:1847-1851 (1980).
70. M. R. Mattern and R. B. Painter, Biochim. Biophys. Acta.563:306 -312 (1979)
71. M. Gellert, Proc. XI Intl. Congress Biochem. p. 11 (1979).
72. R. Colwill and R. Sheinin, Canad. J. Biochem. in press (1982).
73. N. R. Cozzarelli, Science 207:953-960 (1980).
74. L. H. Thompson, R. Mankovitz, R. M. Baker, J. A. Wright, J. E. Till, L. Siminovitch and G. F. Whitmore, J. Cell Physiol. 78:431-440 (1971).
75. G. Setterfield, R. Sheinin, I. Dardick, G. Kiss and M. Dubsky, J. Cell Biol. 77:246-263 (1978).
76. P. R. Ganz, R. Hand and R. Sheinin, manuscript in preparation.
77. P. R. Ganz and R. Sheinin, manuscript in proparation.
78. M. L. Slater and H. L. Ozer, Cell 7:289-295 (1976).
79. P. R. Srinivasan, R. Gupta and L. Siminovitch, Somat. Cell Genet. 6:567-582 (1980).
80. W. N. Hittelman and P. N. Rao, J. Cell Physiol. 95:333-342 (1978).

81. K. K. Jha, M. Siniscalco and H. L. Ozer, Somat. Cell. Genet. 6: 603-614 (1980).
82. M. Rieber and J. Bacalao, Exptl. Cell Res. 85: 334-339 (1974).
83. M. Rieber and J. Bacalao, Cancer Res. 34:3083-3088 (1974).
84. R. Sheinin, G. Settlefield, I. Dardick, G. Kiss and M. Dubsky, Canad. J. Biochem. 58:1359-1369 (1980).
85. T. Nishimoto and C. Basilico, Somat. Cell Genet. 4:323-340 (1978).
86. H. E. Schwartz, G. C. Moser, S. Holmes and H. K. Meiss, Somat. Cell Genet. 5:217-224 (1979).
87. T. Nishimoto, T. Takahaski and C. Basilico, Somat. Cell Genet. 6: 465-476 (1980).

PART III

REPAIR OF DNA DAMAGE

MECHANISMS OF DNA REPAIR AS REVEALED BY ARTIFICIAL INTRODUCTION OF ENZYMES INTO VIABLE CELLS

Mutsuo Sekiguchi, Hiroshi Hayakawa
Katsumi Yamashita and Kenji Shimizu

Department of Biology, Faculty of Science
Kyushu University 33, Fukuoka 812, Japan

ABSTRACT

Procedures for artificial introduction of active protein molecules into living cell systems have been developed and used to analyze cellular mechanisms for DNA repair. Plasmolyzed cells of *Escherichia coli* strains carrying a mutation in one of *uvrA*, *uvrB* and *uvrC* genes acquired ultraviolet (UV) resistance when the cells were exposed to high concentrations of T4 endonuclease V. With increasing concentrations of the enzyme, survival of the plasmolyzed cells after UV irradiation increased while colony-forming ability of unirradiated plasmolyzed cells was not significantly affected. The effect of T4 endonuclease V was specific for *uvr* mutants; wild type strains as well as strains having a mutation in *recA* or *polA* gene were not reactivated. On the other hand, *E. coli* DNA polymerase I was effective for enhancing survival of plasmolyzed cells of *polA* mutant, pre-exposed to UV. Thus, *E. coli* DNA polymerase I (molecular weight, 109,000 daltons) can be taken up into permeable cells and function *in vivo* to replace defective functions of the particular mutants.

T4 endonuclease V was introduced into cultured Xeroderma pigmentosum cells of complementation group A with the aid of UV-inactivated Sendai virus, resulting in a restoration of normal levels of UV-induced unscheduled DNA synthesis. A clear dose response was observed between the level of UV-induced unscheduled DNA synthesis of Xeroderma pigmentosum cells and the amount of T4 endonuclease V activity added. Impaired repair ability of Xeroderma pigmentosum cells belonging to other complementation groups was also restored by the enzyme treatment; cells of group A, B, C, D and E to the normal level and cells of group F to half the normal level. It is suggested that all the studied groups of Xeroderma pigmentosum (A through F) may be defective, at least in part, in the first step of excision repair.

INTRODUCTION

Artificial introduction of active protein molecules into viable cells may provide a useful means for analysis of cellular functions and also for assay of certain proteins whose activity cannot be measured by ordinary biochemical methods. We have explored the possibility of making permeable cells which still retain cell viability and of introducing biologically active proteins into such cells.

Although many methods to permeabilize bacterial cells have been presented[1-6], most of these treatments are too drastic to preserve colony-forming ability of the cells. By examining various procedures we found that plasmolysis is most suitable for insertion of functional enzymes into viable cells of *Escherichia coli*. We have shown that under appropriate conditions *uvr* mutants of *E. coli* acquire considerable degrees of ultraviolet (UV) resistance after treatment of the permeable cells with T4 endonuclease V[7]. We recently obtained evidence indicating that DNA polymerase I could also be inserted into cells by this procedure.

Several attempts have been made to introduce macromolecules into cultured mammalian cells[8-10]. These depended on the interaction of the cell membrane with liposomes or with Sendai virus (hemagglutinating virus from Japan; HVJ). We have shown that T4 endonuclease V can be inserted effectively into human cells in which the enzyme is functional on UV-damaged chromosomal DNA[11]. We have used this technique to analyze molecular defects in Xeroderma pigmentosum (XP) cells[12,13].

The purpose of this article is to review our studies on enzyme insertion and its application to analysis of the cellular DNA repair mechanism. The results will be discussed in the light of recent developments in the enzymology of excision repair.

INSERTION OF ENZYMES INTO VIABLE BACTERIAL CELLS

Experimental Design

It has been shown that UV-induced pyrimidine dimers in DNA are eliminated *in vivo* during the recovery process of the cell after UV irradiation[14,15]. Repair by excision appears to involve a number of steps; (1) a single-stranded break is formed near a pyrimidine dimer in irradiated DNA, (2) a nucleotide fragment containing the dimer is excised from the DNA, (3) new nucleotides complementary to those of the intact opposite strand are inserted into the gap so formed, and (4) the phosphodiester link is finally made between the "new" and "old" nucleotides, to reform the original double-stranded DNA.

Enzymes that catalyze each one of these reactions have been found in E. coli infected with bacteriophage T4. These include endonuclease V[16,17], 5'→3' exonucleases (exonuclease B and C)[18], DNA polymerase[19] and DNA ligase[20]. A possible mechanism of excision repair, deduced from the mode of action of these enzymes, is shown in Fig. 1.

T4 endonuclease V plays an essential role in initiating the repair reactions. The enzyme is coded by denV gene (sometimes called v gene) of T4, and the v mutant, which has a mutation in the denV gene preventing induction of the endonuclease, exhibits increased sensitivity to UV[21,22]. Since the enzyme is small (molecular weight, 16,000) and has a high specificity for UV-irradiated DNA, we have chosen it to test for insertion.

Fig. 2 illustrates the experimental design. Suppose that we add an active repair enzyme into repair defective mutant cells. We may observe reactivation, provided that the inserted enzyme functions in situ to replace the defective function. For instance, the type I

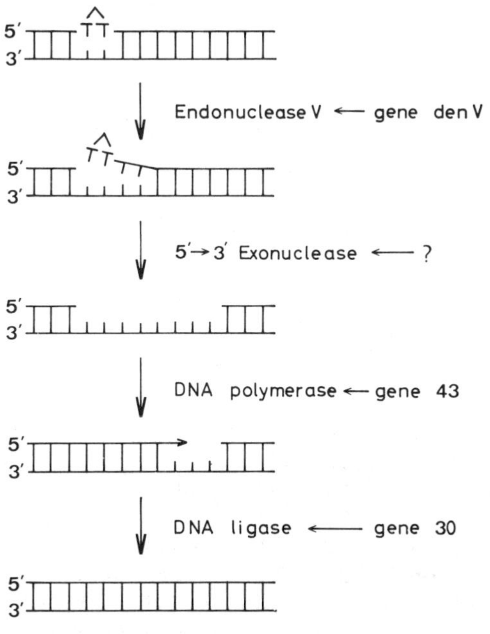

Fig. 1. A possible mechanism of excision repair in T4-infected E. coli.

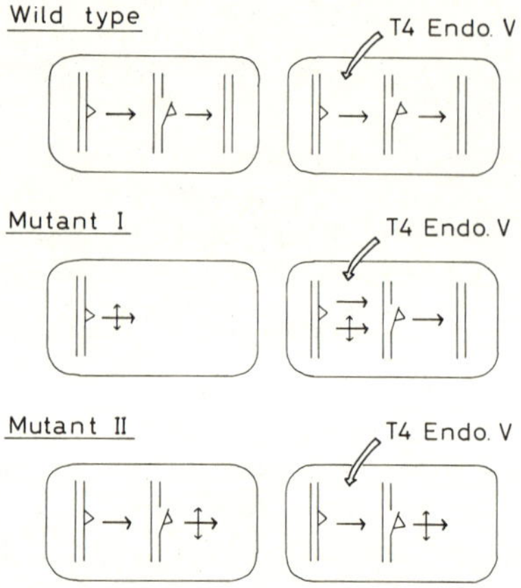

Fig. 2. A model for DNA repair by enzyme insertion.

mutant that is defective in the incision step may be rescued by a supply of T4 endonuclease V. However, the type II mutant which is defective in other step(s) of repair cannot be rescued by such an enzyme. Thus, it is assumed that effective introduction of T4 endocuclease V into certain E. coli mutants should increase survival of the cells after UV irradiation.

Acquisition of UV Resistance by uvr Mutants on Introduction of T4 Endonuclease V

Since the cell membrane of bacteria is usually impermeable to macromolecules, we must permeabilize the cell before enzyme treatment. We have examined various treatments which are known to increase permeability of E. coli cells, and found that plasmolysis is most suitable for this purpose. When we carefully prepare plasmolysed cells under defined conditions, they become permeable to macromolecules while still retaining colony-forming ability.

Plasmolysed cells were prepared by suspending bacteria in ice-cold plasmolysing solution (2M sucrose, 10mM EGTA and 40mM Tris·HCl (pH. 8.0)), and letting them stand at 0^oC for 30 min. The suspension of the plasmolysed cells was diluted 10-fold with a buffer containing T4 endonuclease V, and the mixture was kept at 0^oC for 30 min. After appropriate dilution UV sensitivity of the cells was determined. We calculated the reactivation index by dividing the surviving fraction (ratio of the number of colonies of UV-irradiated sample to that of non-irradiated sample) of enzyme-treated cells by the surviving fraction of non-treated cells.

When plasmolysed cells of E. coli strain N212, which has mutations in both uvrA and recA genes, were treated with various amounts of T4 endonuclease V, the result shown in Fig. 3 was obtained. With increasing concentrations of the enzyme, survivals of UV-irradiated cells increased while numbers of unirradiated cells decreased slightly. Levels of the reactivation were almost proportional to the amount of the enzyme applied, and more than 30-fold of reactivation was obtained at high concentrations of the enzyme. No reactivation took place when the plasmolysed cells were treated with an enzyme preparation which had been inactivated by heat treatment.

The involvement of T4 endonuclease V in the reactivation process was clearly shown in an experiment, in which the effect of enzyme fractions from two types of cells was compared. We prepared extracts from E. coli infected with T4D (wild type) and with v_1 mutant, and the extracts were processed in similar manners. Although elution

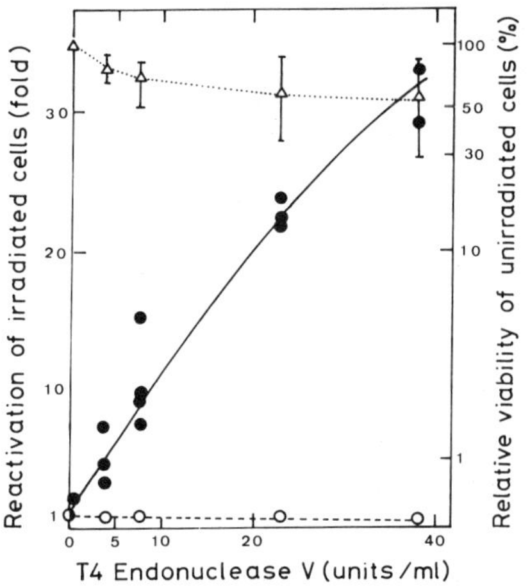

Fig. 3. Effect of T4 endonuclease V on UV resistance of plasmolysed cells[7]. E. coli N212 (uvrA, recA) was grown in L-broth to an optical density (660nm) of 0.9 and plasmolysed by suspending in 2M sucrose-10mM EGTA-40mM Tris·HCl (pH 8.0), and kept at 0°C for 30 min. To 5μl of the permeable cell suspension were added 50μl of a buffer containing various concentrations of T4 endonuclease V, and the mixture was placed in an ice-bath for 30 min. After dilution with medium, UV sensitivity of the cells was measured. (●) plasmolysed cells, treated with enzyme; (o) plasmolysed cells, treated with heated enzyme (45°C for 20 min). Although the heated enzyme possessed no endonuclease activity, the data were plotted against the initial activity before heating. (Δ) relative viability of unirradiated, plasmolysed cells after incubation with T4 endonuclease V.

profiles of protein from a CM-Sephadex C25 column were essentially the same for the two samples, T4 endonuclease V activity, determined in vitro, was found only in the wild type sample. When the enzyme fraction derived from T4D-infected cells was applied to plasmolysed cells of E. coli N212, a significant level of the activity that recovers the cells from UV-induced damages was detected. On the other hand, only slight reactivation was induced by the fraction from T4$\underline{v}_1$-infected cells.

Fig. 4 shows the effect of T4 endonuclease V on UV resistance of three different strains. The enzyme treatment was effective in increasing UV resistance of a uvrA mutant, but not of the wild type strain nor recA mutant. It was further demonstrated that uvrB and uvrC mutants were also reactivated by the enzyme treatment, whereas virtually no reactivation took place in a polA mutant (data not shown). Thus, the action of T4 endonuclease V is specific for the uvr mutants. This is consistent with the notion that the incision step of excision repair in E. coli is controlled by at least three genes, uvrA, uvrB and uvrC[23,24].

These characteristics of the reactivation are consistent with the idea that T4 endonuclease V is taken up by permeable cells. The enzyme must have access to the UV-damaged sites on the chromosomal DNA to exert such effect on irradiated cells. Furthermore, to form a visible colony, the enzyme-treated cells must undergo at least 20 cycles of cell division. Thus, the observed phenomenon can be taken as a strong proof that an active enzyme is inserted into viable cells.

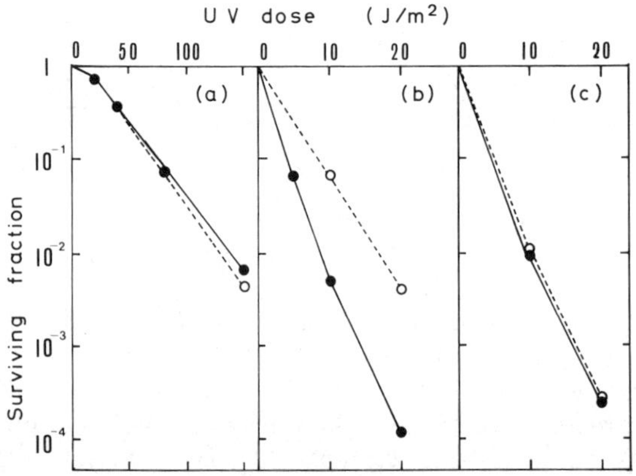

Fig. 4. Specific reactivation of uvrA mutant by treatment with T4 endonuclease V[7]. Plasmolysed cells of E. coli strains were swollen in a buffer with or without T4 endonuclease (25 units/ml) at 0°C for 30 min. The cells were diluted 10-fold with M9S medium and irradiated at various doses. (a) W3623 (Wild type). (b) N17-9 (uvrA). (c) KL16-99 (recA). (•) Treated with T4 endonuclease V; (o) not treated with the enzyme.

Insertion of E. coli DNA Polymerase I into Viable Cells

T4 endonuclease V is a relatively small protein and, unlike most proteins, is positively charged at neutral pH. In order to show the general usefulness of the present system, it was necessary to demonstrate that a large, negatively charged protein can be introduced into plasmolysed cells. For this purpose, we examined DNA polymerase I of *E. coli*, whose molecular weight is 109,000 daltons and which possesses a negative charge at neutral pH.

Fig. 5 shows the effect of DNA polymerase I on the survival of plasmolysed *polA* mutant cells after UV irradiation. With increasing concentrations of the enzyme (up to 30 units/ml), we observed more increase in survival of the cells. The addition of an enzyme preparation which had been inactivated by heat treatment caused no such reactivation.

It was noted, however, that the extent of the reactivation by DNA polymerase I is rather low. Only 8-fold increase was obtained at a concentration as high as 36 units/ml, and the level of reactivation did not increase even though the concentration of the enzyme was increased further. It seems that *E. coli* DNA polymerase I, compared with T4 endonuclease V, is less accessible to the sites of action in permeabilized cells.

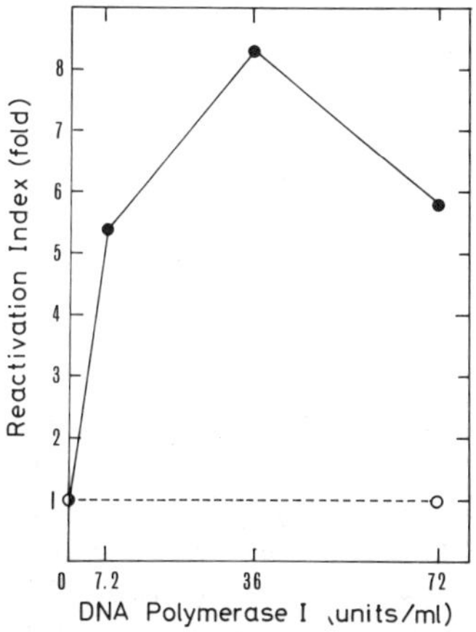

Fig. 5. Acquisition of UV-resistance by *E. coli* *polA* cells on introduction of DNA polymerase I. *E. coli* P3478 (*polA*) cells were plasmolysed and treated with various amounts of DNA polymerase I. (●) Cells treated with *E. coli* DNA polymerase I; (o) cells treated with the heated enzyme (50°C for 20 min).

To be reactivated by DNA polymerase I, the recipient cells must be polA mutants. E. coli strain JG138, N611 and P3478, all of which are defective in the polA gene, were reactivated on application of the polymerase. On the other hand, the uvrA mutants, which could be reactivated by T4 endonuclease V, were not reactivated by the polymerase treatment. Thus, the actions of the enzymes inserted are specific.

An Attempt to Assay the umuC-controlled Activity by the Protein Insertion Technique

The umuC gene is considered to be responsible for a principal step of induction or fixation of mutation after UV irradiation[25]. Mutants defective in the umuC gene are unable to induce mutations while normally inducing other SOS functions such as prophage induction and cell filamentation.

The major difficulty in studying the molecular mechanism for induction of mutation may be due to its rare occurrence. Moreover, there is no sensitive method for assaying in vitro activity involved in the process. We thought that it might be useful to develop a procedure to assay the activity promoting mutation.

Fig. 6 shows the effect of the umuC mutation on survival and induction of mutations of cells after UV irradiation. The umuC mutant is slightly more sensitive to UV than the wild type strain and is almost defective in producing UV-induced mutations. At 10 J/m^2, the mutation frequency of the umuC mutant is 10^{-4} that of the wild type strain. This confirms the results of Kato and Shinoura[25] and provides a basis for the biological assay of the umuC-associated activity.

We have examined whether the impaired ability of $umuC^-$ cells can be restored by externally supplied cell-free extracts. TK501 (umuC, his) cells were irradiated with UV (4 J/m^2), plasmolysed and then treated with an extract of JC3890 ($umuC^+$) which had been irradiated (5 J/m^2) and incubated at 37°C for 45 min in L-broth.

As shown in Table 1, a 10- to 50-fold increase in the number of His^+ revertants was observed. On the other hand, an extract of $umuC^-$ cells which had been treated in the same manner or an extract of $umuC^+$ cells which had not been irradiated showed no such an activity. The activity was heat-labile and was inactivated by phenol treatment. It seems that a protein(s) required for promotion of mutation may be formed under control of the umuC gene after UV irradiation.

It has been shown that SOS functions are induced when spr tif mutant is incubated at elevated temperature[26]. As also shown in

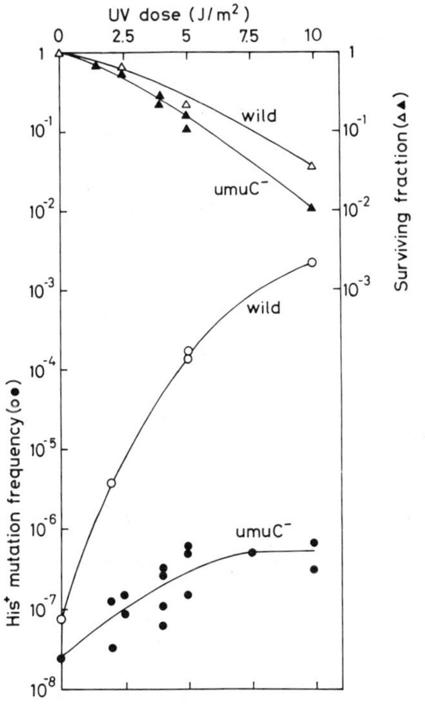

Fig. 6. UV sensitivity and UV-induced mutability of umuC mutant. Overnight cultures of E. coli JC3890 ($umuC^+$, his) and TK501 (umuC, his)[25] were washed twice with a minimal medium and starved for required amino acids by incubating in the minimal medium at 37°C for 1 hr. Bacteria were irradiated with various doses of UV and, after appropriate dilution, plated on two types of plates; semi-enriched agar for counting number of His^+ revertants and nutrient agar for counting number of viable cells.

Table 1, an extract prepared from DM1187 (spr, tif, $umuC^+$) cells which had been incubated at 42°C without irradiation was effective for inducing mutation, whereas an extract from the same cells incubated at 30°C exerted no such effect.

These results are consistent with the current concept for inducible repair[27]. It might be possible to purify the umuC gene product by monitoring the activity by the protein insertion technique.

Table 1. Effect of Various Extracts on Induction of Mutation in UV-irradiated, Plasmolysed Cells of umuC Mutant

Source of extracts		Number of His^+ revertants per 10^7 cells	
Strain	Pre-treatment	Expt. I	Expt. II
None(control)	——	5	4
JC3890($umuC^+$)	+UV, 37°C	69	211
TK501(umuC)	+UV, 37°C	5	-
JC3890($umuC^+$)	-UV, 37°C	5	-
DM1187(spr, tif, $umuC^+$)	-UV, 30°C	4	-
DM1187(spr, tif, $umuC^+$)	-UV, 42°C	20	245

ANALYSIS OF DNA REPAIR MECHANISM IN HUMAN CELLS

Restoration of Impaired Repair Ability of Xeroderma Pigmentosum Cells by Introduction of T4 Endonuclease V

Xeroderma pigmentosum (XP) is a rare hereditary disease caused by recessive mutation of autosomal genes. It is characterized by development of pigmentation abnormalities and numerous malignancies on exposure of the skin to sunlight. Cleaver[28] first demonstrated that cells from XP patients lack an ability to perform repair DNA synthesis after UV irradiation. XP cells might be defective in some early step(s) of excision repair.

We analyzed the defect in XP cells by the protein insertion technique. Fig. 7 shows our experimental plan. If XP cells are defective in the incision step of excision repair, UV-induced unscheduled DNA synthesis may be restored upon introduction of T4 endonuclease V into viable XP cells.

Levels of unscheduled DNA synthesis can be measured autoradiographically, as follows: Cells on a coverslip are irradiated, incubated with [^{3}H]thymidine, and washed. A photographic emulsion is placed over the cells and left in place for one week in cold. When the emulsion is developed, changes produced by the beta particles from disintegrating tritium atoms appear as grains over the nucleus. Thus, the number of grains over a nucleus can be used as a quantitative measure of the amount of [^{3}H]thymidine incorporated into the DNA of that nucleus.

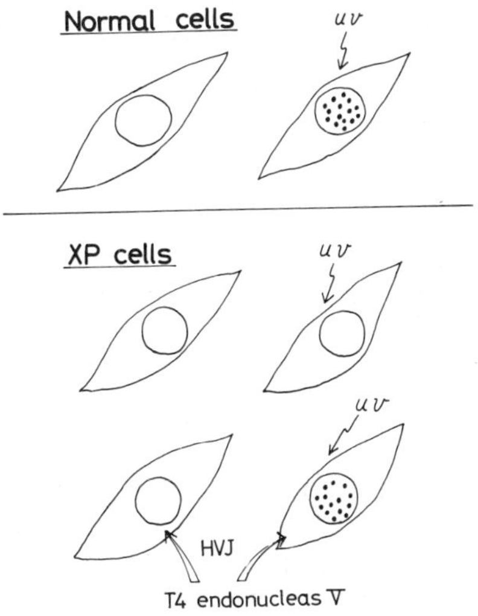

Fig. 7. A model for restoration of repair replication of XP cells by introduction of T4 endonuclease V.

To facilitate the enzyme insertion we used Sendai virus, which is known to have cell fusion activity[29]. At an early stage of the cell fusion reaction, the structure of the cell membrane is partially damaged by the action of the virus and then the structure is restored again. Thus, it seemed possible that macromolecules could be inserted into viable cells from the outside during the interaction between the cell membrane and Sendai virus.

In an experiment, the result of which is shown in Fig. 8, XP30S (complementation group A) was used to determine whether T4 endonuclease V could be inserted effectively into cells by Sendai virus and whether it could function on UV-damaged human chromosomal DNA. The cells were irradiated with UV (30 J/m^2) and exposed to the enzyme and UV-inactivated Sendai virus for 15 min at 0^oC and then for 15 min at 37^oC. Then the cells were cultured in a medium containing $[^3H]$ thymidine and subjected to autoradiography. We counted the number of grains per nucleus for a hundred nuclei and plotted their distribution. From the figure, it is evident that unscheduled DNA synthesis in the XP30S cells increased markedly, reaching the same level as that in normal cells. When the XP cells were exposed to Sendai virus alone, no increase in the level of unscheduled DNA sysnthesis was observed. It has been shown that the level of unscheduled DNA synthesis of irradiated normal cells is not affected by exposure to the enzyme and Sendai virus.

The effect of T4 endonuclease V was specific for UV-damaged DNA. No unscheduled DNA synthesis was seen in non-irradiated cells upon exposure to both the enzyme and Sendai virus.

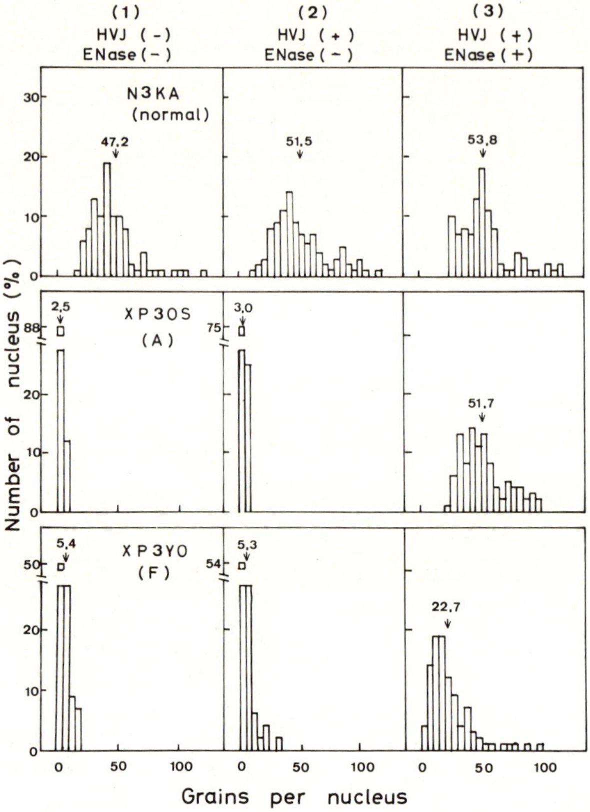

Fig. 8. Restoration of UV-induced unscheduled DNA synthesis in XP cells by the concomitant treatment with T4 endonuclease V and Sendai virus[13]. Cells were irradiated at 30 J/m^2 and treated as follows: (1) without treatment, (2) treated with UV-inactivated Sendai virus (HVJ), (3) treated with both T4 endonuclease V (ENase) and UV-inactivated virus. Arrows indicate mean average grain numbers per nucleus calculated from numbers in 100 nuclei.

A clear relationship was observed between the amount of T4 enzyme added and the level of unscheduled DNA synthesis of the XP cells. The level increased with higher enzyme concentrations (from 1 to 10 units/ml), reaching a plateau at concentrations over 10 units/ml.

To show unambiguously the involvement of T4 endonuclease V in the restoration of unscheduled DNA synthesis in XP cells, we examined the effect of an enzyme fraction derived from T4v_1-infected cells. Extracts from E. coli infected with T4D (a wild type strain) or with T4v_1 (an endonuclease V-defective mutant) were subjected to phase partition in dextran-500 and polyethylene glycol-6000, and the resulting supernatant was dialyzed and applied to a column of CM-Sephadex C-25. Absorbed proteins were eluted from the column with 0.5M KCl, and six 5ml fractions were collected. As shown in the upper half of Fig. 9, elution profiles of protein were essentially the same for the two samples; however, little or no T4 endonuclease V activity was found in the fractions from T4v_1-infected cells.

When the fractions derived from T4D-infected cells were applied to the XP cells with the aid of Sendai virus, significant levels of the activity that restores UV-induced unscheduled DNA synthesis were detected (see the lower half of Fig. 9). On the other hand, only slight restoration was induced by the fractions derived from T4v_1-infected cells. The levels of the in vivo reactivation activity coincided well with the levels of the endonuclease activity in vitro. Thus, it is evident that T4 endonuclease V, and not other components in the enzyme preparation, is responsible for the restoration.

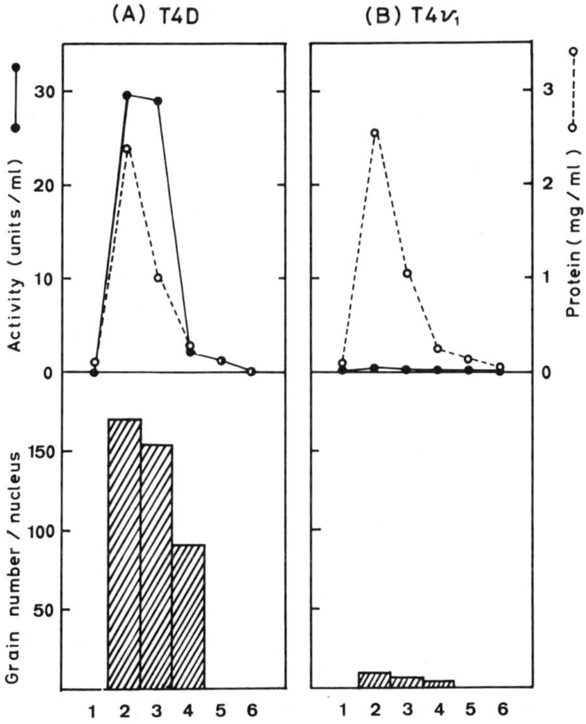

Fig. 9 Effects of various fractions of T4D-infected and T4v_1-infected E. coli on UV-induced unscheduled DNA synthesis of XP100S (group A). The data were taken from reference[12].

Effect of T4 Endonuclease V on different Complementation Groups of XP Cells

DeWeerd-Kastelein et al.[30] first demonstrated that cells of classical XP patients and deSanctis-Cacchione patients were genetically complementary. Both types of cells showed low levels of unscheduled DNA synthesis, but when they were fused by Sendai virus the levels became normal. Using this technique, five complementation groups, designated groups A to E, had been identified[31] by 1975. Recently, two additional groups, F and G, were found[32,33].

The effect of T4 endonuclease V on different kinds of XP cells was studied. Cells were irradiated with UV and then treated with both the enzyme and Sendai virus or with virus alone. The levels of unscheduled DNA synthesis after treatment with Sendai virus alone were 5% of the level in normal cells in group A, 7% in group B, 13% in group C, 20% in group D, and 67% in group E. These levels were in agreement with the standard levels in these cells estimated without treatment with Sendai virus.

As shown in Table 2, the unscheduled DNA synthesis of all these XP cells was restored to almost the normal level on treatment with both the enzyme and Sendai virus; that is, the level increased to 95% of normal level in group A, 91% in group B. 99% in group C, 101% in group D, and 105% in group E. Little or no restoration of unscheduled DNA synthesis was observed when an enzyme fraction derived from $T4v_1$-infected cells was applied.

XP cells belonging to complementation group F possess some unusual properties; they showed a low level of unscheduled DNA synthesis (about 10% of the normal level), despite their relatively high level of repair capacity as measured by both survival of UV-irradiated cells and the host-cell-reactivation of irradiated herpes simplex virus[32]. When T4 endonuclease V was introduced into the group F cells, the impaired ability for unscheduled DNA synthesis was partially restored, the extent of restoration being about half the normal level (see Fig. 8). Since the full recovery of unscheduled DNA synthesis by T4 endonuclease V was achieved in a parallel experiment with group A cells, this can be taken as one characteristic of XP cells of group F.

DISCUSSION

It was found, in these studies, that T4 endonuclease V is capable of restoring UV-induced unscheduled DNA synthesis in all the XP cells tested, among complementation groups A to F, suggesting that at least six genes are involved in controlling the incision step in human cells. It was shown that the incision step in *E. coli* is controlled by three genes, whereas only one gene, the *denV* gene, is

Table 2. Effect of T4 Endonuclease V and the Corresponding Enzyme Fraction Derived from T4$\underline{v}_1$- infected Cells on Repair Replication of Various XP Cells

Expt. No.	Cell line	Number of grains/nucleus		
		Control	+ Endonuclease V	+ Fraction from v
I	HNSF(normal	80	80	——
	CRL1223(A)	4	76	——
	CRL1199(B)	5	73	——
	CRL1170(C)	10	79	——
	CRL1160(D)	16	81	——
II	XP100S(A)	4	171	10
	CRL1160(D)	33	154	25
III	CRL1199(B)	6	69	5
	CRL1161(C)	9	64	11
IV	GM708(E)	57	81	53

Letters in parenthesis indicate complementation groups. The data were taken from reference[11,12].

involved in bacteriphage T4[22,23]. Thus, the processes of excision repair in human cells are more complex than in bacteria and bacteriophage, although the basic reactions may be common to all.

There are several possible explanations for the functions of these six genes: (a) human cells may possess more than one species of enzyme catalyzing the incision step; (b) the enzyme(s) may be composed of subunits, and their formation may be controlled by different genes; (c) some other protein factor(s) besides an enzyme(s) may be required for the reaction in vivo; and (d) there may be a regulator gene, which controls the formation of the enzyme and/or factor. These mechanisms are nonexclusive and the observed phenomena may be due to a combination of several mechanisms.

Recent studies[34-40] have revealed that the incision step of excision repair is much more complex than previously supposed. Based on several lines of evidence with microbial DNA repair enzymes, the following reaction scheme was proposed: a cleavage of the

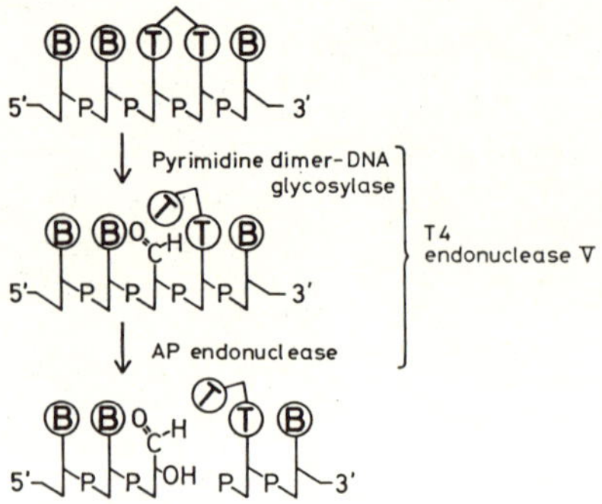

Fig. 10. A model for the initial steps of excision repair. The model was drawn on the basis of the mode of action of T4 endonuclease V. The enzyme catalyses the first two steps in the scheme.

N-glycosyl bond between the 5'-pyrimidime of a dimer and the corresponding sugar first takes place, and then a phosphodiester bond on the 3'-side of the apyrimidinic site is cleaved, Fig 10.

We showed in a recent study that a homogeneous preparation of T4 endonuclease V possesses both pyrimidine dimer-DNA glycosylase and apurinic/apyrimidinic (AP) endonuclease activities[41]. The ratio of the two activities did not change in the last two steps of purification. It was shown, moreover, that introduction of an amber mutation in the denV gene caused a simultaneous loss of the two activities, and suppression of the mutation rendered both activities partially active. These results strongly suggested that the glycosylase and the AP endonuclease reside in a single polypeptide chain.

Although it is uncertain at the present time whether a similar reaction scheme is applicable to other biological systems, this finding raises many interesting questions. Since the incised DNA possesses a free deoxyribose moiety at the 3'-terminus, this must be removed prior to the polymerase reaction. Thus, a 3'⟶5' exonuclease or a specific endonuclease that cleaves the 5'-side of the apyrimidinic site might be involved in the excision repair process. It is also of interest to know whether the two activities, the glycosylase and the AP endonuclease, always share a common polypeptide chain. There is a possibility that the two activities may be present in different protein molecules in mammalian cells. Involvement of many genes in controlling the first step of excision repair in human cells might be related to such a mechanism. To draw a more clear picture, identification and characterization of human incision enzyme(s) is clearly necessary.

ACKNOWLEDGEMENTS

We thank Drs. Y. Okada, H. Takebe, K. Tanaka for collaboration and discussion. The studies presented in this paper were supported by a Grant-in-Aid for Cancer Research and Scientific Research from the Ministry of Education, Science and Culture of Japan.

REFERENCES

1. L. Leive, Proc. Natl. Acad. Sci. USA 53:745-750 (1965).
2. G. Buttin and A. Kornberg, J. Biol. Chem. 241:5419-5427 (1966).
3. R. E. Moses and C. C. Richardson, Proc Natl. Acad. Sci. USA 67: 674-681 (1970).
4. H-P. Vosberg and H. Hoffmann-Berling, J. Mol. Biol. 58:739-753 (1971).
5. R. B. Wickner and J. Hurwitz, Biochem. Biophys. Res. Commun. 47: 202-211 (1972).
6. R. E. Moses, J. Biol. Chem. 247:6031-6038 (1972).
7. K. Shimizu and M. Sekiguchi, Molec. Gen. Genet. 168:37-47 (1979).
8. D. Papahadjopoulos, G. Poste and E. Mayhew, Biochim. Biophys. Acta 363:404-418 (1974).
9. M. Furusawa, T. Nishimura, M. Yamaizumi and Y. Okada, Nature 249: 449-450 (1974).
10. A. Loyter, N. Zakai and R. G. Kulka, J. Cell Biol. 66:292-304 (1975).
11. K. Tanaka, M. Sekiguchi and Y. Okada, Proc. Natl. Acad, Sci. USA 72:4071-4075 (1975).
12. K. Tanaka, H. Hayakawa, M. Sekiguchi and Y. Okada, Proc. Natl. Acad. Sci. USA 74:2958-2962 (1977).
13. H. Hayakawa, K. Ishizaki, M. Inoue, T. Yagi, M. Sekiguchi and H. Takebe, Mutation Res. 80:381-388 (1981).
14. R. B. Setlow and W. L. Carrier, Proc. Natl. Acad. Sci. USA 51:226-231 (1964).
15. R. P. Boyce and P. Howard-Franders, Proc. Natl. Acad. Sci. USA 51:293-300 (1964).
16. S. Yasuda and M. Sekiguchi, Proc. Natl. Acad. Sci. USA 67:1839-1845 (1970).
17. E. C. Friedberg and J. J. King, J. Bacteriol. 106:500-507 (1971).
18. K. Schimizi and M. Sekiguchi, J. Biol. Chem. 251:2613-2619 (1976).
19. M. Goulian, Z. J. Lucas and A. Kornberg, J. Biol. Chem. 243:627-638 (1968).
20. B. Weiss and C. C. Richardson, Proc. Natl. Acad. Sci. USA 57:1021-1028 (1967).
21. W. Harm, Virology 19:66-71 (1963).
22. K. Sato and M. Sekiguchi, J. Mol. Biol. 102:15-26 (1976).
23. P. Howard-Flanders, R. P. Boyce and L. Theriot, Genetics 53:1119-1136 (1966).
24. E. Seeberg, Proc. Natl. Acad. Sci. USA 75:2569-2573 (1978).

25. T. Kato and Y. Shinoura, Molec. Gen. Genet. 156:121-131 (1977).
26. D. W. Mount, Proc. Natl. Acad. Sci. USA 74:300-304 (1977).
27. E. M. Witkin, Bacteriol. Rev. 40:869-907 (1976).
28. J. E. Cleaver, Nature 218:652-656 (1968).
29. Y. Okada and J. Tadokoro, Exp. Cell Res. 26:108-118 (1962).
30. E. A. deWeerd-Kastelein, W. Keijzer and D. Bootsma, Nature New Biol. 238:80-83 (1972).
31. K. H. Kraemer, H. G. Coon, R. A. Petinga, S. F. Barrett, A. E. Rahe and J. H. Robbins, Proc. Natl. Acad. Sci. USA 72: 59-63 (1975).
32. S. Arase, T. Kozuka, K. Tanaka, M. Ikenaga and H. Takebe, Mutation Res. 59:143-146 (1979).
33. W. Keijzer, N. G. J. Jaspers, P. J. Abrahams, A. M. R. Taylor, C. F. Arlett, B. Zelle, H. Takebe, P. D. S. Kinmont and D. Bootsma, Mutation Res. 62:183-190 (1979).
34. L. Grossman, S. Riazzudin, W. A. Haseltine and C. P. Lindan, Cold Spring Harb. Symp. Quant. Biol. 43:947-955 (1978).
35. W. A. Haseltine, 1. k. gordon, C. P. Lindan, R. H. Grafstrom, N. L. Shaper and L. Grossman, Nature 285:634-641 (1980).
36. E. H. Radany, and E. C. Friedberg, Nature 286:182-185 (1980).
37. H. R. Warner, B. F. Demple, W. A. Deutch, C. M. Kane and S. Linn, Proc. Natl. Acad. Sci. USA 77:4602-4606 (1980).
38. B. Demple and S. Linn, Nature 287:303-208 (1980).
39. P. C. Seawell, C. A. Smith and A. K. Ganesan, J. Virology 35: 790-797 (1980).
40. L. K. Gordon and W. A. Haseltine, J. Biol. Chem. 255:12047-12050 (1980).
41. Y. Nakabeppu and M. Sekiguchi, Proc. Natl. Acad. Sci. USA 78 (in press).

DNA POLYMERASES AND DNA REPAIR IN EUKARYOTIC CELLS

Alain Sarasin, Jean-Michel Rossignol[1] and
Michel Philippe[2]

[1]Institut de Recherches Scientifiques sur le Cancer
B.P. 8, 94802 Villejuif Cedex, France
[2]Laboratoire de Biologie Cellulaire ERA-CNRS n° 400
Université Paris-Val de Marne
94010 Créteil Cedex, France

When living organisms are exposed to deleterious conditions, such as physical or chemical agents, their cellular DNAs are subject to damage. In order to survive and to maintain genetic continuity, living organisms have developed very efficient repair processes. If a DNA damage is defined as any modification of the normal chemistry or any change in the sequence of the nitrogenous bases of DNA, DNA repair processes are biological mechanisms by which the damage is reversed or removed from the DNA. An error-free repair mode will restore the correct base sequence while an error-prone repair mode will give rise to a base sequence different from the original one.

DNA repair has been studied most extensively in prokaryotes, using ultraviolet light-induced damages (essentially pyrimidine dimers) as a model system. Many aspects of DNA repair have been recently reviewed[1-6]. DNA repair modes can be classified in different ways. For the purpose of this paper, we have chosen one classification with respect to DNA polymerase involvement. Four DNA repair modes do not need any DNA polymerase: enzymatic photoreactivation; mismatch repair; adaptation; and insertion. Three others involve a DNA polymerase: excision-repair; post-replication repair; and SOS repair. The enzymology of DNA repair processes is now well understood in prokaryotes. Unfortunately, the study of DNA repair mechanisms in higher organisms, particularly in mammalian systems is much less

Present address: [1]Department of Biochemistry and [2]Department of Biological Sciences, Stanford University, Stanford, CA 94305, U.S.A.

advanced. The absence of well-charaterized animal cell mutants in DNA repair pathways and the presence of a highly complex chromatin structure make the experiments more difficult to perform and to interpret. However, the discovery that some cancer-prone diseases are due to a lack of cell DNA repair ability[7] gave rise to a new era in the study of DNA repair and to the crucial relationship between DNA damage and induction of cancer in man.

The purpose of this paper is to study the role of the different DNA polymerases (α, β, γ) in DNA repair processes occurring in eukaryotic systems. As mentioned above, only three major DNA repair modes involve DNA polymerases if one directly transposes DNA repair models from bacteria to mammalian cells. Among these three processes, two of them are used in conjunction with DNA replication (post-replication repair, SOS repair). The third one (excision-repair) acts on non-replicating DNA. After a brief survey of the properties of the eukaryotic DNA polymerases, the role of these enzymes in excision-repair, post-replication repair and SOS repair modes will be discussed.

EUKARYOTIC DNA POLYMERASES

In eukaryotic cells three separate classes of DNA polymerases α, β and γ have been extensively studied[8-10]. Another polymerase, called δ, has been described in two cases[11,12], but has not been well-characterized. The principal characteristics of these enzymes are described in Table 1.

1. DNA Polymerase α

The DNA polymerase α is a multimeric enzyme with a high molecular weight. All the results on the structure of the α enzyme isolated from drosophila embryos, calf thymus and rat liver cells, indicate that the enzyme is composed of a large subunit of about 150 000 daltons which has been found to possess the catalytic activity and which is associated with at least three smaller subunits[13-17]. On the contrary, the DNA polymerase α isolated from human KB cells and mouse myeloma does not possess the large subunit[18,19].

In every case, DNA polymerase α is described as an acidic protein whose activity is sensitive to various agents such as N-ethylmaleimide, aphidicolin or high salt concentrations (Table 1). On the contrary, dideoxythymidine triphosphate which is an analogue of deoxythymidine triphosphate is not inhibitory to the activity when the two components are present. DNA polymerase α seems to be the main enzyme involved in DNA replication, especially because of its presence in dividing cells (and its disappearance in non-dividing differentiated cells[20]), its increased activity during the S phase[21] and its association to the SV40 replicating mini-chromosome[22].

Table 1. Principal Characteristics of Eukaryotic DNA Polymerase

Characteristics	α	β	γ
Molecular weight	>150,000	≃40,000	≃4 x 47,000
Sedimentation coefficient[a]	6-8 S	3-4 S	7-8 S
Isoelectric point	5.2	8.5-9.5	5.4
Cellular compartment	nucleus	nucleus	mitochondria: nucleus
Optimum KCl concentration	20-50mM	100-200mM	100-200mM
Activity with various templates:			
Activated DNA	yes	yes	yes
RNA-primed single-stranded DNA	yes	no	no
Poly A : oligo (dT)	no	yes	yes
Activity in the presence of various inhibitors:			
2mM N-Ethyl Maleimide	total inhibition	no effect	partial inhibition
Aphidicolin	inhibition[b]	no effect[c]	no effect[c]
Dideoxy TTP	no effect	inhibition	inhibition
Phosphonoacetate	inhibition[d]	no effect[e]	no effect[e]

[a]Measured in sucrose gradient containing 0.5M KCl.
[b]50% inhibition with 2μg/ml of Aphidicolin.
[c]With 125μg/ml of Aphidicolin.
[d]50% inhibition with 30μM of Phosphonoacetate.
[e]With 200μM of Phosphonoacetate.

2. DNA Polymerase β

The DNA polymerase β is a low molecular weight enzyme (40 000 daltons) composed of only one basic polypeptide[23,24]. Unlike the DNA polymerase α, N-ethyl-maleimide, aphidicolin and high salt concentrations do not inhibit its activity. However dideoxythimidine has been found to be a strong inhibitor (Table 1).

This enzyme is present in almost all cells including terminal differentiated cells[25,26]. Moreover, its level of activity does not change significantly during the cell cycle[21].

3. DNA Polymerase γ

The molecular weight of this enzyme is not well established (200 000 to 300 000 daltons)[9] but a recent result seems to indicate that the DNA polymerase γ is composed of four subunits of 47 000 daltons each[27]. Its activity is inhibited by N-ethyl-maleimide (like DNA polymerase α) and by dideoxythymidine (like DNA polymerase β) but it is not sensitive to high salt concentrations or aphidicolin (Table 1).

The DNA polymerase γ is mainly located in the mitochondria and is the only polymerase found in this cellular compartment[28,29] suggesting that this enzyme is involved in the replication of mitochondrial DNA[21]. However, evidence for the involvement of DNA polymerase γ in the adenovirus replication has been presented[30]. Like DNA polymerase β, the γ enzyme is also present in terminal differentiated cells[25,26].

As mentioned in the introduction, the mechanisms of DNA repair occurring in eukaryotic cells are not well known and a hypothesis could only be drawn from the considerably better knowledge of DNA repair in Escherichia coli. Since it has been reported[31] that mitochondria were unable to repair their DNA when damaged by UV-light, we shall focus our attention on nuclear DNA repair and more precisely on excision repair, post-replication repair and SOS repair, which all involve the participation of a DNA polymerase activity. As pointed out by Hanawalt et al.[3], excision repair is only possible in a duplex region, while post-replication repair and SOS repair may occur in the area of the replication fork. In consequence, the role of DNA polymerases in the excision repair pathway has been studied preferentially in non-dividing cells. On the contrary, studies with dividing cells have given data both on post-replication repair and SOS repair modes which are active at the replication fork, and on excision repair mode which is active on the double-stranded DNA present before or after the onset of DNA replication.

DNA POLYMERASE INVOLVEMENT IN THE EXCISION REPAIR PATHWAY OF NON-DIVIDING CELLS

1. Excision Repair Pathway

In principle, the excision-resynthesis repair mode is very simple. The following enzymatic events are required (see Fig. 1):

(a) Incision of the DNA strand in a position adjacent to the damage by a specific endonuclease.
(b) Excision of a part of the strand containing the damage by an exonuclease producing a gap in the molecule.

(c) Gap-filling repair synthesis performed by a DNA polymerase using the complementary and non-damaged strand as template.
(d) The phosphodiester bond is sealed by a DNA ligase.

Several endonucleases, specific for various DNA damages, have been isolated and purified from prokaryotic and eukaryotic cells (see references[3,4] for reviews). The studies of a pyrimidine dimer specific endonuclease isolated from Micrococcus luteus or T4 phage-infected E. coli have permitted the elucidation of the incision step at the level of a pyrimidine dimer[32,33]. In mammalian cells, the incision repair pathway seems to involve the same steps as those astablished in bacteria. Particularly, cells isolated from patients affected by Xeroderma pigmentosum (XP) - a human autosomal recessive disease - are deficient in the incision step close to a pyrimidine dimer. The in vitro-addition of bacterial UV-specific endonuclease in UV-damaged XP cells restores a full excision repair ability[34-36].

2. DNA Polymerase Involvement

The biological systems, used to determine which DNA polymerase is involved in the excision-resynthesis pathway, are either non-dividing differentiated cells or cultured cells in which DNA replication is blocked by a specific inhibitor. Since DNA polymerase α is not detectable in some differentiated cells[21], DNA polymerase β and γ are the obvious candidates for performing the resynthesis step.

Among the differentiated cells, adult rat brain neurons have been the most extensively studied[25]. UV-light irradiation gives rise to an important unscheduled DNA synthesis in the neuronal rat nuclei, in which 99.2% of total DNA polymerase activity is due to the DNA polymerase β[20]. This result strongly suggests that the β enzyme play an important role in excision repair after UV-irradiation. One must be very cautious in extending these results to all the non-dividing cells especially because each type of differentiated cell seems to lose a certain number of functions during the differentiated process. This point can be illustrated by some brain cells which have lost their ability to repair some alkylated bases[37,38]. Moreover, no significant increase in DNA polymerase activity is found after UV-irradiation of fully differentiated bull spermatozoa heads (M. Philippe unpublished results), although spermatozoa nuclei, like neuron nuclei, contain the same type of DNA polymerase (DNA polymerase β).

Using permeabilized baby hamster kidney (B.H.K.) cells which are blocked in G_0, Castellot et al.[39] have shown that DNA synthesis takes place after X-ray irradiation but not after UV-light irradiation. This DNA synthesis, like the DNA polymerase β in vitro, is not affected by arabinosyl cytosine triphosphate (an analogue of dATP), N-ethyl-maleimide and KCl. This result suggests the involve-

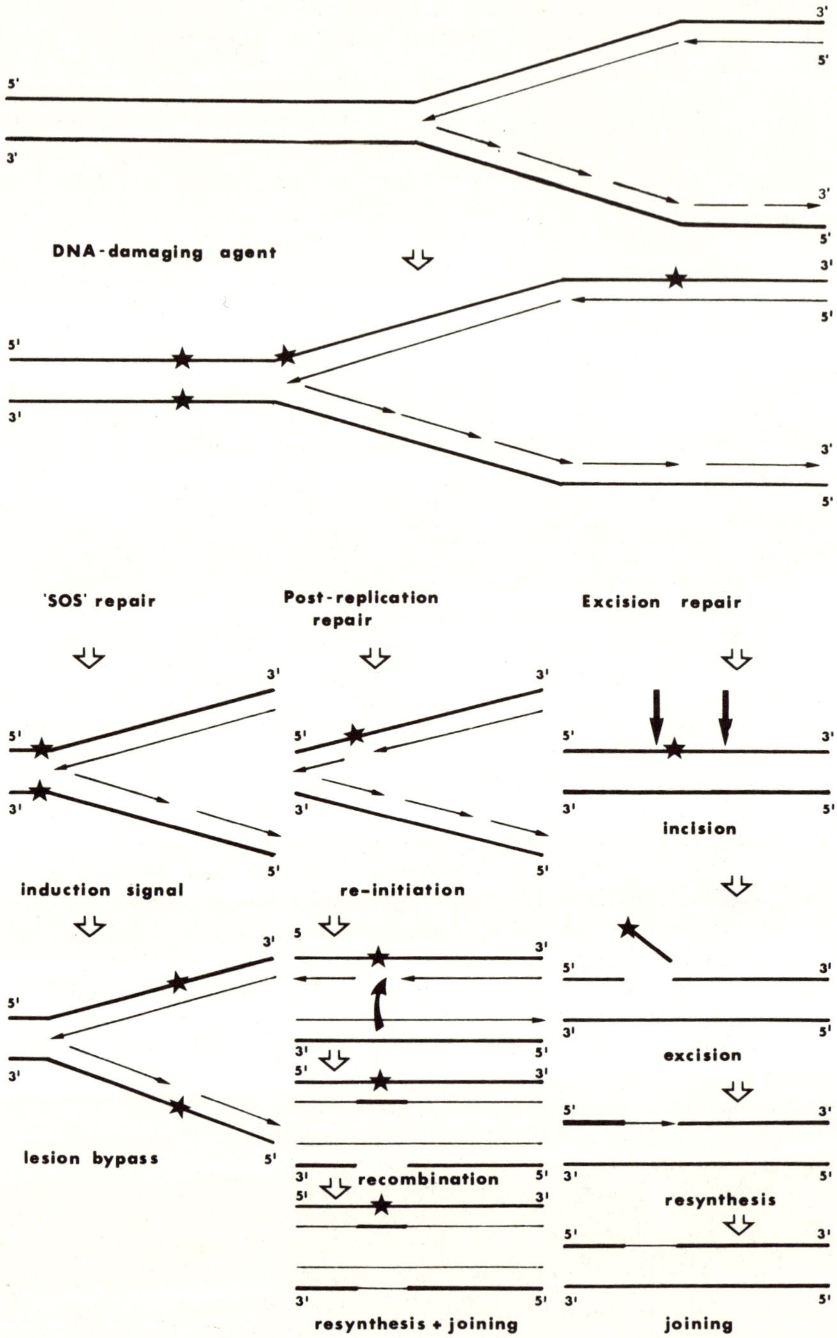

Fig. 1. Schematic overview of repair pathways.

ment of DNA polymerase β in the resynthesis process of X-ray damage repair. Rossignol et al.[40] have, by arginine deprivation, completely inhibited the replicative DNA synthesis in confluent Syrian hamster fibroblasts. After treatment of these cells by N-ethyl-N-methyl-N'-nitro-N-Nitrosoguanidine (MNNG) they found, by autohistoradiography, an unscheduled DNA synthesis whose level was maximum at the end of the treatment. Only DNA polymerase β was detected in cytosol and nuclear extracts prepared from normal and MNNG-treated fibroblasts. This result strongly suggests that the MNNG-induced DNA repair synthesis is performed by DNA polymerase β. Moreover "X-ray like" agents induced only the replacement of 3 to 10 nucleotides. Such small gaps can only be filled in vitro by the DNA polymerase β and not by the polymerase α which is unable to bind at a small gap[41]. Such a synthesis of small fragments of DNA probably requires a small amount of enzyme, which is in good agreement with the fact that the level of DNA polymerase β is quite identical in control cells and in cells treated with X-ray-like compounds.

Using confluent cells in culture, Butt et al.[42] found that only DNA polymerase β was present in isolated nuclei. They have shown that the incorporation of dNTPs was increased after treatment of the nuclei with DNase I. The same result is obtained with isolated bull spermatozoa heads (M. Philippe, unpublished results). These experiments suggest that repair synthesis may occur in these cells if an incision step has been performed.

From all these data where scheduled DNA synthesis does not occur, DNA polymerase β unambiguously participates in the excision repair process, at least after treatment with "X-ray like" agents. When fully differentiated cells are able to repair their DNA after being damaged, DNA polymerase β seems to participate in the repair process, and when these cells are not able to repair their DNA, DNA polymerase β does not appear to be the limiting factor.

In bacteria, repair synthesis is essentially performed by DNA polymerase I or by DNA polymerase III in the mutant lacking the DNA polymerase I activity[3]. These two polymerase activities are associated <u>in vivo</u> with a 5'→3' exonuclease activity responsible for the excision step of the damaged DNA. The excision and resynthesis steps are probably closely linked events. Purified eukaryotic DNA polymerases have been shown to lack any exonuclease activity. Only two reports have described an exonuclease activity associated with a DNA polymerase[12,19]. In consequence, the resynthesis step performed by polymerase β in eukaryotic cells should be preceded by an excision process. In fact, several exonucleases have been described in eukaryotes especially from rabbit tissue, human KB cells or the placenta (see[3] for review).

DNA POLYMERASE INVOLVEMENT IN REPAIR PATHWAYS OF DIVIDING CELLS

1. Post-replication Repair Pathway

Bulky lesions represent a block for the progression of the replication fork. This inhibition has been demonstrated essentially for UV-like compounds (UV light, AAAF, aflatoxin B_1, polycyclic hydrocarbons, ...) both in prokaryotes[3,6] and in eukaryotes[6,43,44]. We have developed an in vitro assay to determine whether the purified rat liver β polymerase is also blocked by pyrimidine dimers. The template poly(dT)-oligo$(dA)_{12-18}$ is irradiated at 254nm at various fluences and then DNA polymerase activity is followed by incorporation of labelled dATP. Figure 2 shows that increasing UV-doses give rise to increasing inhibition of polymerase activities. The experimental curve is identical to the theoretical curve we calculated if DNA polymerases β were blocked by a dimer (Fig. 2). By looking at the kinetics of polymerase β synthesis on unirradiated or UV-irradiated templates, we have deduced that pyrimidine dimers are not simply a pause in the progression of the polymerase β, but really block it (Fig. 3). Since

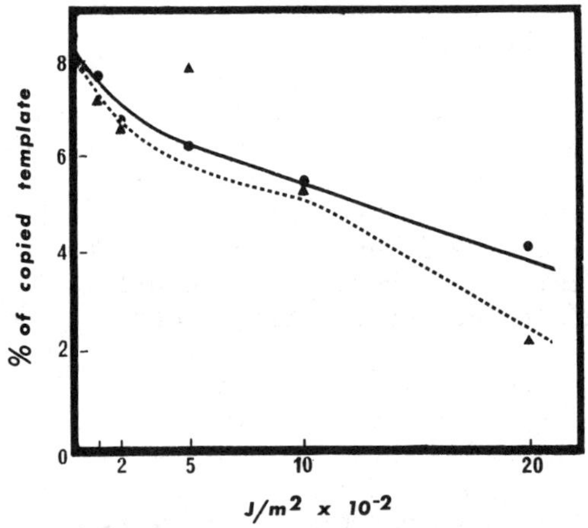

Fig. 2. Activity of rat liver polymerase β on UV-irradiated poly(dT)·oligo$(dA)_{12-18}$. Poly(dT) (about 750 nucleotides) is UV-irradiated at 254nm and then hybridized with oligo(dA) (12-18 nucleotides) at the nucleotide ratio 4:1. The activity of purified polymerase β is measured on that template in the presence of: Amnediol-HCl 50mM, pH 8.6, $MnCl_2$ 0.1mM, 2-mercaptoethanol 1mM, BSA 200μg/ml and H^3-dATP 125μM (200cpm per pM). The three other triphosphate nucleotides do not change the result. The dotted line is the theoretical inhibition of DNA synthesis if pyrimidine dimers were really a block for the polymerase (for experimental details see[69]).

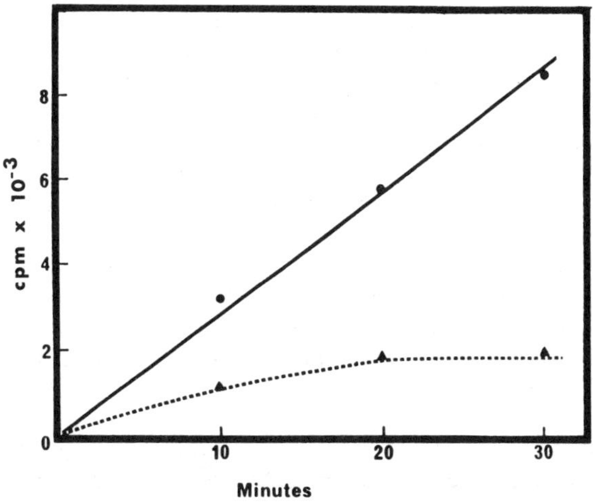

Fig. 3. Kinetics of DNA polymerase β activity on UV-irradiated or unirradiated poly(dT)·oligo(dA). Experimental conditions are identical to the ones described in Figure 2. ●——● unirradiated template; ▲----▲ 2000J/m^2-UV-irradiated template.

our polymerase β preparation was free of any contaminating enzymes, especially exonucleases, the block encountered at the level of a pyrimidine dimer is really intrinsic to the enzyme and is not due to an associated editing function (J.M. Rossignol and A. Sarasin, unpublished results).

The lesions, which inhibit the progression of the replication fork, cannot be repaired by the excision pathway because of the physical separation between the two template strands. To overcome this DNA replication blockage, other kinds of repair pathways are used by the cell. The most popular hypothesis - called post-replication repair or daughter-strand gap repair[3]- is schematized in Figure 1. This model based on DNA repair of E. coli after UV-irradiation, proposes that DNA synthesis resumes beyond the lesion leaving a gap in the newly-synthesized template opposite each lesion. These gaps are the substrates for sister-strand exchanges which will be filled with undamaged DNA from the symmetrical parental strand. Discontinuities in the parental strand will result from this recombination step and will be filled by repair synthesis using the undamaged complementary daughter strand as a template (Fig. 1). This daughter-strand gap repair is a tolerating process rather than a repair process per se, since the lesion is not eliminated but is simply diluted out. In fact, once the replication fork is passed, the excision repair process can eliminate the lesion.

In mammalian cells, experimental protocols identical to that used in bacteria to show the post-replication repair mode, have revealed a cellular response resembling daughter-strand gap repair[45]. This process seems to be deficient in fibroblasts isolated from XP variant patients[46] and is sensitive to caffeine in some cells[47]. After UV-irradiation, the size of newly-synthesized DNA is comparable to the average distance between pyrimidine dimers in the parental DNA[44,45,48,49]. Thereafter, the size of newly-synthesized DNA increases with time, although the pyrimidine dimers seem to remain present on parental DNA[44]. However, the interpretation of these experiments is very difficult, essentially because the eukaryotic chromosomes contain multiple replicons which are replicated at various times during the S-phase.

In conclusion, no clear mechanism could be deduced from the eukaryotic experiments. Several extreme models can be described after the blockage of the replication fork at the dimer:

(1) A gap can be left opposite the lesion and synthesis can resume beyond it, as in bacteria. The gap can be filled by strand exchanges or de novo synthesis.
(2) DNA synthesis can resume after a pause bypassing the lesion.
(3) DNA synthesis is completely blocked and the DNA continuity is assumed by the adjacent replicon[50].

These limitations stress the interest in studying replication on damaged templates from very well-defined DNA molecules such as small DNA viruses[44].

2. DNA Polymerase Involvement

As said before, in dividing cells, post replication repair and excision repair may occur at the same time. For the moment this is not possible without using DNA synthesis inhibitors to separate experimentally these two DNA repair processes. For this reason, in this chapter we will talk in terms of total DNA repair.

In dividing cells, two different approaches have been used to study the possible role of DNA polymerases in the DNA repair process. The first is the measurement of the activity of the three DNA polymerases after DNA-damaging treatment compared to undamaged controls, in order to detect any correlation between a specific increase in one DNA polymerase activity and a specific DNA repair mode. The second approach consists of using specific inhibitors of each DNA polymerase and of following the consequences of a specific polymerase inhibition on various DNA repair pathways.

(a) Variation of DNA polymerase activities during DNA repair processes. Using human lymphocytes stimulated by photohemagglutinin, Bertazzoni[51] reported that the appearance of an activity peak of DNA polymerase α coincided with maximum DNA synthesis. The DNA polymerase β activity is also increased during DNA synthesis, but reaches its maximum at a later time. After UV-irradiation, the capacity of these cells to perform DNA repair shows two maxima. The first correlates with the peak of DNA replication and the second with the peak of DNA polymerase β activity. However, after UV-irradiation of monkey CV-1 cells, Wicker et al.[52] did not detect any significant variations in the level of activity of DNA polymerases α, β and γ, whatever the UV doses and whatever the time after irradiation. Interestingly, in the same treated cells, the level of DNA ligase activity increased significantly after UV-irradiation[53]. Suzuki et al.[54] have studied the variations in DNA polymerase α, β and γ, activities after treatment of proliferating HeLa cells with MNNG. They reported that all three DNA polymerase activities were stimulated by MNNG treatment. For higher concentration of these chemicals, the activities strongly decreased. But the maximum of activity for each DNA polymerase was reached at different concentrations of MNNG.

From all these data it is quite impossible to show unambiguously a specific increase in one DNA polymerase during DNA repair. Moreover, no experiments have shown any significant change in the sedimentation pattern or in the sensitivity to specific inhibitors of DNA polymerases isolated from DNA repairing cells compared to controls. This point has clearly been established by Ciarrochi et al.[55].

(b) Influence of specific inhibitors of DNA polymerases on DNA repair processes. Aphidicolin (see[56] for review), which is a tetracyclic diterpene tetraol obtained from Cephalosporium aphidicola, is a specific direct inhibitor of DNA polymerase α in vitro but has no effect on DNA polymerase β or γ[57].

Almost the same amount of aphidicolin required to inhibit DNA polymerase α in vitro inhibits DNA replication in vivo, suggesting that DNA polymerase α inside the cell nucleus is also the specific target of the drug[58]. This conclusion is confirmed by the isolation of aphidicolin-resistant cell mutants which contain aphidicolin-resistant DNA polymerase α or higher levels of aphidicolin-sensitive DNA polymerase α[59]. In consequence, the use of aphidicolin has been proposed as a good way to investigate the biological role of DNA polymerase α in repair processes.

Using UV-irradiated HeLa cells, Pedrali-Noy and Spadari[60] have reported that DNA repair was unaffected by aphidicolin, and then concluded that only DNA polymerase β was involved in DNA repair, DNA polymerase α being related only to DNA replication. A similar result has been found by Guilotto and Mondello, who studied the repair synthesis in mitotic HeLa cells, in which both α and β polymerases were present[61]. However, several contradictory results have now been reported[55,62-64]. Hanaoka et al.[62] found that UV-induced DNA repair is considerably decreased by aphidicolin treatment in hydroxyurea and arabinosyl cytosine treated HeLa cells. However, if hydroxyurea and arabinosyl cytosine are potent inhibitors of DNA replication[65,66] they are also, as pointed out by Snyder and Regan[63], inhibitors of

DNA repair[67,68], thus making the conclusion of Hanaoka et al. somewhat tenuous. Berger et al.[64] used human lymphocytes stimulated by phytohemagglutinin and treated with MNNG or UV-light. In the permeable cell system they used, they found an inhibitory effect of several substances e.g., arabinosyl cytosine triphosphate, phosphonoacetic acid, aphidicolin, N-ethyl-maleimide, on both DNA replication and repair. In particular, the fact that aphidicolin and N-ethylmaleimide inhibit DNA repair, suggests that DNA polymerase α is implicated in the process. Yet the complete inhibition of repair was never obtained and was always less important than DNA replication inhibition. This fact suggests that the three DNA polymerases may be involved in repair and replication. Ciarrochi et al.[55] measured the DNA repair synthesis in confluent cultured human fibroblasts, after exposure to several UV-like and X-ray like DNA damaging agents. They showed that aphidicolin completely inhibits DNA repair and replicative DNA synthesis in both intact and permeabilized cells. On the contrary, dideoxythymidine triphosphate, a specific inhibitor of DNA polymerases β and γ does not affect either replicative or repair synthesis. The authors concluded that DNA polymerase α is clearly involved in both repair and DNA replication. They excluded the possibility that DNA polymerase β can take over the function of DNA polymerase α when the latter is inhibited[55]. Finally, in a discussion of all these results, Snyder and Regan emphasized that permeabilized lymphocytes usually do not recover from the permeabilization treatment, giving rise to doubt concerning the physiological state of these cells[63]. They also pointed out that several studies in this field have not used specific and direct assays for the analysis of DNA repair. Following DNA repair both by removal of pyrimidine dimers from DNA and by sedimentation of the DNA on alkaline sucrose gradient, Snyder and Regan have shown that aphidicolin clearly inhibits the DNA repair process in UV-irradiated fibroblasts. From these results, they only draw the conclusion that some stage of excision repair is being blocked by aphidicolin[63].

It is clear that the problem of DNA polymerase involvement in repair processes is the subject of controversy, so one must be very wary of drawing any general conclusions from all these data. Discrepancies among different investigations may be due to the use of different methods or to differences between cell types. It is also plausible that the two enzymes (α and β polymerases) are associated to form a kind of repair-replication multienzymatic complex.

DNA POLYMERASE INVOLVEMENT IN INDUCIBLE REPAIR PATHWAYS

1. SOS Repair Process

Most of DNA damaging agents which block the progression of replication forks induce, in bacteria, a series of pleiotropic effects which are called "SOS functions". The expression of these functions is responsible for the lysogenic induction of λ phage in lysogenic bacteria, the reactivation and the mutagenesis of UV-irradiated phage (Weigle's reactivation) and the filamentous growth of bacteria[70-72]. These functions are under the coordinate control of recA and lexA genes and the molecular mechanism of their expression is well understood. Treatment of E. coli with DNA damaging agents such as UV-light, mitomycin C, ionizing radiations, alkylating agents or metabolized aflatoxin B_1 drastically increases the synthesis of the recA protein. This protein, in the presence of single-stranded DNA, due to DNA damage or DNA repair of damages, is partly activated into a protease. This protease activity cleaves the lexA gene product, which is the repressor of several SOS function genes (recA, lexA, uvrA, uvrB, umuC, sfiA ... genes). The cleavage of lexA protein results in the expression of SOS genes. The recA protease activity is also specific for the cleavage of λ repressor whose consequence is lysogenic induction[73] (see[3,5] for review). The expression of umuC gene seems to be responsible for reactivation and mutagenesis of UV-irradiated λ phage.

In mammalian cells, treatment with various physical or chemical agents that damage DNA prior to infection with a UV-irradiated virus enhances virus survival. This phenomenon, which has been called "induced virus reactivation" or "enhanced virus reactivation", was first demonstrated with UV-irradiated Herpes virus infecting lightly UV-irradiated monkey kidney cells[74]. This result has now been confirmed using a variety of UV-irradiated viruses which replicate their DNA in the cell nucleus[75], such as simian adenovirus[75], SV40[75-77] and several DNA parvoviruses[78] utilizing as their host a variety of cell lines such as monkey or rabbit kidney cells, Potoroo cells, HeLa cells, normal or XP human fibroblasts, and rat cells. Interestingly, some normal embryonic cells do not exhibit the phenomenon of radiation enhanced reactivation, although host cell reactivation is normal in these cells[5].

X-ray or γ-irradiation of human or monkey kidney cells also enhances the survival of UV-irradiated Herpes virus, simian adenovirus 7 and SV40[75] or human adenovirus 2[79]. Chemical carcinogen treatment of monkey kidney cells has been shown to be very efficient in enhancing the survival of UV-irradiated SV40[76,80]. This effect was obtained both with compounds producing "UV-like" damage (such as metabolized aflatoxin B_1 or acetoxy-acetyl-aminofluorene) or those producing "X-ray" like damage (such as methyl-methane sulfonate or ethyl-methane sulfonate). These results have been confirmed using UV-irradiated Herpes virus[81] and Lu III parvovirus[82]. SV40 reactivation was also promoted in monkey cells which were treated by drugs inhibiting DNA replication such as: hydroxyurea or cycloheximide[76]. We have hypothesized that the inhibition of scheduled DNA synthesis is a direct or indirect signal inducing a new repair process able to replicate UV-irradiated DNA[76].

With split-dose experiments, d'Ambrosio and Setlow[83] have shown that post-replication DNA synthesis after a UV-challenging dose, was more efficient in Chinese hamster cells or in human cells pretreated with a small dose of UV-light compared to that of DNA synthesis in control cells. This result has been interpreted as a consequence of an induced replication system able to better replicate past pyrimidine dimers.

UV-induced mutations in E. coli are completely dependent upon the recA-lexA genes, indicating that SOS repair is essentially an error-prone pathway[70,71]. Recently, a new gene called umuC has been shown to be responsible for UV-induced mutations[84]. In $umuC^-$ E. coli, no UV-induced mutations are observed, although the other SOS functions are normally expressed. At the same time, Weigle-reactivation of UV-irradiated λ is strongly depressed in $umuC^-$ cells. In mammalian cells, evidence has shown that enhanced virus survival is accompanied by a mutation increase in the repaired virus[80,85].

To analyze the mutagenic activity of DNA-damaging treated cells on UV-irradiated virus repair, we used the biological system composed of cultured monkey kidney cells and thermosensitive simian virus 40. We used either an early mutant of SA40 (tsA58) which is defective in initiation of DNA replication because of a point substitution on the T antigen gene, or a late mutant of SV40 (tsB201) which is defective in virus production because of a mutation on the VP1-protein gene. Unirradiated or 1500 J/m^2 UV-irradiated ts SV40 are used to infect carcinogen-treated monkey kidney cells for one lytic cycle at 33° (72 hours) and the progeny survival was measured both at 33° and 41°. The mutation frequency is determined as the ratio of progeny survival at 41° over progeny survival at 33°. Treatment of monkey kidney cells with the chemical carcinogen acetoxy-acetyl-aminofluorene 24 h before infection with UV-irradiated SV40 mutants strongly increases the mutation frequency in the surviving viruses (Fig. 4B). Unirradiated SV40 mutants did not exhibit a significantly increased mutation frequency in AAAF-treated monkey cells (Fig. 4A). However, because of the very low spontaneous mutation level (10^{-8}), we observed a large standard deviation in these results, and an increase by a small factor could not have been easily detected. This result confirms the increased mutation frequency we have seen with UV-irradiated ts SV40 growing in UV-irradiated monkey cells[80] and the one seen by Das Gupta and Summers using a UV-irradiated Herpes simplex virus growing in UV-irradiated monkey cells[85].

We have analyzed the DNA sequence of one SV40 revertant (R-14-10) obtained during UV-induced reactivation and have shown that it involves a single base-pair substitution, an AT to TA transversion. This transversion occurs opposite a possible thymine dimer site and it is located nine base pairs away from the original tsA mutation, which is a CG to TA transition[86].

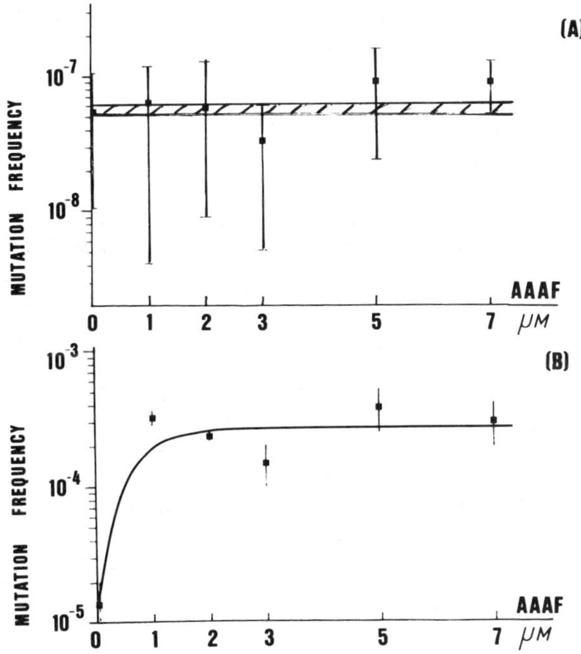

Fig. 4. Mutation frequencies toward wild-type phenotype of tsB201 SV40 mutant in AAAF-treated monkey kidney cells. Control or 1500 J/m^2 UV-irradiated tsB201 SV40 mutants were used to infect control or AAAF-treated CV1-P monkey cells during one lytic cycle at 33° (72h). Mutation frequency is determined as the ratio of progeny survival at 41° over progeny survival at 33°[80]. (A), infection with unirradiated SV40; (B), infection with 1500 J/m^2 UV-irradiated SV40.

2. DNA Polymerase Involvement

The most striking characteristic of the SOS repair pathway in E. coli is its very efficient mutagenic activity[70-72]. To account for this mutagenic effect, Radman and coworkers have put forward a rather elegant hypothesis based on the study of UV-irradiated single stranded ϕX 174 DNA replication[87]. E. coli DNA polymerases are blocked by the first pyrimidine dimer encountered, probably because its 3'-5' exonuclease proof-reading activity. In UV-irradiated *E. coli*, a transdimer synthesis on UV-irradiated ϕX 174 DNA has been detected. A modified DNA polymerase might account for the bypass of dimers which will be intrinsically error-prone. The inhibition of the proof-reading activity could be responsible for SOS repair in UV-irradiated *E. coli*[88]; however, eukaryotic DNA polymerases free of any exonuclease activity are still blocked *in vivo* and *in vitro* by pyrimidine dimers[43,44,48,49] (see Figs. 2 and 3). In consequence, a mechanism other than the inhibition of proof-reading activity could be responsible for SOS repair pathway, although *E. coli* DNA polymerase III seems to be involved in SOS repair pathway. Since the *umuC* gene is necessary for UV-induced mutations in *E. coli*, it is plausible that the *umuC* gene product is a cofactor which will relax the DNA polymerase fidelity, giving rise to mutation during the bypass of lesions.

In mammalian cells, where SOS repair has been clearly established using viral DNAs as molecular probes, there is no data concerning the type of DNA polymerases which may be involved in the SOS repair pathway. With various in vitro assays, it has been reported that replacement of Mn^{++} by Mg^{++} is able to change the specificity of purified DNA polymerases[69,89], although bypass of dimers is not observed. However, the addition of terminal deoxynucleotidyl transferase in an *in vitro* assay has been reported to permit a significant bypass of polymerase α on pyrimidine dimers[90].

It is clear that such in vitro assays represent a way to analyze the effects of various cofactors such as ionic strength, ss-DNA-binding proteins, nucleases..., on the polymerase ability to replicate past-dimer. There exist even fewer experiments which attempt to analyze the molecular mechanism of SOS repair *in vivo*. Some new polymerases or modified polymerases have been described *in vivo* after treatment with various DNA-damaging agents such as chemical carcinogens or mitomycin C. However, no real attempts have been made to determine whether these "new" polymerases are not simply non-specific degradation products of constitutive DNA polymerases.

Since SOS repair seems to be closely linked with DNA replication, it is tempting to speculate that the normal polymerase α, which is blocked by a bulky lesion, will need the help of a relaxing cofactor to permit it to replicate past the lesion. The recent isolation of various DNA repair deficient or "mutator" mutants of eukaryotic cells will permit us to approach the fidelity of the replication or of the repair host cell machinery.

CONCLUSION

It seems obvious that the principal biological role for the DNA polymerase α is the replication of cellular chromosomes, or some viral DNA (SV40); for the DNA polymerase γ, it is the replication of mitochondrial DNA or some viral DNA (adenovirus, parvovirus); for the DNA polymerase β, it is the replication step of the excision-resynthesis repair process. For the other DNA repair modes, it is more difficult to differentiate the role of the DNA polymerases, especially of DNA polymerases α and β. A possible explanation for this difficulty is that these two polymerases might be working together in some processes, such as post-replication repair or SOS repair. In the future, the isolation and the study of mammalian cell mutant or of human cells isolated from patients, in which some steps of DNA replication or DNA repair were impaired, will give rise to a possible way to follow the exact role of DNA polymerases and accessory proteins in these processes. Finally, the use of simple models, such as DNA replication or DNA repair of virus-containing small DNA molecules, will offer an easy way to analyze the exact role of eukaryotic DNA polymerases *in vivo* and *in vitro*.

ACKNOWLEDGEMENTS

We are grateful to Dr. L. Daya-Grosjean and to Dr. A. Gentil for their critical reading of this paper.

REFERENCES

1. P. C. Hanawalt, E. C. Friedberg and C. F. Fox, eds., DNA Repair Mechanisms, Academic Press, New York (1978).
2. T. Lindhal, *Prog. Nucl. Acid Res. Mol. Biol*. 22:135-192 (1979).
3. P. C. Hanawalt, P. K. Cooper, A. K. Ganesan and C. A. Smith, *Ann. Rev. Biochem*. 48:783-836 (1979).
4. J. Laval and F. Laval, *in*:"Molecular and Cellular Aspects of Carcinogen Screening Tests", R. Montesano, H. Bartsch and L. Tomatis, eds., Lyon IARC Scientific Publications (1980).
5. M. Defais-Villani, P.C. Hanawalt and A. Sarasin, *Adv. in Radiat. Biol*., in press (1982).
6. J. D. Hall and D. W. Mount, *Prog. Nucl. Acid Res. Mol. Biol*.25: 53-126 (1981).
7. J. Cleaver, *Nature* 218:652-656 (1968).
8. A. Weissbach, *Ann. Rev. Biochem*. 46:25-47 (1977).
9. M. G. Sarngadharan, M. Robert-Guroff and R. C. Gallo, *Biochem. Biophys. Acta* 516:419-487 (1978).
10. A. Weissbach, *Arch. Biochem. Biophys*. 198:386-396 (1979).
11. J. J. Byrnes, K. M. Downey, V. L. Black and A. G. So, *Biochemistry* 15:2817-2823 (1976).
12. M. Y. W. Tsang-Lee, C. K. Tan, A. G. So and K. M. Downey, *Biochemistry* 19:2096-2101 (1980).
13. G. R. Banks, J. A. Boezi and I. R. Lehman, *J. Biol. Chem*. 254: 9886-9892 (1979).
14. K. Mac Kune and A. M. Holmes, *Nucl. Acids. Res*. 6: 3341-3352 (1979).

15. M. Mechali, J. Abadiedebat and A. M. de Recondo, J. Biol. Chem. 255:2114-2122 (1980).
16. G. Villani, B. Sauer and I. R. Lehman, J. Biol. Chem. 255:9479-9483 (1980).
17. A. Spanos, S.G. Sedgwick, G.T. Yarranton, U. Hubscher and G. R. Banks, Nucl. Acids Res. 9:1825-1839 (1981).
18. P. A. Fisher and D. Korn, J. Biol. Chem. 252:6528-6535 (1977).
19. Y. C. Chen, E. Bohn, S. R. Planck and S. H. Wilson, J. Biol. Chem. 254:11678-11687 (1979).
20. U. Hubscher, C. C. Kuenzle and S. Spadari, Proc. Natl. Acad. Sci. U.S.A.76:2316-2320 (1979).
21. L. M. S. Chang and F. J. Bollum, J. Biol. Chem. 247:7948-7950 (1972).
22. M. Mechali, M. Girard and A. M. de Recondo, J. Virology 23:117-125 (1977).
23. L. M. S. Chang, J. Biol. Chem. 248:3789-3795 (1973).
24. M. Yamaguchi, K. Tanabe, Y. N. Taguchi, M. Nishizawa, T. Takahashi and A. Matsukage, J. Biol. Chem. 255:9942-9948 (1980).
25. J. Waser, U. Hubscher, C. C. Kuenzle and S. Spadari, Eur. J. Biochem. 97:361-368 (1979).
26. M. Philippe and P. Chevaillier, Biochem. J. 175:595-600 (1978).
27. M. Yamaguchi, A. Matsukage and T. Takahashi, J. Biol. Chem. 255: 7002-7009 (1980).
28. A. I. Scovassi, R. Wicker and U. Bertazzoni, Eur. J. Biochem. 100:491-496 (1979).
29. W. Zimmerman, S. M. Chen, A. Bolden and A weissbach, J. Biol. Chem. 255:11847-11852 (1980).
30. M. M. Kwant and P. C. van der Vliet, Nucl. Acids Res. 8:3993-4007 (1980).
31. D. A. Clayton, J. M. Doda and E. C. Friedberg, Proc. Natl. Acad. Sci. U.S.A. 71:2777-2781 (1974).
32. W. A. Haseltine, L. K. Gordon, C. P. Lindan, R. H. Grafstrom, N. L. Shaper and L. Grossman, Nature 285:634-641 (1980).
33. P. C. Seawell, C. A. Smith and A. K. Ganesan, J. Virology 35: 790-797 (1980).
34. K. Tanaka, M. Sekiguchi and Y. Okada, Proc. Natl. Acad. Sci. U.S.A. 72:4071-4075 (1975).
35. G. Ciarrochi and S. Linn, Proc. Natl. Acad. Sci. U.S.A. 72:1887-1891 (1978).
36. C. A. Smith and P. C. Hanawalt, Proc. Natl. Acad. Sci. U.S.A. 75: 2598-2602 (1978).
37. G. P. Margison and P. Kleihues, Biochem. J. 148:521-525 (1975).
38. P. Kleihues and J. Bucheler, Nature 269:625-626 (1977).
39. J. J. Castellot, M. R. Miller, D. M. Lehtomaki and A. B. Pardee, J. Biol. Chem. 254:6904-6908 (1979).
40. J. M. Rossignol, A. Gentil, J. Lacharpagne and A. M. de Recondo, Biochem. Intern. 1:253-261 (1980).
41. T. S. F. Wang and D. Korn, Biochemistry 19:1782-1790 (1980).
42. T. R. Butt, W. M. Wood, E. L. McKay and P. L. R. Adams, Biochem. J. 173:309-314 (1978).

43. P. Moore and B. S. Strauss, Nature 278:664-666 (1979).
44. A. Sarasin and P. C. Hanawalt, J. Mol. Biol. 138:299-319 (1980).
45. A. R. Lehmann, J. Mol. Biol. 66:319-337 (1972).
46. A. R. Lehmann, S. Kirk-Bell, C. F. Arlett, M. C. Paterson, P. H. Lohman, E. A. de Weerd-Kastelein and D. Bootsma, Proc. Natl. Acad. Sci. U.S.A. 72:219-223 (1975).
47. Y. Fujiwara and M. Tatsumi, Mutation Res. 37:91-110 (1976).
48. H. J. Edenberg, Biophys. J. 16:849-860 (1976).
49. R. Meneghini, Biochim. Biophys. Acta. 425:419-427 (1976).
50. W. K. Kauman and J. E. Cleaver, J. Supramol. Struct. Cell Biochem. 5:188 (1981)
51. U. Bertazzoni, M. Stefanini, G. Pedrali-Noy, E. Guilotto, F. Nuzzo, A. Falaschi and S. Spadari, Proc. Natl. Acad. Sci. U.S.A. 73:785-789 (1976).
52. R. Wicker, A. I. Scovassi and S. Nocentini, Nucl. Acids Res. 6: 1591-1605 (1979).
53. M. Mezzina and S. Nocentini, Nucl. Acids Res. 5:4317-4328 (1978).
54. K. Suzuki, M. Miyaki, N. Akamatsu and T. Ono, FEBS Lett. 119:150-154 (1980).
55. G. Ciarocchi, J. G. Jose and S. Linn, Nucl. Acids Res. 7:1205-1219 (1979).
56. J. A. Huberman, Cell 23:647-648 (1981).
57. S. Ikegami, T. Taguchi and M. Ohashi, Nature 275:458-460 (1978).
58. G. Pedrali-Noy and S. Spadari, J. Virol. 36:457-464 (1980).
59. A. Sugino and K. Nakayama, Proc. Natl. Acad. Sci. U.S.A. 77:7049-7053 (1980).
60. G. Pedrali-Noy and S. Spadari, Mutation Res. 70:389-394 (1980).
61. E. Guilotto and C. Mondello, Biochem, Biophys. Res. Commun. 99: 1287-1294 (1981).
62. F. Hanaoka, M. Kato, S. Ikegami, M. Ohashi and M. Yamada, Biochem. Biophys. Res. Commun. 87:575-580 (1979).
63. R. D. Snyder and J. D. Regan, Biochem. Biophys, Res. Commun. 99: 1088-1094 (1981).
64. N. A. Berger, K. K. Kurohara, S. J. Petzold and G. W. Sikorski, Biochem. Biophys. Res. Commun. 89:218-225 (1979).
65. J. E. Cleaver, Radiat. Res. 37:334-340 (1969).
66. N. R. Cozarelli, Ann. Rev. Biochem. 46:641-668 (1977).
67. A. R. S. Collins, Biochim. Biophys. Acta 478:461-473 (1977).
68. W. Dunn and J. Regan, Molec. Pharmacol. 15:367-374 (1979).
69. J. M. Rossignol, Ph.D. Thesis, University of Paris (1980).
70. M. Radman, in: "Molecular Mechanisms for Repair of DNA," P. C. Hanawalt and R. B. Setlow, eds., Plenum Press, New York 355-367 (1975).
71. E. M. Witkin, Bacteriol. Rev. 40:869-907 (1976).
72. R. Devoret, A. Goze, Y. Moulé and A. Sarasin, in: "Mécanismes d'altération et de réparatiod du DNA : relations avec la mutagénèse et la cancérogénèse chimique, "Colloques Internationaux du C.N.R.S., R. Daudel, Y. Moulé and F. Zajdela, eds., Paris, 256:283-291 (1977).
73. P. L. Moreau, M. Fanica and R. Devoret, Biochimie 62:687-694 (1980).

74. L. E. Bockstahler and C. D. Lytle, Biochem. Biophys Res. Commun. 41:184-189 (1970).
75. L. E. Bockstahler and C. D. Lytle, Photochem. Photobiol. 25:477-482 (1977).
76. A. Sarasin and P. C. Hanawalt, Proc. Natl. Acad. Sci. U.S.A. 75: 346-350 (1978).
77. A. Sarasin, Biochimie 60:1141-1144 (1978).
78. C. D. Lytle, Natl. Cancer Inst. Monograph. 50:145-149 (1978).
79. W. P. Jeeves and A. J. Rainbow, Mutation Res. 60:33-41 (1979).
80. A. Sarasin and A. Benoit, Mutation Res. 70:71-81 (1980).
81. C. D. Lytle, J. Copey and W. D. Taylor, Nature 272:60-62 (1978).
82. M. Gunther, R. Wicker, S. Tiravy and J. Copey, in: "Chromosome Damage and Repair", E. Seeberg, ed., 605-610 (1981).
83. S. M. D'Ambrosio and R. B. Setlow, Proc. Natl. Acad. Sci. U.S.A. 73:2396-2400 (1976).
84. T. Kato and Y. Shinoura, Mol. Gen. Genet. 156:121-131 (1977).
85. U. B. Das Gupta and W. C. Summers, Proc. Natl. Acad. Sci. U.S.A. 75:2378-2381 (1978).
86. A. Sarasin, C. Gaillard and J. Feunteun, in:"Induced Mutagenesis : Molecular mechanisms and their implications for environmental protection", C. W. Lawrence, ed., Plenum Press, New York, in press (1981).
87. P. Caillet-Fauquet, M. Defais and M. Radman, J. Mol. Biol.117: 95-112 (1977).
88. G. Villani, S. Boiteux and M. Radman, Proc. Natl. Acad. Sci. U.S.A. 75:3037-3041 (1978).
89. B. Strauss, P. Moore and S. Rabkin, J. Supramol. Struct. Cell. Biochem. 5:191 (1981).
90. S. Yoshida, Biochim. Biophys. Acta 652:324-333 (1981).

STIMULATION OF recA PROTEIN DEPENDENT STRAND ASSIMILATION AND DNA COMPLEX FORMATION BY SINGLE-STRANDED DNA BINDING PROTEINS*

George M. Weinstock,** Kevin McEntee,*** and
I. Robert Lehman

Department of Biochemistry
Stanford University School of Medicine
Stanford, California 94305

ABSTRACT

RecA protein promotes the homology dependent hybridization of single-stranded DNA (ssDNA) with linear or circular duplex DNA (dsDNA) [strand-assimilatin]. Single-stranded DNA binding protein of E. coli (SSB) or the gene 32 protein of bacteriophage T4 stimulates the rate and extent of strand assimilation by a mechanism that does not appear to involve a recA protein·SSB complex. Furthermore, SSB stimulates pairing of circular ssDNA with RFII DNA indicating that free ends on the ssDNA partner are not necessary for SSB action. SSB also stimulates formation of recA protein·DNA complexes under conditions where it enhances strand assimilation, suggesting that these complexes are intermediates in the assimilation reaction. Finally, the recA629 mutant protein, which is cold labile for ssDNA annealing, promotes complex formation at 37°C, but not at 30°C, indicating that the mechanisms of strand pairing in assimilation and ssDNA annealing are closely related.

*This work was supported by grants from the National Institutes of Health (GM06196) and the National Science Foundation (PCM74-00865). K. McEntee was a recipient of an American Cancer Society Senior Fellowship. G. M. Weinstock was supported by the Bank of America-Giannini Foundation.

**Present address: Cancer Biology Program, NCI Frederick Cancer Research Center, Frederick, Maryland 21701.

***Present address: Department of Biological Chemistry, UCLA School of Medicine Los Angeles, CA 90024.

INTRODUCTION

The recA protein of Escherichia coli is required for homologous recombination and repair of DNA damage in vivo. Highly purified recA protein catalyzes the pairing and annealing of DNA molecules in vitro, activities which presumably reflect the role of this protein in the cell. When recA protein is purified from a conditional cold-sensitive recA mutant, the mutant protein is cold labile for ssDNA annealing activity[1].

Although recA protein promotes the ATP dependent annealing of complementary ssDNA as well as the ATP dependent annealing of ssDNA molecules with homologous duplex DNA (strand assimilation), the requirements for these reactions are quite different. The rate of ssDNA annealing increases linearly with recA protein concentration while the rate of strand assimilation is sensitive to the ratio of recA protein to ssDNA in the reaction and decreases sharply when the ratio of recA protein exceeds 5 nucleotides/recA monomer[2,3]. Hydrolysis of ATP and ADP and P_i that accompanies both ssDNA annealing and strand assimilation is blocked by inhibitors of ATP analogs that inhibit the annealing function, indicating a tight coupling of these activities[1,2].

Strand assimilation promoted by recA protein is markedly stimulated by ssDNA binding protein (SSB) whereas recA protein catalyzed strand annealing is inhibited by SSB[4]. Although stimulating strand assimilation, SSB significantly inhibits the ssDNA dependent ATPase activity of recA protein, greatly reducing the amount of ATP hydrolysis accompanying strand assimilation.

In the presence of ssDNA and the ATP analog adenosine 5'O-3 thiotriphosphate (ATPγS), recA protein binds and partially unwinds duplex DNA[3,5]. These recA protein·DNA complexes* are extremely stable and require no homology between the duplex and ssDNA for their formation. It is thought that these complexes represent an intermediate in strand assimilation which, in the presence of ATP, is resolved in a subsequent step requiring ATP hydrolysis. Excess ssDNA inhibits formation of these recA protein·DNA complexes, and we have shown that an additional effect of SSB is to relieve this inhibition[4]. One means by which SSB might stimulate both recA protein-dependent strand

*In this paper we refer to recA protein·duplex DNA complexes formed in the presence of ssDNA as recA protein·DNA complexes. These complexes are identical to "ternary complexes" described by Shibata et al.[3] in their requirements for formation. However, direct evidence that these complexes contain ssDNA, dsDNA and recA protein and the stoichiometry of the substrates in these complexes is still lacking. Because of these considerations we prefer to use the term recA protein·DNA complexes.

assimilation and formation of recA protein·DNA complexes is by covering the free single-stranded regions to prevent their competition with dsDNA for binding to recA protein. In this scheme, other DNA binding proteins should stimulate strand assimilation, a prediction that has been verified using the phage T4 ssDNA binding protein, gene 32 protein (P32). *In vivo* this protein is essential for phage DNA replication, recombination and repair[6,7].

In addition, we have investigated the formation of recA protein·DNA complexes and demonstrate formation of these complexes under conditions of strand assimilation. When a cold labile mutant recA protein is substituted for the wild-type recA protein, recA protein·DNA complex formation becomes cold labile, providing strong evidence that strand annealing and strand assimilation activities of recA protein are mechanistically related.

RESULTS

In Figure 1 the rate and extent of strand assimilation into linear P22 DNA is compared in the presence and absence of SSB. In the absence of SSB (180 pmol recA protein) only 30% of the duplex molecules are retained on nitrocellulose filters after 30min when 410 pmol of ss P22 DNA are included in the reaction (2.3 nt/recA monomer). Increasing the ssDNA concentration (7-8 nt/recA monomer) inhibits the reaction, consistent with earlier results[4]. In the presence of SSB the reaction continues until 80-90% of the duplex DNA

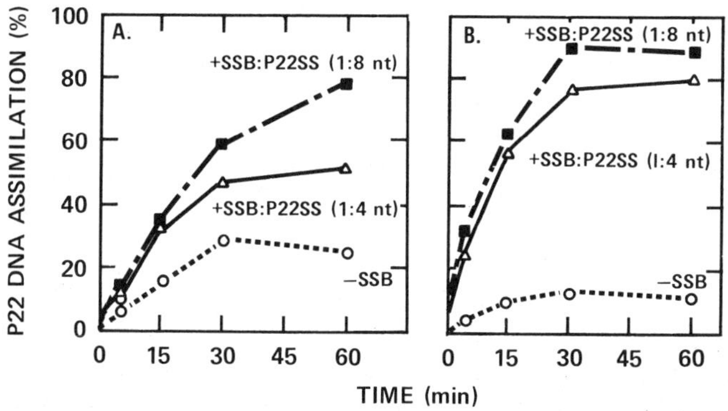

Fig. 1. Kinetics of recA protein dependent strand assimilation in the presence and absence of SSB. Reaction mixtures (200µl) contained 1.05 nmol ^{3}H-labelled P22 dsDNA, 180 pmol recA protein and 900 M ATP. SSB was added to the indicated ratio, and samples (40µl) were taken and assayed as decribed[4]. The amount of heat denatured ss P22 DNA was 410 pmol in (A) and 1230 pmol in (B).

is retained on filters. Furthermore, comparing the rates of the reactions in the presence of SSB indicates that the rate of strand assimilation is approximately proportional to the ssDNA concentration (at saturating SSB) in the reaction. Increasing the SSB beyond saturation by 2-fold slightly inhibits the reactions but increasing SSB to 4 times saturation sharply inhibits strand assimilation (data not shown). Considerable stimulation of recA protein dependent strand assimilation is observed when the T4 gene 32 protein (P32) replaces SSB in the reaction. The stimulation shown in Figure 2 is somewhat less than that obtained with SSB, at approximately the same DNA·binding protein ratio (8-10 nt/monomer). Neither binding protein alone promotes detectable levels of strand assimilation and the P32 stimulated reaction is ATP dependent (data not shown). In addition,

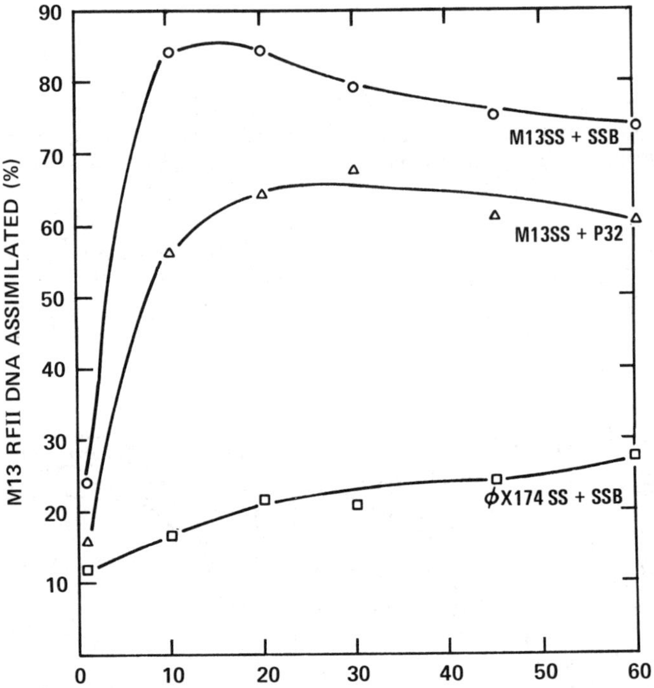

Fig. 2. Kinetics of recA protein-dependent strand assimilation in the presence of T_4 gene 32 protein. Reaction mixtures (200μl) contained 200 pmol recA protein, 1mM ATP, 2000 pmol ^{3}H-M13 RFII DNA, 2100 pmol M13 ssDNA fragments or 2100 pmol ϕX174 ssDNA fragments, 5.2μg E. coli SSB or 4.8μg phage T_4 gene 32 protein (P32). Samples (30μl) were taken at the indicated times and assayed for assimilation as described[4]. Little or no assimilation is detected in the absence of SSB (<5%) because of excess ssDNA in the reaction (10 nt/recA monomer).

stimulation of strand assimilation requires that the ssDNA be homologous to the duplex, as demonstrated in Figure 2 by the failure of ϕX174ss to interact with M13 RFII. The tight, cooperative binding of P32 to ssDNA is likely to prevent nonproductive recA protein-ssDNA interactions in a manner analogous to SSB. We imagine that P32 inhibits the ssDNA dependent ATPase activity although this has not been tested directly. However, it is important to note that masking ssDNA and inhibiting ATP hydrolysis by recA protein are not sufficient to stimulate strand assimilation since SSB purified from a lexC mutant (the lexC mutation is located within or very near the SSB structural gene) binds tightly to ssDNA and inhibits the ATPase activity of recA protein but fails to efficiently stimulate strand assimilation[4] (and unpublished results).

One possible effect of SSB binding to ssDNA is to localize recA protein at sites or regions where SSB binds less tightly. For example, cooperative interactions between SSB protomers are less favorable at the ends of ssDNA molecules where an SSB molecule has only a single neighbor. Binding of recA protein to these molecules could occur preferentially at the ends of the SSB coated ssDNA chain. We have tested this possibility by using circular ss M13 DNA and M13 RFII molecules. RecA protein promotes pairing of these DNAs presumably by causing the interrupted strand of the duplex to "invade" the complementary circular ssDNA. As shown in Table 1, SSB stimulates strand assimilation of circular ss M13 DNA into M13 RFII duplex molecules. This experiment indicates that SSB stimulation of strand assimilation has no requirement for ends on the free ssDNA partner.

Table 1. SSB Stimulation of Circular ss M13 Strand Assimilation into M13 RFII DNA

Description	% RFII DNA retained
Complete*	100
-SSB	4.6
-ATP + ATP S (100μM)	54
-ATP + UPT (1.5mM)	11

*The complete reaction (100μl) contained 220 pmol recA protein, 1800 pmol ^{3}H-labelled M13 RFII, 2100 pmol circular M13 ssDNA, 1mM ATP and 5μg SSB. All incubations were for 30min at 37°C and assimilation was assayed as described[4].

The SSB stimulated assimilation reaction requires ATP[4]. However, unlike the annealing reaction, this reaction will also utilize ATPγS (reference 4; Table 1), although this analog is not hydrolyzed by *recA* protein. This indicates that extensive ATP hydrolysis is not necessary for D-loop formation. However, in view of the relatively high SS/DS ratio in the assay, the ATPγS reaction may be considerably less efficient than the reaction with ATP. UTP, which is hydrolyzed by *recA* protein, does not stimulate assimilation (Table. 1).

The *E. coli* DNA binding protein, SSB, has been shown to complex with DNA polymerase II and exonuclease I and to stimulate the activities of these proteins *in vitro*[8]. We investigated whether SSB and *recA* protein form a complex by incubating the proteins together in the absence of DNA but otherwise under the conditions of the strand assimilation assay and sedimenting the protein through a linear 5-30% sucrose gradient. No evidence was obtained from this experiment that *recA* protein and SSB form a complex or interact (data not shown). However, we cannot rule out the possibility that a small amount of *recA* protein associates with SSB since under these conditions, *recA* protein sediments heterogeneously and overlaps the peak of SSB protein (which sediments as a tetramer) in the sucrose gradient. We feel justified in concluding that recA protein and SSB do not form a tight complex in the absence of DNA.

The stimulation of *recA* protein-promoted strand assimilation by SSB is in part due to the ability of SSB to prevent nonproductive *recA* protein-ssDNA interactions which compete for binding of duplex DNA. However, this masking effect cannot be the only mechanism by which strand assimilation is stimulated. We have pointed out that ADP inhibits strand assimilation and SSB, by inhibiting the ATPase activity of *recA* protein, retards the generation of ADP in reaction[4]. Cunningham and coworkers demonstrated that in the presence of ATPγS, ssDNA stimulates the binding of *recA* protein to dsDNA[5]. The duplex DNA in these complexes is partly unwound presumably to receive the ssDNA chain in the subsequent annealing step[5]. *RecA* protein·DNA complexes, although likely intermediates in strand assimilation, have been demonstrated only when the nonhydrolyzable ATPγS analog is used instead of ATP in the reaction[5]. In the presence of SSB, we have found that *recA* protein·DNA complexes are readily formed with ATP and ssDNA (Fig. 3). In the absence of SSB, *recA* protein binding to dsDNA is inhibited by increasing the concentration of ssDNA. When SSB is added to the reaction, *recA* protein DNA complexes increase in proportion to the ssDNA:SSB added. These results suggest that SSB retards dissociation of *recA* protein from DNA. By stabilizing the binding of *recA* protein to ssDNA in this way, SSB may stimulate strand assimilation by making the reaction more processive: *recA* protein would remain bound to the DNA substrates for a longer time before dissociating. In this bound state we presume that, as in the case when ATPγS is used instead of ATP, the duplex is partially unwound and available for pairing with the exogenous single strand.

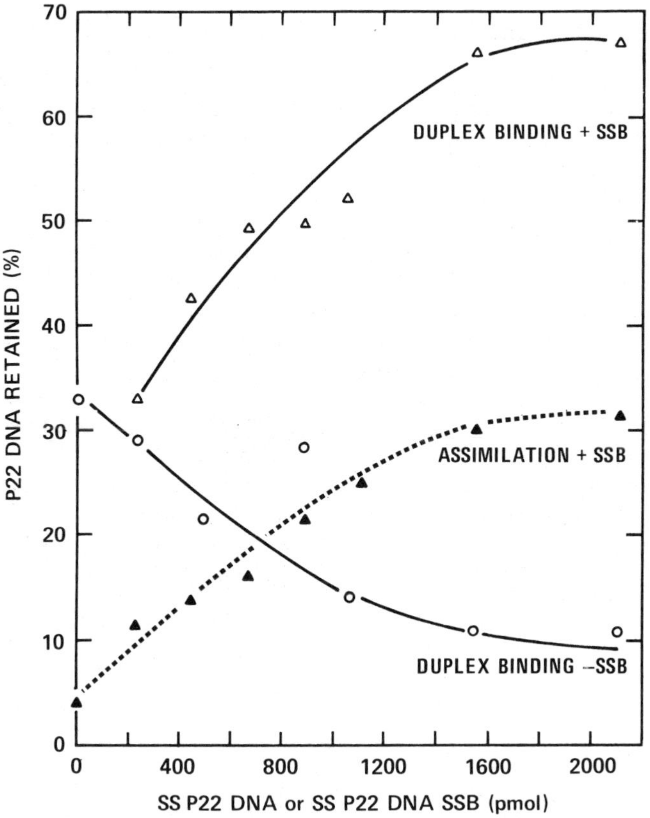

Fig. 3. SSB stimulation of recA protein·DNA complexes in the presence of ATP. Reaction mixtures (200μl) containing 30 pmol recA protein, 1.05 nmol ^{3}H-P22 duplex DNA, 0.9mM ATP and the indicated amount of ss P22 DNA or ss P22 DNA:SSB (8 nt/SSB monomer), were incubated at 37°C for 20min. RecA protein·DNA complexes and strand assimilation were assayed as described[4]. In the absence of recA protein, less than 3% of the ^{3}H-labelled dsP22 DNA bound to filters.

Binding proteins or helix destabilizing proteins stimulate DNA replication in part by making the polymerase more processive during elongation of the growing chain. Enhanced binding of recA protein to DNA in the presence of SSB may be due to a favourable recA protein·SSB interaction or alternatively, by binding to ssDNA, SSB may alter the DNA conformation in such a way as to make recA protein·DNA interaction more favourable. Sedimentation analysis of the two proteins in a sucrose gradient indicates that recA protein and SSB do not strongly interact in solution; however, important protein·protein interactions may occur when the two polypeptides are bound to the same DNA molecule.

RecA Protein·DNA Complexes are Intermediates in Strand Assimilation

The demonstration by Shibata et al.[3] and Cunningham et al.[5] that in the presence of ATPγS, ssDNA stimulated recA protein to bind and partially unwind duplex DNA suggests that these recA protein·DNA complexes represent intermediates in the overall assimilation reaction. We have shown that formation of recA protein·DNA complexes, like the overall assimilation reaction, is inhibited by excess ssDNA and addition of SSB relieves this inhibition[4]. SSB also stimulates formation of recA protein·DNA complexes when ATPγS is replaced by ATP in the reaction (Fig. 3). The amount of strand assimilation into linear P22 DNA increases linear with the concentration of ssDNA·SSB (8 nucleotides/SB monomer) added. RecA protein·DNA complex formation displays an identical response to ssDNA·SSB but the amount of ^{3}H-labelled-dsP22 DNA retained in recA protein DNA complexes is approximately 2.5 times the amount in the assimilated molecules, suggesting that approximately 40% of the recA protein·DNA complexes contain correctly base paired ssDNA and dsDNA molecules. In the absence of SSB, increasing the concentration of ssDNA inhibits the retention of duplex DNA on nitrocellulose filters. This result is completely

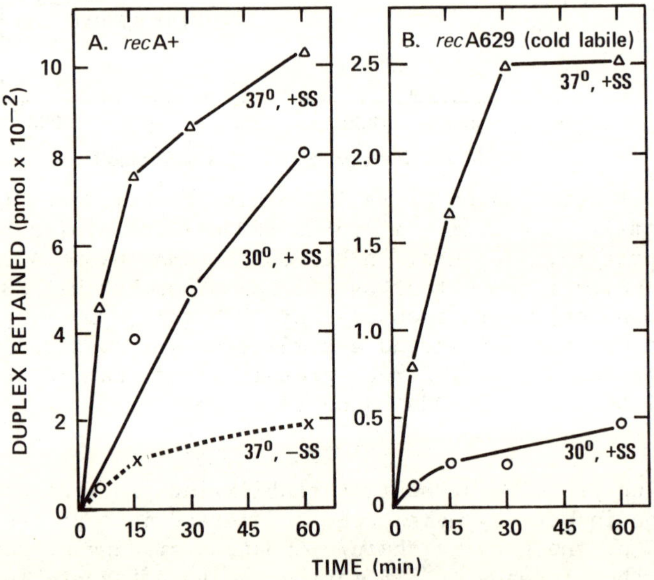

Fig. 4. Cold labile recA protein·DNA complex formation promoted by mutant recA629 protein. Reaction mixtures (200μl) containing 180 pmol wild-type recA protein or approximately 200 pmol recA629 protein, 1.05 nmol ^{3}H-labelled ds P22 DNA, with or without 420 pmol ϕX174 ss DNA and 100μM ATPγS were incubated at 30°C or 37°C for the indicated times and assayed for recA protein·DNA complexes as described[4].

consistent with the idea that ssDNA competes with dsDNA for binding to *recA* protein.

The cold labile *recA*629 protein catalyzes ATP-dependent renaturation of ssDNA at 37°C (in the presence of ATPγS) but not at 30°C (Fig. 4). The wild-type *recA* protein efficiently promotes complex formation at both 37°C and 30°C. At 37°C, the *recA*629 protein is only about 25% as active as the wild-type protein in the reaction. We have also found that the *recA*629 protein has a lower specific activity in the strand renaturation and ATP hydrolysis reactions. We conclude that *recA* protein-dependent strand assimilation and strand renaturation reactions proceed through formation of complexes containing *recA* protein and the DNA molecules that are to be paired. Our results with the cold labile recA629 protein are consistent with this notion and suggest that the enzymatic mechanism of pairing ssDNA molecules is closely related to the pairing of ssDNA chains with duplex molecules.

REFERENCES

1. G. M. Weinstock, K. McEntee and I. R. Lehman, *Proc. Natl. Acad. Sci. U.S.A.* 76:126-130(1979).
2. K. McEntee, G. M. Weinstock and I. R. Lehman, *Proc. Natl. Acad. Sci. U.S.A.* 76:2615-2619 (1979).
3. T. Shibata, R. P. Cunningham, C. DasGputa and C. M. Radding, *Proc. Natl. Adac. Sci. U.S.A.* 76:5100-5104 (1979).
4. T. McEntee, G. M. Weinstock and I. R. Lehman, *Proc. Natl. Acad. Sci. U.S.A.* 77:857-861 (1980).
5. R. P. Cunningham, T. Shibata, C. DasGupta and C. M. Radding, *Nature (London)* 281:191-195 (1979).
6. B. M. Alberts, F. J. Amodio, M. Jenkins, E. D. Gutmann and F. L. Ferris, *Cold Spring Harbor Symp. Quant. Biol.* 33:289-305 (1968).
7. J. Tomizawa, N. Anraku and T. Iwama, *J. Mol. Biol.* 21:247-253 (1966).
8. I. J. Molineux and M. L. Gefter, *J. Mol. Biol.* 98:811-825 (1975).

INDEX